This item must be returned or renewed by the last date shown below. The loan period may be shortened if it is reserved by another reader. A fine will be due if it is not returned on time.

DATE OF RETURN

2 9 APR 2008

UNIVERSITY COLLEGE LONDON
Gower Street London WC1E 6BT

Predicting *the* Performance *of* Multistage Separation Processes

Second Edition

Predicting *the* Performance *of* Multistage Separation Processes

Second Edition

Fouad M. Khoury

Dynamic Optimization Technology Products, Inc.
Houston, Texas

CRC Press

Boca Raton London New York Washington, D.C.

Library of Congress Cataloging-in-Publication Data

Khoury, Fouad M.
    Predicting the performance of multistage separation processes / Fouad M. Khoury. — 2nd. ed.
        p.   cm.
    Originally published: Houston: Gulf Pub. Co., c1995.
    Includes bibliographical references and index.
    ISBN 0-8493-1495-X (alk. paper)
    1. Separation (Technology) I. Title. II. Title: Performance of multistage separation processes.
TP156.S45K48 1999
660′.2842—dc21
                                                                                                99-15552
                                                                                                CIP

About the Author

Fouad M. Khoury, Ph.D., P.E., is a specialist in multistage separation processes, their modeling and optimization. He is a registered professional engineer in Texas and a member of the American Institute of Chemical Engineers. He received his Ph.D. degree in chemical engineering from Rice University and is the author of numerous articles on multistage separation processes, thermodynamics, and transport phenomena.

Dedication

To my wife Yola
for her encouragement and inspiration,
and countless other contributions,

and to our children, Sami and Nadia,
for their understanding and support.

Contents

Preface

Multistage separation processes are the heart of the petroleum, petrochemical, and chemical industries. These industries yield important products as common as gasoline and plastics and as specialized as medical-grade pharmaceuticals.

This book is aimed at performance prediction of multistage separation processes that is essential for their efficient design and operation. It is distinguished by its emphasis on computer modeling, expert interpretation of models, and discussion of modern simulation techniques. It is also unique in that it relates fundamental concepts to intuitive understanding of processes. A generous number of examples are provided in a wide variety of applications to demonstrate the performance of processes under varying conditions and the relationships among the different operating variables. The book is of value as a reference for practicing engineers in the process industry and for advanced level students of engineering process design.

Improved accuracy in predicting thermodynamic and physical properties has occurred simultaneously with major advances in the development of computational techniques for solving complex multistage separation equations. The result has been the emergence of a variety of simulation programs for accurate and efficient prediction of multistage separation processes. This has provided engineers with valuable tools that can help them make more reliable qualitative as well as quantitative decisions in plant design and operation. Frequently, however, effective use of such programs has been hampered by lack of understanding of fundamentals and limitations of prediction techniques. Improper use of simulators can be costly in time and money, which tends to defeat the purpose of computer-aided engineering. These problems are addressed here, and a strategy is pursued that decouples the discussion of conceptual analysis of the material and the computational techniques.

Although rigorous mathematical methods are presented with a good degree of detail, special attention is given throughout the book to keep practical interpretation of the models in focus, emphasizing intuitive understanding. Graphical techniques and shortcut methods are applied wherever possible to gain a handle on evaluating performance trends, limitations, and bottlenecks. Also included are industrial practice heuristics about what ranges of operating variables will work. The student of this book should come away with an enhanced intuitive grasp of the material as well as a basic understanding of the calculational techniques.

The book may be used for a methodical study of the subject or as a reference for solving day-to-day problems. It follows a logical flow of ideas within each chapter and from one chapter to the next, yet each chapter is quite self-contained for quick reference. The discussion starts with fundamental principles, prediction of thermodynamic properties, the equilibrium stage, and moves on to the different types of multistage and complex multistage and multicolumn processes and batch distillation. Although computer simulation is a central theme of this book, no previous experience in the use of simulation software is required.

Earlier chapters use simplified and binary models to analyze in a very informative way some fundamentals such as the effect of reflux ratio and feed tray location, and to delineate the differences between absorption and distillation. Following chapters concentrate on specific areas such as complex distillation, with detailed analyses of various features such as pumparounds and side-strippers and when they should be used. Also discussed are azeotropic, extractive, and three-phase distillation operations, liquid–liquid extraction, and reactive distillation. The applications are clearly explained with many practical examples.

Shortcut computational methods, including modular techniques for on-line, real-time applications, are discussed, followed by a discourse on the major rigorous algorithms in use for solving multicomponent separations.

An understanding of column hydraulics in both trayed and packed columns is essential for a complete performance analysis and design of such devices. The reader will find instructional coverage of these topics, as well as of tray efficiency, in subsequent chapters.

Finally, in a departure from steady-state processes that characterize the rest of the book, the subject of batch distillation is discussed. This process, important for separating pharmaceuticals and specialty chemicals, is presented, including shortcut and rigorous computational methods, along with various optimization techniques.

Fouad M. Khoury, Ph.D., P.E.

Preface to the Second Edition

The favorable reception of the first edition in the industry and academic fields has resulted in the publication of this second, enhanced edition of *Predicting the Performance of Multistage Separation Processes.*

The second edition includes several improvements and updates, and, most important, the addition of about 100 application problems and solutions. These exercises expound on the material throughout the book, and can serve both as teaching material and an applications-oriented extension of the book.

The problems cover three major aspects of the learning process: theory and derivation of application equations, engineering and problem-solving cases, and numerical and graphical exercises. The numerical problems require an algorithm definition and computations which may be done manually. For computer-oriented courses, these problems provide excellent material for program writing exercises.

Fouad M. Khoury, Ph.D., P.E.
Houston, Texas
September 1999

CHAPTER 1

Thermodynamics and Phase Equilibria

The separation processes discussed in this book involve interactions between vapor and liquid phases, or between two liquid phases, or between a vapor phase and two liquid phases. The thermodynamic principles that govern these interactions are introduced in this chapter. Because this chapter is not intended as a full treatise on thermodynamics, only those aspects of the subject that have a direct bearing on phase separation processes are covered. To this end, theory is developed from basic principles and carried through to the formulation of practical methods for correlating relevant thermodynamic properties, such as fugacity and enthalpy. These properties are essential for carrying out heat and material balance calculations in the separation processes described in this book.

When theoretical principles are applied to solve practical problems, various conditions of complexity are encountered. The general approach is to apply the theory to "ideal" systems and then to account for nonidealities by developing models such as equations of state and activity coefficient equations. Models are judged on their consistency with thermodynamic principles and their accuracy in representing actual data. One model may be appropriate for a given system but totally inadequate for another. Hence, the importance of properly selecting a thermodynamic model when attempting to deal with a given separation problem cannot be overemphasized.

1.1 THERMODYNAMIC FUNDAMENTALS

Thermodynamics is a science that relates properties of substances, such as their internal energy, to measurable quantities, such as their temperature, pressure, density, and composition. Thermodynamics also deals with the transformation of energy from one form to another, such as the transformation of internal energy to useful

1

work. In this regard, thermodynamics is more general than mechanics in its formulation of the law of conservation of energy. The energy forms that mechanics is concerned with are entirely convertible to work. Thermodynamics, on the other hand, considers the conversion of thermal energy to work, recognizing that only a fraction of the energy is convertible.

The principles that form the foundation of thermodynamics are embodied in several laws referred to as the *laws of thermodynamics*. In addition, *thermodynamic functions*, which interrelate the various properties of a system, are derived on the basis of these laws. A *system* refers to a part of space under consideration through whose boundaries energy in its different forms, as well as mass, may be transferred.

Within the context of its application to solving practical problems, thermodynamics is primarily concerned with systems at *equilibrium*. From an observational viewpoint, a system is at equilibrium if its properties do not change with time when it is isolated from its surroundings. The concept of equilibrium is a unifying principle that determines energy–work relationships as well as phase relationships.

The principles developed in Section 1.1 are fundamental in the sense that the system considered is not limited to any particular fluid type.

1.1.1 Laws of Thermodynamics

The *first law of thermodynamics* is a formulation of the principle of conservation of energy. It states that the increase in the internal energy of a system equals the heat absorbed by the system from its surroundings minus the work done by the system on its surroundings. For infinitesimal changes, the first law is expressed mathematically by the equation

$$dU = dQ - dW \tag{1.1}$$

where dU is the increase in internal energy, dQ is the heat absorbed, and dW is the work done by the system. The dimensional units of U, Q, and W are energy units.

The internal energy, U, is a basic property that represents the energy stored in a system. It is related to other properties such as work and heat. In an *adiabatic* process, for instance, where no heat crosses the system boundaries, the work done by the system equals the change in its internal energy. This follows from the first law by setting dQ = 0. In a process where no work is done by or on the system, dW = 0, and the change in internal energy equals the heat absorbed or rejected by the system. Heat is considered positive when absorbed by the system, and work is considered positive when done by the system on its surroundings.

The *second law of thermodynamics* relates to the availability of energy in a system for conversion to useful work. In order for a system to perform work, it must have the capacity for spontaneous change toward equilibrium. For instance, a system comprising a hot subsystem and a colder subsystem is capable of performing work as heat passes from the hot to the cold subsystem. Part of the heat is converted to work, while the rest is rejected to the cold subsystem.

The Carnot Engine

An example of a process that can deliver work by absorbing heat from a hot reservoir and rejecting heat to a cold reservoir is the Carnot engine. This is an idealized model consisting of a sequence of processes, each of which is assumed to be *reversible*. A reversible process is one that can be reversed by an infinitesimal change in the external conditions. For instance, in order to compress a gas reversibly, the external pressure at any moment should be $P + \Delta P$, where P is the gas pressure at that moment and ΔP is a small pressure increment. The reversible compression can be changed to a reversible expansion by changing the external pressure to $P - \Delta P$. A reversible process consists of steps in which the system is at equilibrium. In a reversible process there are no losses due to friction or other factors.

Assume that the system used to carry out the Carnot cycle is an amount of *ideal gas* contained in a cylinder fitted with a frictionless piston. The concept of an ideal gas is introduced in Section 1.2. Of consequence at this point is the premise that for an ideal gas the internal energy, U, is a function of temperature only. The Carnot cycle consists of reversible *isothermal* and *adiabatic* processes. An isothermal process is one in which the system temperature is kept constant. An adiabatic process requires that no heat be transferred between the system and its surroundings. The steps are as follows:

1. Reversible adiabatic compression in which the gas temperature changes from T_1, the temperature of the cold reservoir, to T_2, the temperature of the hot reservoir. Since this is an adiabatic process, $dQ = 0$, and, from the first law, $-dW = dU$. The work done on the gas in this step is, therefore,

$$-W_{12} = U_2 - U_1$$

   where U_1 and U_2 are the values of the internal energy at temperatures T_1 and T_2, respectively.

2. Reversible isothermal expansion at temperature T_2, in which an amount of heat, Q_2, is absorbed by the gas from the hot reservoir. For an ideal gas, $dU = 0$ at constant temperature. Therefore, the work done by the gas in this step is

$$W_2 = Q_2$$

3. Reversible adiabatic expansion in which the gas temperature changes from T_2 to T_1. As in step 1, the work done by the gas is

$$-W_{21} = U_1 - U_2$$

4. Reversible isothermal compression to the original state in which an amount of heat, Q_1, is rejected to the cold reservoir at constant temperature T_1. The work done on the system is

$$W_1 = -Q_1$$

The net work done by the system is the sum of the work associated with these steps:

$$W = U_2 - U_1 + U_1 - U_2 + Q_2 - Q_1$$

$$= Q_2 - Q_1$$

The Carnot efficiency is defined as

$$\eta = \frac{W}{Q_2} = \frac{Q_2 - Q_1}{Q_2}$$

The significance of the reversibility of the above processes is that at the end of the cycle the system is brought back to its starting point with no losses incurred due to friction or other causes. Thus, as the system temperature is restored to its starting point, T_1, its pressure is also restored to its starting level of P_1. If the processes are not reversible, additional work would have to be done on the system to bring the pressure back to P_1. The net work would be less than $Q_2 - Q_1$, and the efficiency would be lower than the Carnot efficiency.

It can be shown that the Carnot efficiency may also be expressed as

$$\eta = \frac{T_2 - T_1}{T_2}$$

where the temperatures are on the absolute scale. This is the maximum efficiency attainable by an engine operating between a hot and a cold reservoir. Although developed for an ideal gas model, the efficiency of an engine operating with any medium between temperatures T_1 and T_2 will never exceed the above value.

Entropy

The two equivalent expressions of the efficiency imply that

$$\frac{Q_1}{T_1} = -\frac{Q_2}{T_2} \quad \text{or} \quad \frac{Q_1}{T_1} + \frac{Q_2}{T_2} = 0$$

This equation states that the sum of the Q/T ratio along a Carnot cycle is 0. In general, any closed cycle, starting at some point and moving along reversible paths and returning to the starting point, can be represented by many small isothermal and adiabatic steps. The heat transferred in the adiabatic steps is, by definition, 0. For all the isotherms contained in the loop, the summation of the heat absorbed in each isotherm divided by its absolute temperature is 0:

$$\sum_i \left(Q_i / T_i \right) = 0$$

As the isotherms become infinitely small, the summation may be written as an integral over the closed reversible cycle:

$$\oint dQ/T = 0$$

If any two points A and B are chosen along the reversible cycle, the cycle may be broken into two reversible paths: 1 and 2. The closed cycle integral may be written as the sum of two integrals:

$$\oint \frac{dQ}{T} = \int_A^B \left(\frac{dQ}{T} \right)_1 + \int_B^A \left(\frac{dQ}{T} \right)_2 = 0$$

Subscripts 1 and 2 designate the integration paths. Reversing the integration limits and sign of the second integral, the above equation is rewritten as

$$\int_A^B \left(\frac{dQ}{T} \right)_1 = \int_A^B \left(\frac{dQ}{T} \right)_2$$

Thus, the value of the integral of dQ/T from point A to point B is independent of the path between A and B. A function S, the *entropy*, is now defined such that the change in its value from point A to point B is given by

$$S_B - S_A = \int_A^B \frac{dQ}{T}$$

Since the integral on the right-hand side is independent of the path as long as it is reversible, the change in entropy is independent of the path, and the entropy itself is a function of the thermodynamic coordinates of the system, such as its temperature and pressure. The above equation is next written for an infinitesimal change in entropy:

$$dS = dQ/T \tag{1.2}$$

where dQ is transferred reversibly. Equation 1.2 is a consequence of the second law of thermodynamics and is thus considered the mathematical formulation of this law.

1.1.2 Thermodynamic Functions

The first and second laws of thermodynamics are the basis on which thermody-namic relationships are derived. The two laws, represented by Equations 1.1 and 1.2, are combined in the following statement:

$$dU = TdS - dW$$

The work done by a system as a result of an infinitesimal volume change, dV, against a pressure, P, is dW = PdV. If this is the only form of work, the above equation may be written as

$$dU = TdS - PdV \qquad (1.3)$$

The two above expressions for dU apply only to *closed systems*. A closed system is a fixed-mass body that cannot exchange matter with its surroundings, although it may exchange energy and work.

It was shown that the entropy, S, of a system at equilibrium is a function only of its thermodynamic coordinates such as its temperature and pressure. Such prop-erties are said to be *functions of state*. The internal energy, U, is also a function of state. The internal energy and entropy, along with the temperature, pressure, and volume, are all that is needed to describe the thermodynamics of a system. Additional functions are defined in terms of these five properties to represent other properties that might have practical significance for various applications. These properties, also functions of state, are defined as follows:

Enthalpy

$$H = U + PV$$

Helmholtz function

$$A = U - TS$$

Gibbs free energy

$$G = U + PV - TS$$

$$= H - TS$$

$$= A + PV$$

The differential of the free energy is given by

$$dG = dU + PdV + VdP - TdS - SdT$$

Combining this with Equation 1.3 gives:

$$dG = -SdT + VdP \qquad (1.4)$$

Remember that this equation was derived for a closed system. Because such a system cannot exchange mass with its surroundings, its composition is fixed. Under these circumstances the free energy is a function of only two variables, T and P, and can therefore be expressed as a total differential as follows:

$$dG = \left(\frac{\partial G}{\partial T}\right)_P dT + \left(\frac{\partial G}{\partial P}\right)_T dP$$

where subscripts P and T indicate that the partial derivatives are taken at constant P and T, respectively.

If the system is a homogeneous phase with a variable composition, its free energy is a function of temperature, pressure, and composition. If the system is a solution containing C components with n_1 moles of component 1, n_2 moles of component 2, etc., then G is a function of T, P, n_1, n_2, ..., n_c, and the total differential of G is

$$dG = \left(\frac{\partial G}{\partial T}\right)_{P,n_i} dT + \left(\frac{\partial G}{\partial P}\right)_{T,n_i} dP + \sum_{i=1}^{C}\left(\frac{\partial G}{\partial n_i}\right)_{T,P,n_j(j\neq i)} dn_i \qquad (1.5)$$

The partial derivatives with respect to temperature and pressure are carried out at constant n_i for all the components; that is, at constant composition. For each term within the summation sign, the partial derivative is, with respect to the number of moles of one component, keeping constant the number of moles of all the other components as well as the temperature and pressure.

Equation 1.4 is now applied to a fixed composition system to evaluate the partial derivatives of G at constant and T and P:

$$\left(\frac{\partial G}{\partial T}\right)_{P,n_i} = -S$$

$$\left(\frac{\partial G}{\partial P}\right)_{T,n_i} = V$$

The partial derivative of G with respect to the number of moles of a component i is defined as μ_i, the chemical potential of that component in the phase under consideration:

$$\mu_i = \left(\frac{\partial G}{\partial n_i}\right)_{T,P,n_j} \quad (j \neq i)$$

Equation 1.5 may thus be written in the form:

$$dG = -SdT + VdP + \sum_i \mu_i dn_i \qquad (1.6)$$

At constant temperature and pressure, Equation 1.6 reduces to

$$dG = \sum_i \mu_i dn_i$$

where dG is the differential free energy that results from mixing differential amounts of each one of the components to form a differential amount of solution. The chemical potential of each component is its contribution to the solution free energy, and is, therefore, the partial molar free energy. If many differential amounts of solution with identical temperature, pressure, and composition are mixed together, the total free energy is a simple summation of the differentials. Thus

$$G = \sum_i \mu_i n_i$$

The total differential of G, resulting from the variation of either μ_i or n_i, is then

$$dG = \sum_i \mu_i dn_i + \sum_i n_i d\mu_i$$

This differential is equated to the expression given by Equation 1.6 to show that

$$-SdT + VdP = \sum_i n_i d\mu_i \qquad (1.7)$$

This is the Gibbs–Duhem equation, which relates the variation of temperature, pressure, and chemical potentials of the C components in the solution. Of these C + 2 variables, only C + 1 can vary independently. The Gibbs–Duhem equation has many applications, one of which is providing the basis for developing phase equilibrium relationships.

1.1.3 Conditions for Equilibrium

A system is considered at equilibrium when it has no tendency to move away from its existing conditions. Conversely, a system not at equilibrium tends to undergo spontaneous change toward equilibrium. As a heterogeneous nonequilibrium system moves toward equilibrium, heat flows across regions within the system to equalize their temperatures, work may be generated as the pressures are equalized, and component concentrations tend to vary so as to cause the chemical potential of each component to be equal in all the regions (or phases). These processes can occur concurrently and will, in general, interact with each other.

One theoretical criterion for equilibrium is the expected change in the entropy of a system. Equation 1.2 states that

$$dS = \frac{dQ}{T}$$

for a reversible process. If the process is also adiabatic, $dQ = 0$ and, therefore, $dS = 0$. In an irreversible, or spontaneous, process, $dS > 0$ even if $dQ = 0$. In general, the second law is stated as

$$dS \geq \frac{dQ}{T} \tag{1.8}$$

where the equality sign applies to reversible processes and the inequality sign applies to irreversible processes. Thus, a system is at equilibrium if its entropy is at a maximum.

This fundamental criterion of equilibrium may be applied to practical situations such as determining phase equilibrium conditions. The relationship 1.8 is combined with the first law to arrive at the following expression:

$$dG \leq -SdT + VdP$$

It follows that for a closed system at constant T and P,

$$dG \leq 0$$

This statement dictates that spontaneous changes within a system at constant temperature and pressure tend to lower the free energy of the system. At equilibrium and at constant temperature and pressure,

$$dG = 0$$

The system may be heterogeneous, consisting of a number of homogeneous phases. Each phase is a solution containing a number of components. Assume, for simplicity, that only two phases, α and β, make up the system. Referring to Equation 1.6, a change in the free energy of the system at constant T and P is written as

$$dG = \left(\sum_i \mu_i dn_i \right)^\alpha + \left(\sum_i \mu_i dn_i \right)^\beta$$

The molecules of each component will migrate between the phases so as to minimize the total free energy. At equilibrium, $dG = 0$ and

$$\left(\sum_i \mu_i dn_i \right)^\alpha + \left(\sum_i \mu_i dn_i \right)^\beta = 0$$

Assuming no chemical reactions take place, this relationship must apply to each component independently. Therefore,

$$(\mu_i dn_i)^\alpha + (\mu_i dn_i)^\beta = 0$$

Also, in the absence of chemical reactions, a decrease of one mole of component i in phase α causes an increase of one mole of the same component in phase β. Thus,

$$(dn_i)^\alpha = -(dn_i)^\beta$$

Hence, the condition for phase equilibrium, along with the requirement of uniformity of temperature and pressure, is the equality of the chemical potential for each component in both phases:

$$(\mu_i)^\alpha = (\mu_i)^\beta \tag{1.9}$$

1.2 PVT BEHAVIOR OF FLUIDS

The thermodynamic principles discussed in Section 1.1 are rigorous and general. The "system" for which thermodynamic functions were derived need not be any particular substance. The only requirement is that it satisfy the definition of a system in the thermodynamic context. In order to apply these principles to a given fluid for calculating such properties as enthalpy, free energy, etc., the fluid behavior must be known.

The fluid behavior is commonly represented by its pressure–volume–temperature (or PVT) relationship, expressed in general as

$$f(P, V, T) = 0$$

This relationship might be available in the form of experimental data, or it could be represented by a *model*. Models are usually based on experimental data, but they also possess *predictive* capabilities. That is, they are expected not only to reproduce the correlated data, but also to generate data over reasonable ranges of conditions. Although many PVT models are semiempirical, some are based on theoretical principles such as molecular thermodynamics and statistical thermodynamics. No single PVT correlation exists that can accurately predict all properties for diverse substances over wide ranges of temperature, pressure, density, and composition. Nevertheless, a number of models have demonstrated their usefulness for many applications.

Once the PVT relationship has been proven satisfactory for representing a fluid, it may be used, together with the thermodynamic functions, to derive expressions for other properties such as vapor pressure, vapor–liquid equilibrium relationships, enthalpy departure from ideality, etc.

1.2.1 The Ideal Gas

Perhaps one of the earliest attempts at representing fluid properties centered around the concept of the ideal gas. Experiments on gases at low pressures and densities had led to the following observations: at a given temperature, the volume of a gas is inversely proportional to its absolute pressure and, at a given pressure, the volume of a gas is directly proportional to its temperature, if the latter is measured on an appropriate scale. Later work showed that this scale coincides with the absolute temperature scale associated with the Carnot engine efficiency (Section 1.1.1). The two observations were combined to form the ideal gas equation of state. An equation of state is a fluid behavior model that relates the temperature, pressure, and volume of the fluid in an equation form. The ideal gas equation of state takes the form

$$PV = nRT \tag{1.10}$$

where V is the volume of n moles of gas, P is its absolute pressure, and T is its absolute temperature. The proportionality constant, R, is the universal gas constant. Another characteristic of an ideal gas, which was discussed with the Carnot engine (Section 1.1.1), states that the internal energy of an ideal gas is a function of temperature only and is independent of pressure and gas composition. This aspect of ideal gases is discussed further in Section 1.4 on enthalpy.

Certain ideal gas characteristics follow from the definition of the ideal gas. For instance, the molar volumes of all ideal gases are the same at the same temperature and pressure since R is a universal constant. Also, it follows that the volume of a mixture of ideal gases at a given temperature and pressure is equal to the sum of the volumes of the individual gases at the same temperature and pressure. From the molecular standpoint, an ideal gas consists of molecules, each of which occupies 0 volume and between which no forces of attraction or repulsion exist.

Real gases approach ideal gas behavior at low pressures. Deviations from ideality increase at higher pressures and lower temperatures, that is, as the density goes up. The ideal gas equation cannot predict the transition from gas to liquid since, according to the equation, the volume at a fixed pressure would decrease continuously and proportionately to the absolute temperature as the temperature is decreased. This, of course, is not what is observed in reality.

1.2.2 Real Fluids

The ideal gas equation of state cannot describe real fluids in most situations because the fluid molecules themselves occupy a finite volume and because they exert forces of attraction and repulsion on each other. As the gas is cooled, and assuming its pressure is below the critical point, a temperature is reached where the intermolecular interactions result in a transition from the gas phase to the liquid phase. The ultimate fluid model would be one that could describe this transition as well as the fluid behavior over the entire range of temperature and pressure. Such a model would also be capable of representing mixtures as well as pure components.

Numerous attempts have been made to develop fluid models on the basis of molecular thermodynamics, taking into account the intermolecular forces. It is beyond the scope of this book to review these theories, and, in any case, the theoretical models are not necessarily the ones that are most widely used. The success of a model rests on its ability to represent real fluids. The *principle of corresponding states* is another approach that provides the foundation for some of these models. A number of models, or equations of state, that have proven their practical usefulness for phase equilibrium and enthalpy departure calculations are presented in this section.

Qualitative PVT Behavior of Pure Substances

The phase behavior of a pure substance may be depicted schematically on a pressure–temperature diagram as shown in Figure 1.1. The curve OC, the vapor pressure curve, separates the vapor and liquid phases. At any point on this curve, the two phases can coexist at equilibrium, both phases having the same temperature and pressure. Phase transition takes place as the curve is crossed along any path. Figure 1.1 shows two possible paths: at constant pressure (path AB) and at constant temperature (path DE). At the critical point, C, the properties of the two phases are indistinguishable and no phase transition takes place. In the entire region above the critical temperature or above the critical pressure, only one phase can exist.

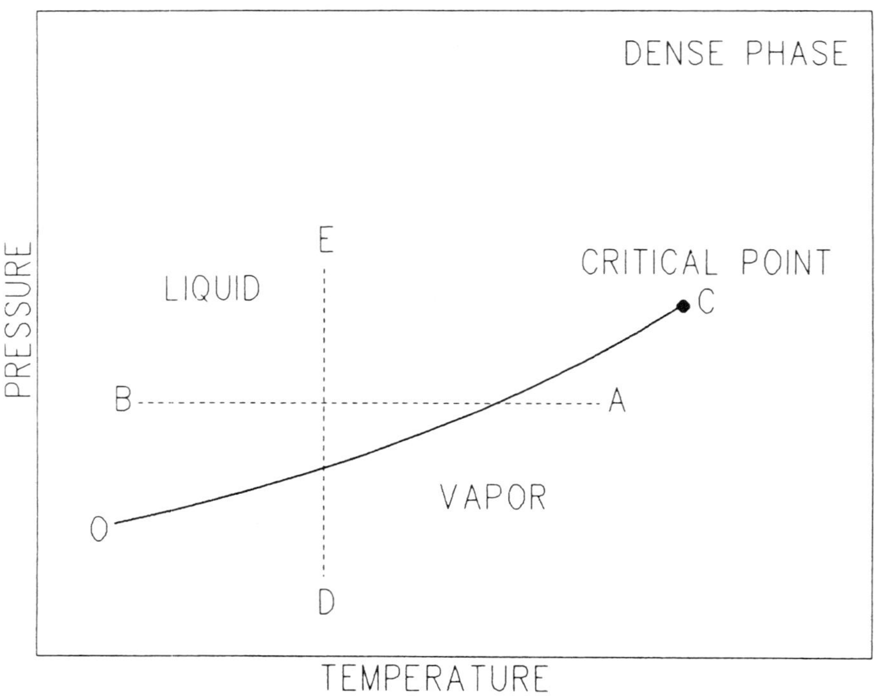

Figure 1.1   P–T diagram for a pure substance.

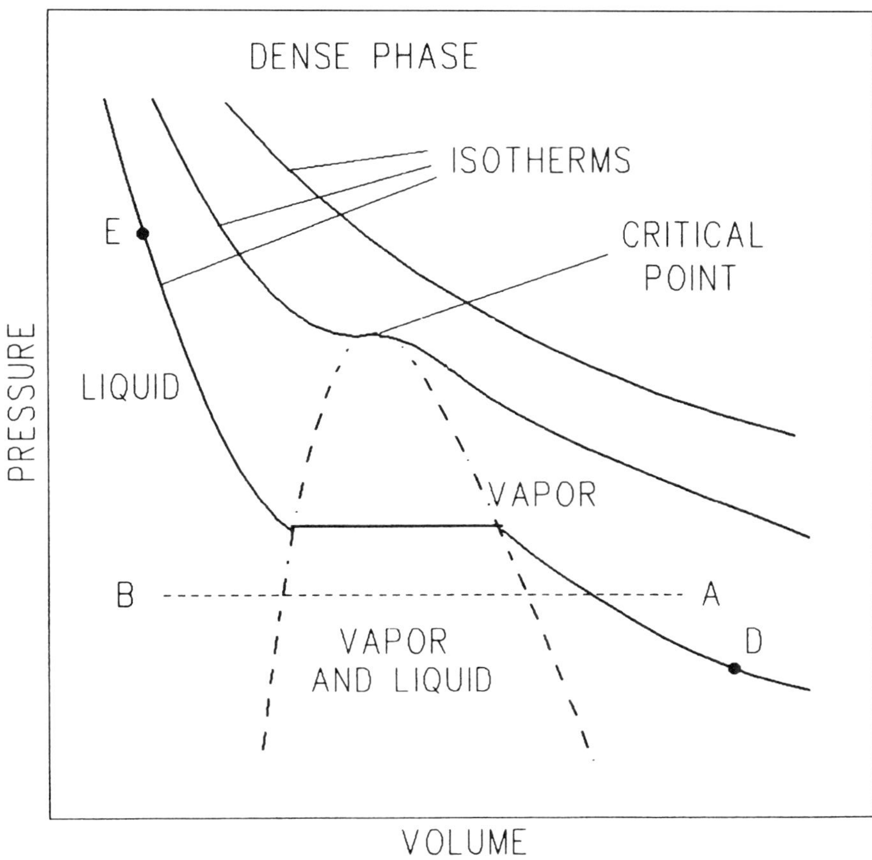

Figure 1.2   P–V diagram for a pure substance.

The PVT behavior of a pure substance may be described on a pressure–volume diagram, as shown in Figure 1.2. The variation of volume with pressure at various fixed temperatures is represented by the *isotherms*. If the isotherm temperature is above the critical, the pressure decreases continuously as the volume increases and no phase change takes place. The critical temperature isotherm is also continuous but has an inflection point at the critical pressure. On subcritical isotherms, the pressure of a liquid drops steeply with small increases in the volume until the liquid starts to vaporize. At this point the pressure remains constant as the total volume increases, as long as both vapor and liquid coexist at equilibrium. When all the liquid has vaporized, the pressure starts falling again as the volume is further increased. The constant pressure and constant temperature processes described by paths AB and DE in Figure 1.1 are shown again in Figure 1.2.

1.2.3 The Principle of Corresponding States

The qualitative observation of PVT behavior of pure substances indicates a continuity in the isotherm at the critical point on a PV diagram. The existence of an inflection point on the critical isotherm at the critical pressure may be shown to imply that the first and second derivatives of the pressure with respect to the volume are equal to 0 at the critical point:

$$\left(\frac{\partial P}{\partial V}\right)_{T_C} = \left(\frac{\partial^2 P}{\partial V^2}\right)_{T_C} = 0 \tag{1.11}$$

These mathematical conditions at the critical point may be applied to various equations of state to determine their parameters in terms of the critical constants. The equations themselves may then be written in terms of the critical constants which are characteristic of each substance. The temperature, pressure, and volume may be replaced in the equation of state by the reduced properties, defined as

$$T_r = \frac{T}{T_C}; \quad P_r = \frac{P}{P_C}; \quad V_r = \frac{V}{V_C}$$

An equation of state written in terms of the reduced properties is a generalized equation that could be applied to any substance. It follows that, if two substances are at the same reduced temperature and pressure, they would have the same reduced volume. This is the concept of the principle of corresponding states, which has been used to correlate PVT properties of similar components.

An equation of state based on the principle of corresponding states as described above takes the form

$$f(P_r, T_r, V_r) = 0$$

Once this function is determined, it could be applied to any substance, provided its P_C, T_C, and V_C are known. One way of applying this principle is to choose a reference substance for which accurate *PVT* data are available. The properties of other substances are then related to it, based on the assumption of comparable reduced properties. This straightforward application of the principle is valid for components having similar chemical structure. In order to broaden its applicability to disparate substances, additional characterizing parameters have been introduced such as *shape factors*, the *acentric factor*, and the *critical compressibility factor*. Another difficulty that must be overcome before the principle of corresponding states can successfully be applied to real fluids is the handling of mixtures. The problem concerns the definition of P_C, T_C, and V_C for a mixture. It is evident that mixing rules of some sort need to be formulated.

Ideas derived from the principle of corresponding states have been incorporated into the development of equations of state, some of which are discussed next.

1.2.4 Equations of State

New equations of state are constantly being published, and it is not the intention here to present a complete overview of these equations. The ones discussed here were selected because they enjoy widespread use or, in some instances, because they possess some historical or theoretical significance.

The van der Waals Equation

This is one of the oldest and most famous equations and one that was based on theoretical reasoning. It represents one of the earlier attempts at representing both the vapor phase and the liquid phase with the same equation although its success is mainly limited to the vapor phase. It is still used for the vapor phase portion of certain phase equilibrium calculations. The equation is given as

$$(P + a/V^2)\ (V - b) = RT \tag{1.12}$$

It is a logical requirement that equations of state approach the ideal gas equation at the limit of low pressures. As the pressure decreases, the volume increases so that at very low pressures $a/V^2 \ll P$ and $b \ll V$. If these terms are dropped from the van der Waals equation, it reduces to the ideal gas form. The terms a/V^2 and b account for intermolecular forces and molecular volume. The parameters a and b are called the attraction and repulsion parameters, respectively. Parameter b is also referred to as the effective molecular volume.

Equation 1.11 may be used to evaluate a and b in terms of the critical constants, resulting in the following expressions:

$$a = 3P_C V_C^2 = 27R^2\, T_C^2 / 64 P_C$$

$$b = V_C / 3 = RT_C / 8P_C$$

If these expressions are substituted in the van der Waals equation, it takes the following form:

$$\left(P_r + 3/V_r^2\right)\left(3V_r - 1\right) = 8T_r$$

The equation in this form is applicable to any substance, provided the critical constants are known. Correlating data in this manner is one illustration of utilizing the principle of corresponding states.

The Virial Equation

The ideal gas equation may be written in the form $Pv/RT = 1$, where v is the molar volume ($v = V/n$). In general, for real fluids, $Pv/RT = z$, where z is known as the compressibility factor. The compressibility factor may be expressed in terms of the pressure in a power series (or *virial* expansion):

$$z = 1 + BP + CP^2 + \ldots$$

The coefficients B, C, etc., are called the second virial coefficient, the third virial coefficient, etc. The first virial coefficient is unity. The virial coefficients (except the first) depend on the temperature and the nature of the fluid.

The virial equation is usually truncated after the linear pressure term that contains the second virial coefficient. Data are readily available for pure component and binary interaction second virial coefficients for a large number of components and binaries. Binary interaction coefficients are required for extending the equation to mixtures. The simplicity of the equation, the availability of coefficient data, and its ability to represent mixtures are some of the reasons the virial equation of state is a viable option for representing gases at densities up to about 70% of the critical density. It may be used for calculating vapor phase properties at these conditions but is not applicable to dense gases or liquids.

The Redlich–Kwong Equation

This equation is an improvement over the van der Waals equation in that it introduces a temperature dependency for the attraction parameter by dividing it by the term $T^{0.5}$. The equation also has a special quadratic term in the volume:

$$\left(P + \frac{a}{T^{0.5}V(V + b)}\right)(V - b) = RT \qquad (1.13)$$

The parameters a and b are evaluated in terms of the critical constants by applying the critical point conditions (Equation 1.11). The results are:

$$a = 0.42748R^2\, T_C^{2.5}/P_C$$

$$b = 0.08664R\, T_C/P_C$$

An alternative method for determining the parameters is by regression of experimental data with the objective of minimizing errors between the experimental and predicted properties. For the regression, either direct PVT data or properties that can be derived from the equation of state, such as vapor–liquid equilibrium or enthalpy departure data, may be used.

Later work on Redlich–Kwong-type equations (known as cubic equations because they are of the third degree in volume when expressed explicitly in the pressure) has shown that greater accuracy is achieved if the equation parameters are correlated as a function of temperature. The original Redlich–Kwong equation has therefore been dropped for the most part from practical applications in favor of the more recent cubic equations.

The Soave Equation

In an effort to improve the equation of state representation of the effect of temperature, the Soave equation (Soave, 1972) replaces the $a/T^{0.5}$ term in the Redlich–Kwong equation with a more general temperature-dependent parameter, $a(T)$:

$$P = \frac{RT}{V - b} - \frac{a(T)}{V(V + b)}$$

The development of the equation was targeted primarily at improving the accuracy of vapor–liquid equilibrium calculations. The underlying reasoning in developing the equations was that a necessary condition for an equation of state to predict mixture vapor–liquid equilibrium properties was that it accurately predict pure component vapor–liquid equilibrium properties, namely pure component vapor pressures.

The Soave equation uses an additional parameter, the acentric factor, to correlate vapor pressure data. The acentric factor, ω, had been defined in earlier work as a parameter that correlates the deviation of the reduced vapor pressure of a particular compound from that of simple molecules. The reduced vapor pressure is correlated with the reduced temperature as follows:

$$\log P_r^0 = C_1 - C_2/T_r$$

It was observed that the reduced vapor pressure of noble gases (simple molecules) at a reduced temperature of 0.7 is about 0.1. It is also known that, at the critical point, $T_r = 1$ and $P_r = 1$. These two conditions were applied to the above equation to evaluate C_1 and C_2. The resulting equation for simple substances is

$$\log(10P_r^0) = 10/3 - 7/3T_r$$

For substances in general, the reduced vapor pressure tends to deviate to varying degrees from this equation. The deviation is accounted for by including the acentric factor in the equation:

$$\log(10^{1+\omega} P_r^0) = 10/3 - 7/3T_r \tag{1.14}$$

The acentric factor is a characterizing parameter for each substance and is defined as

$$\omega = -\log(10P_r^0) \quad \text{at} \quad T_r = 0.7$$

For simple compounds, $P_r^0 = 0.1$ at $T_r = 0.7$ and, therefore, $\omega = 0$.

Equation 1.14 incorporates the definition of the acentric factor and may also be used to predict the vapor pressure, once the acentric factor has been determined. Another route for calculating the vapor pressure is via an equation of state, as described below. In the Soave equation, ω is used in formulating the temperature dependency of the parameter a, which may be considered a function of both T and ω. The function $a(T, \omega)$ was determined with the objective of fitting vapor pressures calculated by the equation of state to experimental pure-component vapor pressure data.

The first step of calculating the vapor pressure from an equation of state is the familiar evaluation of parameters a and b using the critical point conditions, Equation 1.11, with the results

$$a' = 0.42747 R^2 \, T_C^2 / P_C$$

$$b = 0.08664 R \, T_C / P_C$$

Here a', a constant, is the value of the parameter at the critical temperature. At other temperatures, the parameter at (T, ω) is defined as

$$a(T, \omega) = a'\alpha(T, \omega)$$

At the critical temperature, $\alpha = 1$. At other temperatures, α is determined from the vapor pressure fit, leading to the relationship

$$\alpha^{0.5} = 1 + \left(1 - T_r^{0.5}\right)\left(0.480 + 1.574\omega - 0.176\omega^2\right)$$

Vapor pressures are calculated with the Soave equation by equating the pure component fugacity coefficients in the vapor and liquid. It is significant that the same equation of state is applied to both the vapor and liquid phases.

Like other cubic equations of state, the Soave equation can have three real roots for the molar volume (or density or compressibility, depending on the form in which the equation is written). When the composition, temperature, and pressure are such that three roots exist, the largest volume root is used if the system is a vapor, and the smallest volume root is used if the system is a liquid.

Equations of state are extended to mixtures by replacing pure component parameters with mixture parameters. These are expressed in terms of the mixture composition and the parameters of the pure components, using certain mixing rules. The Soave equation uses the following mixing rules:

$$a = \sum_i \sum_j X_i X_j a_i^{0.5} a_j^{0.5}\left(1 - k_{ij}\right)$$

$$b = \sum_i X_i b_i \tag{1.15}$$

where X is the mole fraction. The summations are carried over all the components in the mixture. Parameters a_i (or a_j) and b_i are pure component parameters. The binary interaction parameters k_{ij} are equal to zero for $i = j$ and for similar components. For binaries of dissimilar molecules, such as those between carbon dioxide, hydrogen sulfide, and hydrocarbons, k_{ij} is usually in a range from 0 to 0.2. The parameter k_{ij} may be assumed to be independent of temperature, pressure, and composition, although better representation is obtained with temperature-dependent interaction coefficients. The coefficients are determined from binary experimental data, particularly vapor–liquid equilibrium data.

The Soave equation is widely used for hydrocarbons and related components over broad ranges of temperature and pressure. It is accurate enough for calculating enthalpy and entropy departures, vapor–liquid equilibria, and vapor density in natural gas processing and many petroleum-related operations. The equation is not very accurate in the critical region and for liquid density calculations.

The Peng–Robinson Equation

This equation (Peng and Robinson, 1976) was developed with the goal of overcoming some of the deficiencies of the Soave equation, namely its inaccuracy in the critical region and in predicting liquid densities. The equation is similar to the Soave equation in that it is cubic in the volume; expresses its parameters in terms of the critical temperature, critical pressure, and acentric factor; and is based on correlating pure-component vapor pressure data. The equation is written as

$$P = \frac{RT}{V - b} - \frac{a(T)}{V(V + b) + b(V - b)} \tag{1.16}$$

The parameters are given by

$$a(T, \omega) = 0.45724 \left(R^2 \, T_C^2 / P_C \right) \alpha(T, \omega)$$

$$\alpha^{0.5} = 1 + \left(1 - T_r^{0.5} \right) \left(0.37464 + 1.5422\omega - 0.26992\omega^2 \right)$$

$$b = 0.07780 R \, T_C / P_C$$

Mixtures are handled by calculating mixture parameters, using the same mixing rules used with the Soave equation, given by Equation 1.15.

The Peng–Robinson equation is widely used for the same applications as the Soave equation. Although it may be more accurate than the Soave equation in the critical region and for calculating liquid densities, it is not generally recommended for the latter since better methods for predicting liquid densities are available.

The BWR Equation

One approach towards improving the accuracy of equations of state is to produce a better fit to experimental data by including many adjustable parameters in the equation. The equation of Benedict, Webb, and Rubin (1951) was first introduced with eight parameters. In later development the accuracy of the equation was improved by modifying it to include 13 parameters (Starling, 1973). Since the parameters must be determined individually for each substance, the equation applicability is limited to those substances for which parameters are available. Mixtures are handled using mixing rules that require interaction coefficients for several of the parameters. Interaction coefficients are available for only a limited number of components, which further restricts the equation's usage. The equation itself and the parameter mixing rules are not presented here because of space limitation. The reader is referred to the original sources for a detailed description of the equation.

Attempts have been made to correlate the BWR parameters in terms of the critical temperature, critical pressure, and acentric factor. The equation is thus reduced essentially to a three–parameter function, not necessarily more accurate than the cubic equations.

The Lee–Kesler–Plocker Equation

By applying the corresponding states principle, the deviations of the properties of a substance from those of a "simple" fluid may be correlated in terms of the acentric factor, as described above for vapor pressures (Equation 1.14). The compressibility factor has also been correlated in terms of the acentric factor in the form of a polynomial

$$z = z^{(0)} + \omega z^{(1)} + \omega^2 z^{(2)} + \ldots$$

which is usually truncated after the linear term. The compressibility factor of a fluid, z, is thus expressed in terms of the compressibility factor of a *simple* fluid, $z^{(0)}$, and that of a *reference* fluid, $z^{(1)}$. Lee and Kesler (1975) developed equations for $z^{(0)}$ and $z^{(1)}$ using data for argon, krypton, and methane to represent the simple fluid, and n-octane data to represent the reference fluid. Plocker et al. (1978) modified the parameter mixing rules and determined binary interaction parameters for many component pairs based on vapor–liquid equilibrium data. With these improvements, the Lee–Kesler–Plocker equation has gained broad acceptance as a reliable generalized equation of state. Like other generalized equations, such as Soave and Peng–Robinson, the Lee–Kesler–Plocker equation is not adapted to polar compounds. The equation details, parameter values, and mixing rules are not provided here but may be found in the referenced sources.

1.3 PHASE EQUILIBRIA

The conditions for equilibrium discussed in Section 1.1.3 are applied here to the problem of phase equilibria. These conditions are that, in order for two or more phases to coexist at equilibrium, they must have the same temperature and pressure and the chemical potential of each component must be equal in all the phases. The chemical potential is not a measurable quantity and is not intuitively related to observable physical properties. Applying the conditions of equilibrium to real fluids involves a transformation to more practical terms and the utilization of fluid models such as equations of state.

1.3.1 Fugacity

The chemical potential of component i in a fluid may be derived on the basis of Equation 1.6. By holding T and n_i constant, then T, P and n_j (j ≠ i) constant, the following relationships may be written:

$$V = \left(\frac{\partial G}{\partial P} \right)_{T,n_i}$$

$$\mu_i = \left(\frac{\partial G}{\partial n_i} \right)_{T,P,n_j}$$

If μ_i is differentiated with respect to P and the reciprocity rule is applied, the following is obtained:

$$\left[\frac{\partial}{\partial P} \left(\frac{\partial G}{\partial n_i} \right)_{T,P,n_j} \right]_{T,n_i} = \left[\frac{\partial}{\partial n_i} \left(\frac{\partial G}{\partial P} \right)_{T,n_i} \right]_{T,P,n_j}$$

or

$$\left(\frac{\partial \mu_i}{\partial P} \right)_{T,n_i} = \left(\frac{\partial V}{\partial n_i} \right)_{T,P,n_j}$$

Defining the partial pressure of component i as

$$p_i = \left(\frac{n_i}{n} \right) P$$

and combining this definition with the ideal gas equation to eliminate P, the ideal gas equation for component i may be written as

$$p_i V = n_i RT$$

Thus, for an ideal gas, the derivative of the volume with respect to the number of moles of component i is

$$\left(\frac{\partial V}{\partial n_i}\right)_{T,P,n_j} = \frac{RT}{P_i}$$

Substitution in the above equation gives

$$\left(\frac{\partial \mu_i}{\partial p_i}\right)_{T,n_j} = \frac{RT}{P_i}$$

In its integrated form, this equation becomes

$$\mu_i = \mu_i^0 + RT \ln p_i \qquad (1.17)$$

where μ_i^0 is an integration constant, a function of temperature only.

For real fluids the partial pressure is replaced by the fugacity, a defined property, maintaining the same form of Equation 1.17:

$$\mu_i = \mu_i^0 + RT \ln f_i \qquad (1.18)$$

The fugacity bears the same relationship to the chemical potential for real fluids as does the partial pressure for ideal gases. Because of the direct relationship between chemical potential and fugacity, the condition for equilibrium expressed by Equation 1.9 is equivalent to the equality of component fugacities in the phases:

$$f_i^L = f_i^V \qquad (1.19)$$

This expression of the condition for equilibrium is used in phase equilibrium calculations more frequently than the equality of the chemical potentials because the fugacity is more closely related to observable properties than the chemical potential.

The ratio

$$\phi_i = \frac{f_i}{P_i}$$

is called the fugacity coefficient. As the pressure approaches zero, the fluid approaches ideal gas behavior; hence,

$$\lim_{p_i \to 0} \phi_i = 1$$

Pure Substances

To evaluate the fugacity of a pure component, Equation 1.4, which is valid for a closed system with a fixed composition, can be applied to the special case of a pure component. At constant temperature this equation becomes

$$dG = vdP$$

For an ideal gas $v = RT/P$ so that

$$dG = RTdP/P = RTdlnP$$

For real fluids the fugacity is substituted for the pressure. At constant temperature,

$$dG = vdP = RTdlnf \qquad (1.20)$$

or

$$dlnf = vdP/RT$$

The fugacity of a pure component is calculated by integrating this equation between 0 pressure and system pressure. At $P = 0$, $f = P$ and $\Phi = 1$. Also, as the pressure approaches 0, the ideal gas law applies. The integration results in the following equation:

$$\ln \frac{f}{P} = \ln \phi = \int_{0}^{P} \frac{z-1}{P} dP \qquad (1.21)$$

The compressibility factor, z, must be available as a function of pressure, either through an equation of state or directly from experimental data.

Mixtures

The fugacity of a component i in a homogeneous phase is derived in a manner analogous to pure substances. The change in the partial free energy of component i in an ideal gas mixture at constant temperature resulting from a change in its partial pressure is

$$d\hat{G}_i = RTd \ln p_i = \hat{V}_i dP$$

where the caret designates partial quantities, or quantities associated with a component in solution. For nonideal solutions, vapor or liquid, the partial pressure is replaced by the fugacity:

$$RTd \ln \hat{f}_i = \hat{V}_i dP$$

The change in component fugacity resulting from a change in pressure from P_1 to P_2 is evaluated by integrating the partial molar volume at constant temperature:

$$\ln\left(\frac{\hat{f}_{i2}}{\hat{f}_{i1}}\right) = \frac{1}{RT}\int_{P_1}^{P_2}\hat{V}_i\,dP \tag{1.22}$$

The fugacity coefficient of component i in a mixture is defined as

$$\hat{\phi}_i = \frac{\hat{f}_i}{P_i} = \frac{\hat{f}_i}{PX_i}$$

where p_i, the partial pressure of component i, is defined as the total pressure multiplied by X_i, the mole fraction of component *i*. As the pressure approaches 0, the fugacity approaches the partial pressure so that

$$\lim_{p\to 0}\hat{\phi}_i = 1$$

The lower limit of the integral is defined at $P = 0$, where $\hat{\phi}_i = 1$, resulting in the following expression for the fugacity coefficient:

$$\ln\hat{\phi}_i = \frac{1}{RT}\int_0^P\left(\hat{V}_i - \frac{RT}{P}\right)dP$$

$$= \int_0^P\left(\frac{\hat{z}_i - 1}{P}\right)dP \tag{1.23}$$

The partial molar volume or compressibility factor is evaluated from an equation of state or from experimental data.

The condition for vapor–liquid equilibrium, Equation 1.19, may be stated in terms of fugacity coefficients:

$$Y_i\hat{\phi}_i^V P = X_i\hat{\phi}_i^L P \tag{1.24}$$

In this equation, X_i and Y_i are the mole fractions of component i in the liquid and vapor, respectively. The vapor–liquid distribution coefficient, or K-value, defined as $K_i = Y_i/X_i$, is therefore

$$K_i = \frac{\hat{\phi}_i^L}{\hat{\phi}_i^V} \tag{1.25}$$

For liquids that cannot be represented by equations of state, the liquid fugacities are expressed in terms of activity coefficients, discussed in Section 1.3.3. For ideal solutions, Equation 1.24 reduces to Raoult's law.

1.3.2 Phase Equilibrium in an Ideal System

Mixture properties are related to constituent component properties by means of the general relationship

$$\{\text{Mixture Property}\} = \Sigma \,\{\text{Mole Fraction of i}\} \times \{\text{Partial Property of i}\}$$

This relationship was incorporated, for example, in the derivation of the equation for the free energy of a mixture. The volume of a mixture is

$$V = \sum_i X_i \hat{V}_i$$

where $\hat{V}_i$ is the partial volume of component i. The volume of an ideal solution is equal to the sum of the constituent component volumes at the same temperature and pressure (Section 1.2.1):

$$V = \sum_i X_i V_i$$

It follows that, in an ideal solution,

$$\hat{V}_i = V_i$$

It also follows from Equation 1.20 and the definition of fugacity and fugacity coefficient that, for ideal solutions,

$$\hat{\phi}_i = \phi_i$$

and

$$\hat{f}_i = X_i f_i \tag{1.26}$$

where f_i is the fugacity of the pure component. Equation 1.26 is known as the Lewis–Randall rule.

Raoult's Law

The above characterization of ideal solutions does not require the fluid to behave as an ideal gas. In fact, certain liquid mixtures behave as ideal solutions, but they obviously do not obey the ideal gas law. In a mixture forming a vapor phase and a liquid phase at equilibrium with each other, either one of the phases, or both phases, may approach ideal solution behavior. Ideal solution behavior is approached at low pressures and usually with mixtures of chemically similar components. Referring to Equation 1.24, if the vapor phase is assumed to behave as an ideal solution, $\hat{\phi}_i^V = 1$, and the left-hand side of the equation reduces to Y_iP, where P is the total pressure. Moreover, for ideal liquid solutions,

$$\phi_i^L = \frac{p_i^0}{P}$$

where p_i^0 is the vapor pressure of component i. Thus, if both phases are assumed to behave as ideal solutions, the following relationship, known as Raoult's law, holds:

$$Y_iP = X_ip_i^0 \tag{1.27}$$

The vapor–liquid distribution coefficient in an ideal solution is therefore

$$K_i = \frac{Y_i}{X_i} = \frac{p_i^0}{P}$$

Since p_i^0 is a function of temperature only, the vapor–liquid distribution coefficient, or K-value, is also a function of temperature only at constant total pressure. It bears a simple relationship to the total pressure and is independent of the composition.

Binary Ideal Solutions

Equation 1.27 is applied to a binary ideal solution to compute the partial pressure of each component and the total pressure:

$$p_1 = Y_1P = X_1p_1^0$$

$$p_2 = Y_2P = X_2p_2^0$$

$$P = p_1 + p_2 = X_1p_1^0 + \left(1 - X_1\right)p_2^0$$

$$= X_1\left(p_1^0 - p_2^0\right) + p_2^0$$

At constant temperature, p_1^0 and p_2^0 are constant; hence, for an ideal binary solution at constant temperature, the partial pressures and the total pressure are linear functions of the molar composition (X_1 or X_2). Example 1.1 in Section 1.3.4 illustrates Raoult's law behavior and deviations from it.

Henry's Law

In deriving Raoult's law (Equation 1.27), the implied assumption was that all components are condensable; that is, their critical temperature is higher than the mixture temperature. If a component's critical temperature is below the mixture temperature, it does not condense at that temperature as a pure component. There would be no vapor pressure for that component at the system temperature and Raoult's law would not apply. The component could nonetheless exist as a solute in the liquid phase. Henry's law states that the concentration of the gas component dissolved in the liquid is directly proportional to the partial pressure of that component in the vapor phase at equilibrium with the solution:

$$p_i = k_{Hi} X_i \tag{1.28}$$

where k_{Hi} is Henry's law constant. The constant is a function of the temperature and, to a lesser degree, of the pressure. Its value is specific to the particular solute and solvent.

Henry's law is mainly valid at low concentrations of the solute, typically below 1%, and is therefore applicable to low miscibility solutes. By extension, the solutes are not limited to noncondensables but could include condensables with low miscibility.

1.3.3 Phase Equilibrium in Nonideal Systems

For many mixtures, the interactions between molecules in the liquid phase are too strong to permit adequate representation by equations of state. Such mixtures generally involve components with chemically dissimilar molecules. In the vapor phase, intermolecular forces become more significant at higher pressures, but the vapor phase fugacity coefficient could still be adequately calculated from PVT behavior information such as equations of state. Vapor fugacity methods have been developed for special situations such as those where molecules associate in the vapor phase. The liquid phase deviation from ideal solution behavior is quantified by introducing the *liquid activity coefficient.*

Activity Coefficients

For a component i in the liquid, the activity coefficient γ_i, is defined as

$$\gamma_i = \frac{\hat{f}_i^L}{f_i^L X_i}$$

where $\hat{f}_i^L$ is the fugacity of component i in the liquid solution and f_i^L is the *standard state* fugacity of component i. The standard state is defined as pure liquid component i at system temperature and pressure. In an ideal liquid solution, $\gamma_i = 1$ and

$$\hat{f}_i^L = f_i^L X_i$$

In the general case, the definitions of fugacity coefficient and activity coefficient are combined with Equation 1.24 to give

$$Y_i \hat{\phi}_i^V P = \hat{f}_i^L = X_i \gamma_i f_i^L$$

The distribution coefficient is, therefore, given by the expression

$$K_i = \frac{\gamma_i f_i^L}{\hat{\phi}_i^V P} \tag{1.29}$$

The vapor phase fugacity coefficient, $\hat{\phi}_i^V$, may be calculated, as before, from Equation 1.23, using for instance an equation of state. The pure component fugacity in the liquid state is equal to its fugacity in the vapor at equilibrium with the liquid:

$$f_i^L \left(\text{at } p_i^0 \right) = f_i^V \left(\text{at } p_i^0 \right)$$

Since the system pressure is not, in general, equal to the vapor pressure, the effect of pressure on the fugacity of the liquid must be taken into account in calculating f_i^L. Using the basic definition of fugacity, f_i^L is calculated by carrying out the integration over two pressure steps: from 0 pressure to the vapor pressure and from the vapor pressure to the system pressure:

$$\ln f_i^L = \ln p_i^0 + \int_0^{p_i^0} \left(V_i^V - \frac{RT}{P} \right) dP + \int_{p_i^0}^{P} \frac{V_i^L}{RT} dP$$

The first integral involves the vapor volume, V_i^V, and may be evaluated using an equation of state. The second integral, known as the Poynting correction factor, takes into account the effect of pressure on the fugacity in the liquid. The volume in this integral, V_i^L, is the liquid molar volume and may be obtained from experimental data or from liquid density estimation methods (API, 1978; Rackett, 1970).

In order to calculate the distribution coefficient by Equation 1.29, the activity coefficient γ_i must be evaluated. Activated coefficients are generally determined from experimental data and correlated on the basis of thermodynamic phase equilibrium principles. The relationship most often used for this purpose is the Gibbs–Duhem equation (Equation 1.7). At constant temperature and pressure, this equation becomes

$$\sum n_i d\mu_i = 0$$

or, in terms of mole fractions,

$$\sum X_i d\mu_i = 0$$

The chemical potential is related to the fugacity by Equation 1.18, which may be written in differential form:

$$d\mu_i = RTd \ln \hat{f}_i = RTd \ln \left(f_i \gamma_i X_i\right)$$

The last expression is based on the definition of the activity coefficient. Substituting this expression in the Gibbs–Duhem equation at constant T and P gives the following:

$$\sum X_i RTd \ln \left(f_i \gamma_i X_i\right) = 0$$

or

$$\sum X_i d \ln \gamma_i + \sum dX_i = 0$$

Note that $d\ln f_i = 0$ at constant temperature and pressure. Since the total number of moles is constant, $\Sigma dX_i = 0$ and, hence,

$$\sum X_i d \ln \gamma_i = 0 \qquad\qquad (1.30)$$

Since Equation 1.30 was derived at constant temperature and pressure, it may be used to correlate vapor–liquid equilibrium data rigorously only if such data were obtained at constant temperature and pressure. A binary vapor–liquid equilibrium system can exist only at one composition at a fixed temperature and pressure since it has only 2 degrees of freedom. (Degrees of freedom are discussed in more detail in Chapter 2.) Nevertheless, binary data are those that are most commonly used for correlating activity coefficient data because they are the most commonly available. Moreover, multicomponent mixtures can be represented adequately in terms of binary coefficients.

If Equation 1.30 were to be applied to binary isothermal data, the term that represents the effect of composition on system pressure would be neglected. It may be shown (Walas, 1985) that the error incurred by neglecting this term is small since it is a fraction of the change in liquid volume due to mixing, which is usually a negligible quantity.

If, on the other hand, Equation 1.30 were to be applied to isobaric binary data, the term that represents the effect of composition on the temperature would be neglected. The error incurred by neglecting this term is related to the heat of mixing, which can be significant for components with widely differing boiling points or chemical structure.

It is, therefore, more accurate to correlate isothermal than isobaric binary vapor–liquid equilibrium data for predicting liquid activity coefficients on the basis of Equation 1.30.

Thermodynamic Consistency of VLE Data

Experimental vapor–liquid equilibrium data, namely equilibrium temperature, pressure, and vapor and liquid compositions, can be used to calculate the activity coefficients from Equation 1.29. The K-values are calculated from the composition data:

$$K_i = \frac{Y_i}{X_i}$$

The fugacities and fugacity coefficients, f_i^L, $\hat{\phi}_i^V$, are calculated from the compositions, temperature, and pressure, using, for instance, an equation of state and liquid density data as described earlier. The activity coefficients are then calculated by rearranging Equation 1.29:

$$\gamma_i = \frac{K_i \hat{\phi}_i^V P}{f_i^L}$$

An equation that satisfies the Gibbs–Duhem equation in the form of Equation 1.30 is then used to correlate, or fit, the experimentally determined activity coefficients. Normally, binary VLE data are correlated, generating binary interaction coefficients. The coefficients may then be used to reproduce the binary VLE data and also to predict multicomponent VLE behavior as described in the discussion of some of the activity coefficient equations later in this section.

Before the activity coefficients are represented with an equation, it is important to check the VLE data for thermodynamic consistency against Equation 1.30. As concluded earlier, the error introduced by applying Equation 1.30 to binary isothermal data is usually negligible. The consistency check is described for this type of data, which is the most commonly used for equation development. Equation 1.30 is written for a binary as

$$X_1 d\ln\gamma_1 + X_2 d\ln\gamma_2 = 0$$

By substituting $X_2 = 1 - X_1$ and rearranging, the following equation is obtained:

$$d\ln\gamma_2 = -X_1(d\ln\gamma_1 - d\ln\gamma_2)$$

This equation is combined with the expression for the derivative of a product,

$$d[X_1(\ln\gamma_1 - \ln\gamma_2)] = (\ln\gamma_1 - \ln\gamma_2)dX_1 + X_1(d\ln\gamma_1 - d\ln\gamma_2)$$

to give the equation

$$\ln \frac{\gamma_1}{\gamma_2} dX_1 = d \ln \gamma_2 + d\left[X_1\left(\ln \gamma_1 - \ln \gamma_2\right)\right]$$

This equation is now integrated between $X_1 = 0$ and $X_1 = 1$. The limit at $X_1 = 0$ corresponds to pure component 2 with activity coefficient $\gamma_2 = 1$. Similarly, at $X_1 = 1$, $\gamma_1 = 1$. Therefore, at $X_1 = 0$, $\ln \gamma_2 = 0$, and at $X_1 = 1$, $\ln \gamma_1 = 0$. The integration leads to the following:

$$\int_{X_1=0}^{X_1=1} \ln \frac{\gamma_1}{\gamma_2} dX_1 = 0 \tag{1.31}$$

The consistency test is carried out by plotting $\ln (\gamma_1/\gamma_2)$ vs. X_1 and graphically evaluating the integral in Equation 1.31. The curve consists of a positive part and a negative part above and below the line $\ln (\gamma_1/\gamma_2) = 0$. The data points are considered thermodynamically consistent, and the assumption of negligible effect of variable pressure is deemed valid if the areas above and below the line $\ln (\gamma_1/\gamma_2) = 0$ are equal.

The actual representation of activity coefficient data involves solving Equation 1.30. Several integrated forms of this equation for binary systems are in use, the most common of which are presented here.

The Margules Equation

Different forms of this equation have been proposed (Wohl, 1946). One of these is the three-suffix, or third-degree form:

$$\log \gamma_1 = \left(2A_{21} - A_{12}\right)X_2^2 + 2\left(A_{12} - A_{21}\right)X_2^3$$

$$\log \gamma_2 = \left(2A_{12} - A_{21}\right)X_1^2 + 2\left(A_{21} - A_{12}\right)X_1^3 \tag{1.32}$$

It can be proved that these equations are solutions to the Gibbs–Duhem equation by taking the differentials $d\log\gamma_1$ and $d\log\gamma_2$ and substituting them in Equation 1.30 written for a binary. It should also be recalled that $X_1 + X_2 = 1$ and that $dX_1 = -dX_2$.

The binary constants A_{12} and A_{21} are usually determined by regressing binary isothermal vapor–liquid equilibrium data. If these data are accurate in the vicinities around $X_1 = 0$ and $X_1 = 1$, the binary constants may be determined from the limiting conditions as follows:

$$\lim_{X_1 \to 0} \log \gamma_1 = A_{12}$$

$$\lim_{X_2 \to 0} \log \gamma_2 = A_{21}$$

The constants may, in fact, be calculated from a single equilibrium data point (vapor–liquid or liquid–liquid) although the results would be heavily weighted to that point, probably resulting in less accuracy for the remaining range of compositions.

The Margules equation is generalized in its four-suffix form to multicomponent mixtures as follows (Wohl, 1946):

$$\ln \gamma_i = \left(1 - X_i\right)^2 \left[A_i + 2X_i\left(B_i - A_i - D_i\right) + 3D_i X_i^2\right] \tag{1.33}$$

where

$$A_i = \sum_j X_j A_{ij} / \left(1 - X_i\right)$$

$$B_i = \sum_j X_j A_{ji} / \left(1 - X_i\right)$$

$$D_i = \sum_j X_j D_{ij} / \left(1 - X_i\right)$$

The binary interaction constants, A_{ij}, A_{ji}, and D_{ij} ($= D_{ji}$), are determined from binary phase equilibrium data.

Though one of the oldest activity coefficient equations, the Margules correlation is still commonly used for a wide range of nonideal systems. Its accuracy diminishes, however, as the molecules of a binary are more and more dissimilar in size or chemical structure.

The van Laar Equation

The binary form of this equation is as follows:

$$\ln \gamma_1 = A_{12}\left[\frac{A_{21}X_2}{A_{12}X_1 + A_{21}X_2}\right]^2$$

$$\ln \gamma_2 = A_{21}\left[\frac{A_{12}X_1}{A_{12}X_1 + A_{21}X_2}\right]^2 \tag{1.34}$$

This equation is also a solution to the Gibbs–Duhem equation. The constants may be calculated from infinite dilution activity coefficients or from a single vapor–liquid equilibrium data point.

The multicomponent van Laar equation is as follows (Carlson et al., 1942):

$$\ln \gamma_i = \sum_1 A_{i1}Z_1 - \sum_j A_{ij}Z_iZ_j - \frac{1}{2}\sum_{j \neq i}\sum_{k \neq i}\left(A_{jk} A_{ij} / A_{ji}\right)Z_jZ_k \tag{1.35}$$

where

$$Z_1 = \frac{X_1\, A_{1i}/A_{i1}}{\displaystyle\sum_j X_j\, A_{ji}/A_{ij}}$$

The binary constants A_{ij} and A_{ji} are determined from binary vapor–liquid equilibrium data.

The modified van Laar equation (Black, 1958) extends the applicability of the van Laar equation to many systems, including nonsymmetrical and hydrogen-bonding binaries. The equation, which is generalized to multicomponent mixtures, is expressed in terms of three binary interaction parameters: A_{ij}, A_{ji}, and C_{ij} ($= C_{ji}$). The parameters must be determined from binary phase equilibrium data. The modified van Laar equation reduces to the van Laar equation if $C_{ij} = 0$. Refer to the source (Black, 1958) for detailed mathematical formulation of the equation.

Other activity coefficient equations have been proposed since the earlier equations of Margules and van Laar. It is outside the scope of this book to present the detailed development of these equations, which may be found in the indicated references. The objective here is to present the basis for each of the more commonly used equations, their features, and applicability.

The Wilson Equation

This equation (Wilson, 1964) was developed on the basis of a liquid molecular model which assumes that interactions between molecules of two different components depend on the volume fraction of each component in the vicinity of a molecule of a given component. The volume fractions are determined from the probability of finding a molecule of one component or another in the vicinity of a molecule of the given component. The probabilities, and hence the volume fractions, are expressed through Boltzmann distribution functions of energy in terms of binary interaction parameters. The result for volume fractions is

$$Z_i = X_i/(X_i + A_{ij}X_j)$$

Parameters A_{ij} are temperature dependent, given as

$$A_{ij} = \frac{V_j}{V_i}\exp\!\left[-\left(\lambda_{ij} - \lambda_{ii}\right)/RT\right]$$

where V_i and V_j are molar volumes and $(\lambda_{ij} - \lambda_{ii})$ are energy terms and are binary constants in the Wilson equation. The activity coefficients are related to volume fractions rather than directly to mole fractions. The expression for a binary is

$$\ln\gamma_i = -\ln\,(X_i + A_{ij}X_j) + \beta_{ij}X_j \qquad (1.36)$$

where

$$\beta_{ij} = \frac{A_{ij}}{X_i + A_{ij}X_j} - \frac{A_{ji}}{X_j + A_{ji}X_i}$$

The equation is generalized to multicomponent mixtures in terms of binary parameters only, either A_{ij} and A_{ji} or $(\lambda_{ij} - \lambda_{ii})$ and $(\lambda_{ji} - \lambda_{jj})$. Either set of parameters must be determined to fit binary vapor–liquid equilibrium data. Using the parameters $(\lambda_{ij} - \lambda_{ii})$ and $(\lambda_{ji} - \lambda_{jj})$ allows for a certain degree of temperature dependence through the equations that relate A_{ij} to $(\lambda_{ij} - \lambda_{ii})$.

The Wilson equation is capable of representing both polar and nonpolar molecules in multicomponent mixtures using only binary parameters. It cannot, however, represent liquid–liquid equilibrium systems.

The NRTL (Renon) Equation

The model for the NRTL (nonrandom two-liquid) equation (Renon et al., 1968) is similar to the Wilson model but includes a nonrandomness constant, α_{ij} $(= \alpha_{ji})$, that is characteristic of the types of components in each binary. In addition to this constant, the equation, which is generalized to multicomponent mixtures, utilizes four interaction parameters for each binary: a_{ij}, a_{ji}, b_{ij}, and b_{ji}. Parameters b_{ij} and b_{ji} include a temperature dependency similar to the Wilson coefficients. Parameters a_{ij} and a_{ji} may be added to improve the ability to represent the effect of temperature. The equation may thus be used either in its three-parameter form or in its five-parameter form.

All parameters are determined by regression of binary phase equilibrium data although α_{ij} is usually fixed at some estimated value between 0.2 and 0.5.

The NRTL equation is one of the more successful equations for representing phase equilibrium data, including liquid–liquid equilibrium. It is applicable to multicomponent mixtures, which may include nonsymmetrical binaries. It also has built-in temperature dependency over moderate ranges.

The UNIQUAC Equation

The UNIQUAC (universal quasichemical) equation (Abrams et al., 1975) is based on the two-liquid model in which the excess Gibbs energy is assumed to result from differences in molecular sizes and structures and from the energy of interaction between the molecules.

The equation, which is generalized to multicomponent mixtures, requires pure component data for the van der Waals area and volume parameters, q_i and r_i. Additionally, binary interaction parameters $(u_{ji} - u_{ii})$ and $(u_{ij} - u_{jj})$ are also required and are generally determined from binary phase equilibrium data. Temperature dependency is incorporated in the equation, similar to the Wilson and NRTL equations. The UNIQUAC equation is applicable to many classes of components, including mixtures containing considerably dissimilar molecules, and is also applicable to liquid–liquid equilibrium systems. It can represent temperature dependency over moderate ranges but is not necessarily more accurate than simpler equations in spite of its theoretical foundation.

The UNIQUAC equation is the basis for the development of the group contribution method, the UNIFAC equation, which predicts liquid activity coefficients from component structures on the basis of interactions between chemical functional groups.

1.3.4 Vapor—Liquid Equilibria—Applications

Actual phase equilibrium behavior is best described by examining phase diagrams for binary systems. Since it is generally not feasible to represent multicomponent mixtures graphically, phase diagrams of the constituent binaries may be used as an aid in understanding multicomponent phase behavior. The basis for this is the fact that activity coefficients in multicomponent mixtures can be related to activity coefficients in binaries. Ternary diagrams are commonly used for describing immiscible liquids (Section 1.3.5).

Binary vapor–liquid equilibrium data may be represented by various types of diagrams such as pressure–composition, activity coefficient–composition, and vapor composition–liquid composition plots. Curves representing nonideal systems can have different characteristics, as shown in the examples that follow. The representation of phase behavior may be based on direct experimental data or on predictive calculations using equations such as those presented in the preceding sections. These equations have their roots in thermodynamic principles but also rely on experimental data to determine the various constants. The calculations require evaluating the different variables in Equation 1.29. These are highly complex and tedious calculations but can be performed rapidly and accurately using computer programs. The phase diagrams for the following examples are based on computer simulation. Some of the activity coefficient parameters used in the calculations are given in Table 1.1.

Assuming the vapor phase does not deviate too strongly from ideal solution behavior, the phase equilibrium characteristics are primarily controlled by the liquid activity coefficient's behavior. Under these conditions, the only term that is composition-dependent in Equation 1.29 is the activity coefficient, γ_i.

Example 1.1 Pentane–Hexane and Methanol–Water

The phase behaviors of two binaries, pentane–hexane and methanol–water, are checked against ideal solution behavior. The partial pressure and total pressure isotherms calculated according to Raoult's law at 100°F are compared to expected actual isotherms. The component vapor pressures at 100°F are as follows:

Pentane	15.6 psia	Methanol	4.6 psia
Hexane	5.0 psia	Water	0.9 psia

Table 1.1   Selected Activity Coefficient Binary Interaction

Diethyl Ether (1) – Ethanol (2) at 25°C

	A_{12}	A_{21}
Margules	1.0535	0.9679
Wilson	47.5508	646.2295
Uniquac	362.5405	-41.9356

Ethanol (1) – Benzene (2) at 45°C

	A_{12}	A_{21}
Margules	2.1412	1.3856
Wilson	1594.1685	142.1319
Uniquac	-147.1802	876.8192

Methyl Acetate (1) – Chloroform (2) at 50°C

	A_{12}	A_{21}
Margules	-0.8695	-0.5955
Wilson	19.0577	-422.5855
Uniquac	1349.3593	-740.5781

Water (1) – 3-Hydroxy-2-Butanone (2) at 750 MMHG

	A_{12}	A_{21}
Margules	-0.2977	1.5637
Wilson	642.9837	1476.1815
Uniquac	1018.5457	-569.7499

From Gmehling, J. and U. Onken, *Vapor–Liquid Equilibrium Data Collection*, Dechema Chemistry Data Series, Port Washington, NY, Scholium International, Inc., 1977. With permission.

The isotherms are plotted on pressure–composition diagrams (Figures 1.3A and 1.3B). The curves calculated by Raoult's law are the broken straight lines. They are obtained simply by joining the pure-component vapor pressure points to the zero pressure point on the opposite side to get the partial pressure curves and by joining the two vapor pressure points to get the total pressure curve. The solid lines are obtained by simulation and represent expected behavior of the mixtures. The pentane–hexane binary is predicted by the Soave equation of state, and the methanol–water binary is predicted from activity coefficients calculated by the NRTL equation.

The deviation of the pentane–hexane binary from Raoult's law is negligible because the two components are nonpolar and chemically similar and the system pressure is quite low. This is not the case with the methanol–water binary, where the deviations from Raoult's law are significant.

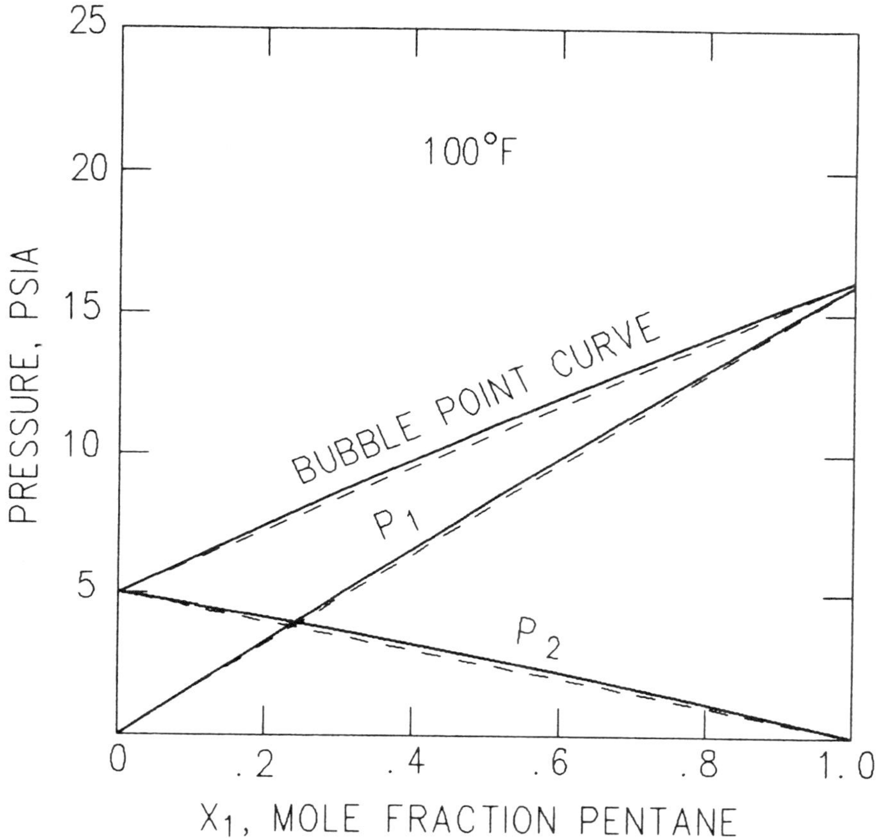

Figure 1.3A Pressure–composition isotherms for pentane–hexane (Example 1.1).

The deviation from ideality for the methanol–water system is considered positive because the actual pressure curves are higher than those calculated by Raoult's law. There are systems for which the deviation is negative; that is, actual pressures are lower than ideally calculated pressures.

Example 1.2 Diethyl Ether–Ethanol

Over the entire composition range in this binary at 25°C, the partial pressures of the individual components are greater than those predicted by Raoult's law, as is the total pressure (Figure 1.4). That is, the system exhibits positive deviation from ideality. The activity coefficients are consequently greater than unity for both components over the entire composition range (Figure 1.5). Moreover, the lower boiling component, diethyl ether, has the higher concentration in the vapor phase over the entire composition range (Figure 1.6).

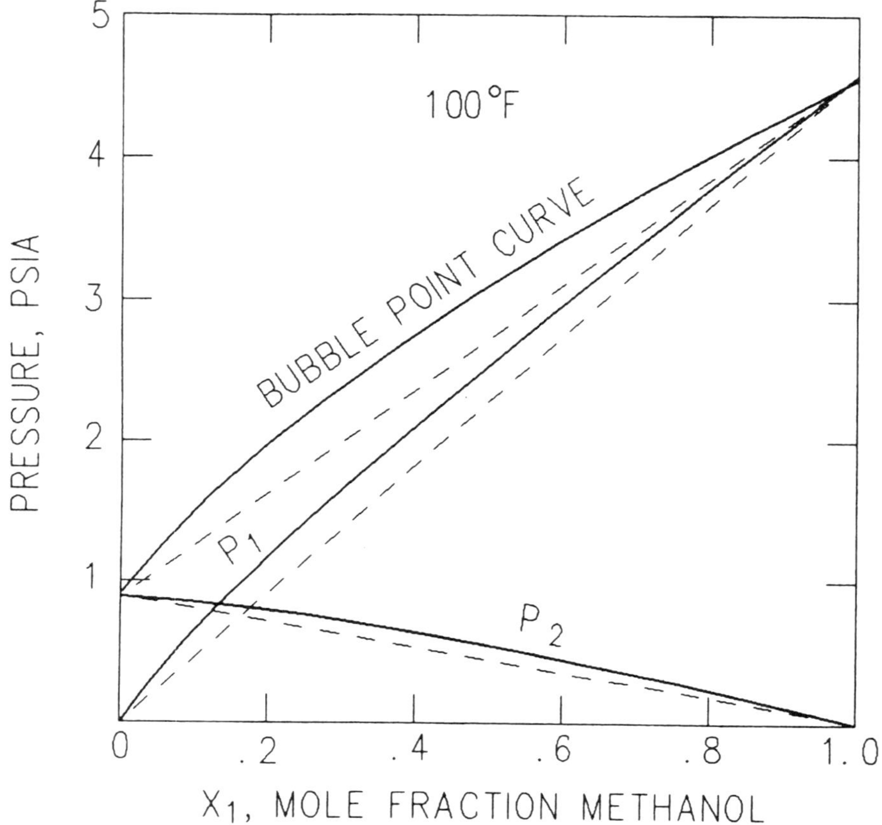

Figure 1.3B Pressure–composition isotherms for methanol–water (Example 1.1).

Example 1.3 Ethanol–Benzene

Here, too, the deviations from Raoult's law are positive for both components over the entire composition range at 45°C (Figures 1.7 and 1.8). The partial pressure deviations are large enough to result in a maximum in the total pressure (Figure 1.7). The binary thus forms a maximum pressure, or minimum temperature, *azeotrope*, a phenomenon discussed in more detail later in this section. In this binary, the concentration of ethanol is higher in the vapor than in the liquid below the azeotrope and is lower in the vapor than in the liquid above the azeotrope. At the azeotropic point, the concentrations in the liquid and vapor are identical (Figure 1.9).

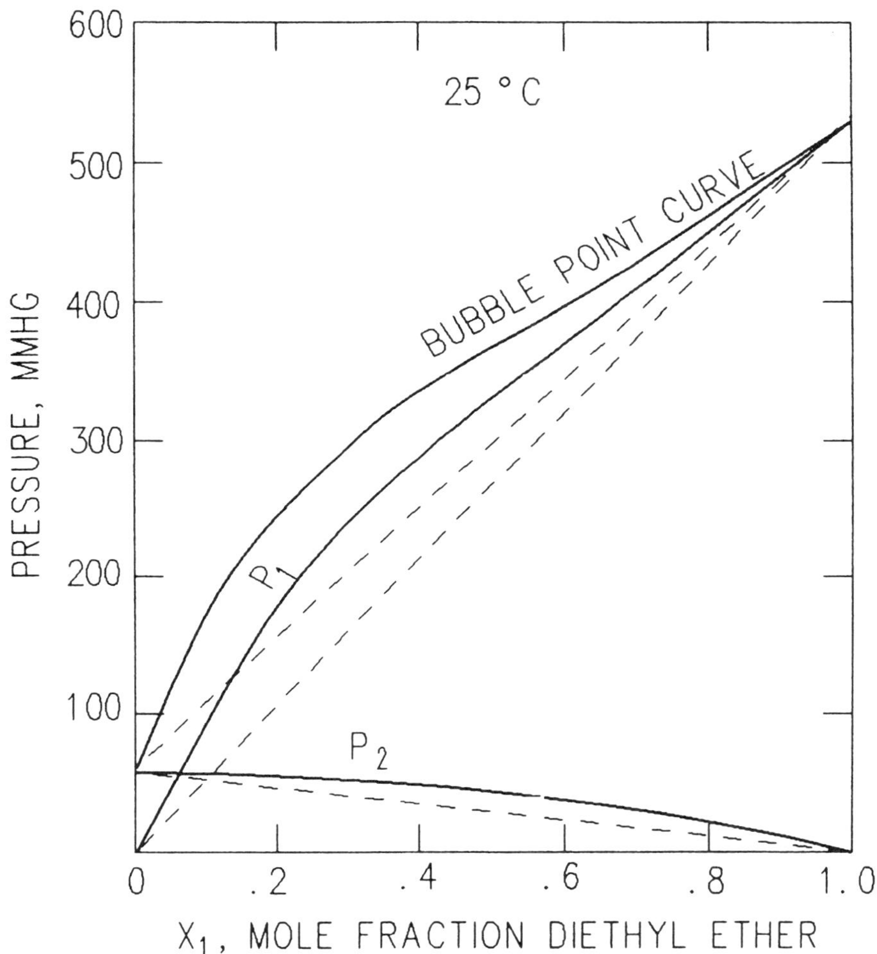

Figure 1.4   Pressure–composition diagram for diethyl ether–ethanol (Example 1.2).

Example 1.4 Methyl Acetate–Chloroform

This system is characterized by negative deviations from Raoult's law over the entire composition range at 50°C (Figures 1.10 and 1.11). The binary forms a minimum pressure, or maximum temperature, azeotrope (Figures 1.10 and 1.12).

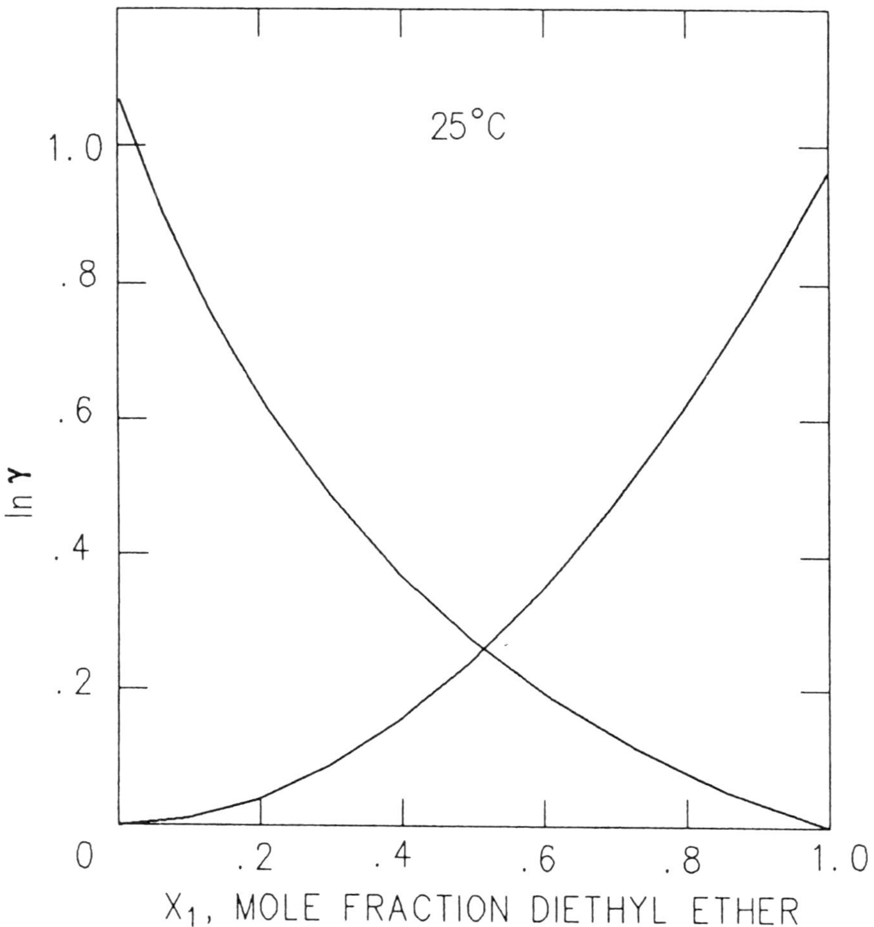

Figure 1.5   Activity coefficients vs. composition for diethyl ether–ethanol (Example 1.2).

Example 1.5 Water–3-Hydroxy-2-Butanone

At 120°C, both components in this binary indicate positive deviation in one composition range and negative deviation in another (Figure 1.13). The system also indicates a slight maximum pressure azeotrope near a water mole fraction of 0.95 (Figures 1.13 and 1.15). Additionally, a maximum is observed for the activity coefficient of water and a minimum for the activity coefficient of 3-hydroxy-2-butanone (Figure 1.14). These phenomena are best represented by the Margules equation.

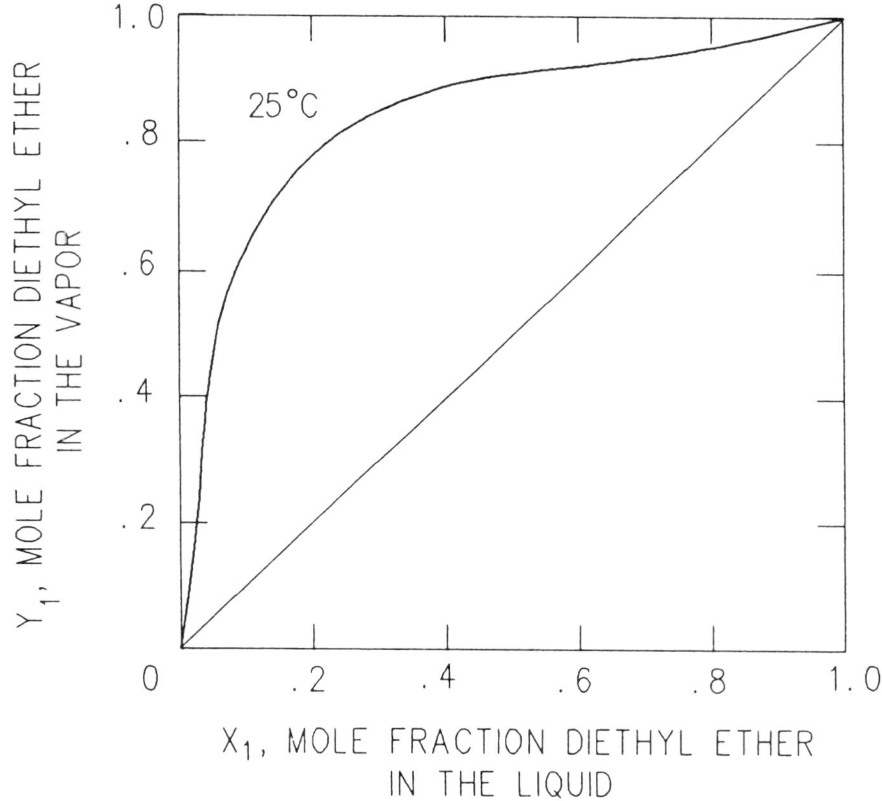

Figure 1.6   Y–X Diagram for diethyl ether–ethanol (Example 1.2).

Azeotropes

The above examples show that, for certain systems, deviations from Raoult's law can cause a maximum or a minimum in the vapor pressure to appear at a certain temperature and composition. At constant pressure, the boiling point or bubble point temperature curve could have a maximum or a minimum. Liquid mixtures whose vapor pressure curve or surface exhibits a maximum or a minimum are said to form azeotropes. The composition at which the azeotrope occurs is the azeotropic composition. Binaries are likely to form azeotropes if they deviate from Raoult's law and if their boiling points are not too far apart (within about 45°F). Azeotropes caused by positive deviations from Raoult's law are minimum boiling, i.e., the azeotrope boils at a temperature below the boiling point of either pure component.

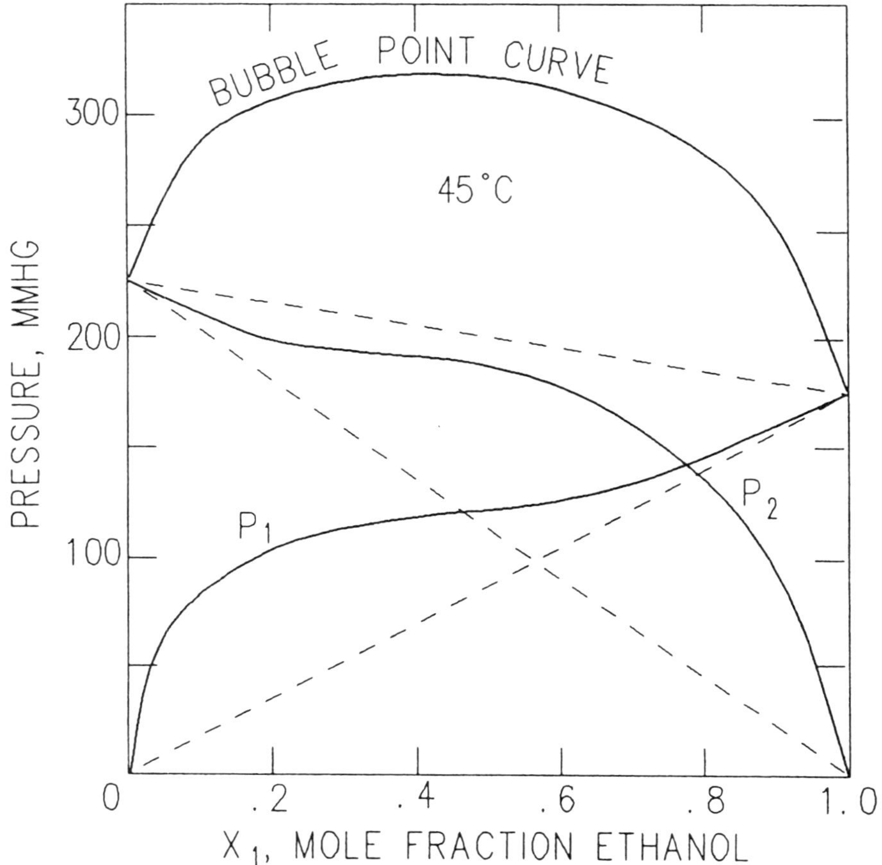

Figure 1.7   Pressure–composition diagram for ethanol–benzene (Example 1.3).

Conversely, azeotropes caused by negative deviations from Raoult's law are maximum boiling.

A binary azeotrope may be described on a vapor–liquid composition plot such as Figures 1.9, 1.12, and 1.15. On each side of the azeotropic point, a different component appears more volatile, having a higher concentration in the vapor than in the liquid. At the azeotropic point, the compositions of the two phases are identical.

The existence of an azeotrope in a given binary may be predicted on the basis of the chemical nature of the components, their boiling points, etc. A quantitative prediction of the azeotropic composition, and the temperature and pressure at which it occurs, can be predicted with an appropriate activity coefficient equation.

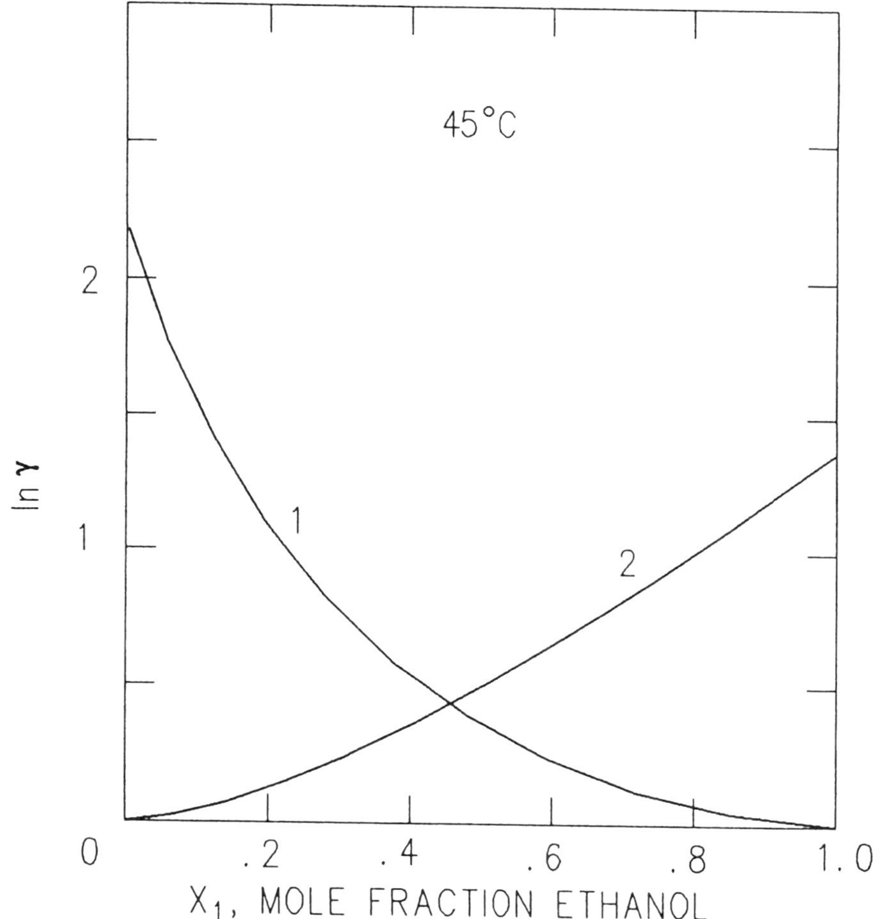

Figure 1.8   Activity coefficients vs. composition for ethanol–benzene (Example 1.3).

The interaction parameters of such an equation would have to be determined from experimental data in the first place. However, once the parameters are available, the effect on azeotropic composition of the temperature and pressure, within reasonable ranges, may be calculated from the activity coefficient equation and Equation 1.29.

The separation of mixtures that form azeotropes often requires the addition of a third component called a solvent or an entrainer. These processes, described in Chapters 2 and 10, may result in the formation of *multicomponent azeotropes* or *heterogeneous azeotropes*. Multicomponent azeotropes, usually ternary, are mix-

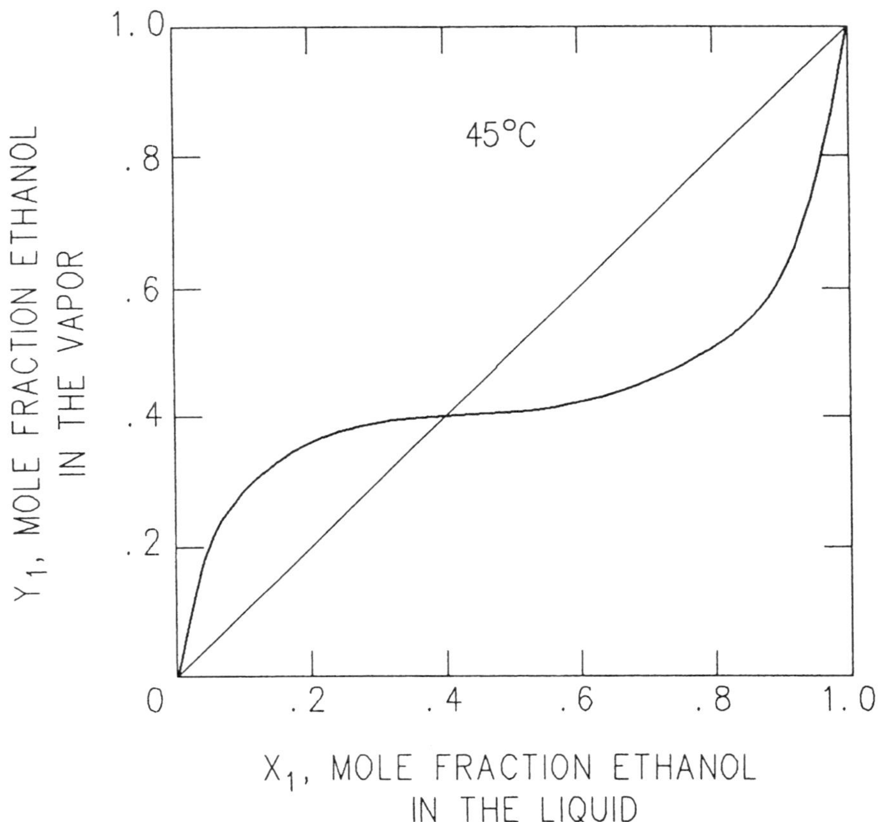

Figure 1.9   Y–X Diagram for ethanol–benzene (Example 1.3).

tures whose equilibrium vapor and liquid phases have the same composition. An activity coefficient equation that can predict binary azeotropes could, in its generalized multicomponent form, predict ternary azeotropes. Heterogeneous azeotropes are those forming two equilibrium liquid phases and are discussed in the next section.

1.3.5 Liquid–Liquid and Vapor–Liquid–Liquid Equilibria

The condition for equilibrium between vapor and liquid phases, expressed by Equation 1.19 as the equality of each component fugacity in the two phases, applies to equilibrium between two liquid phases or any number of phases such as two

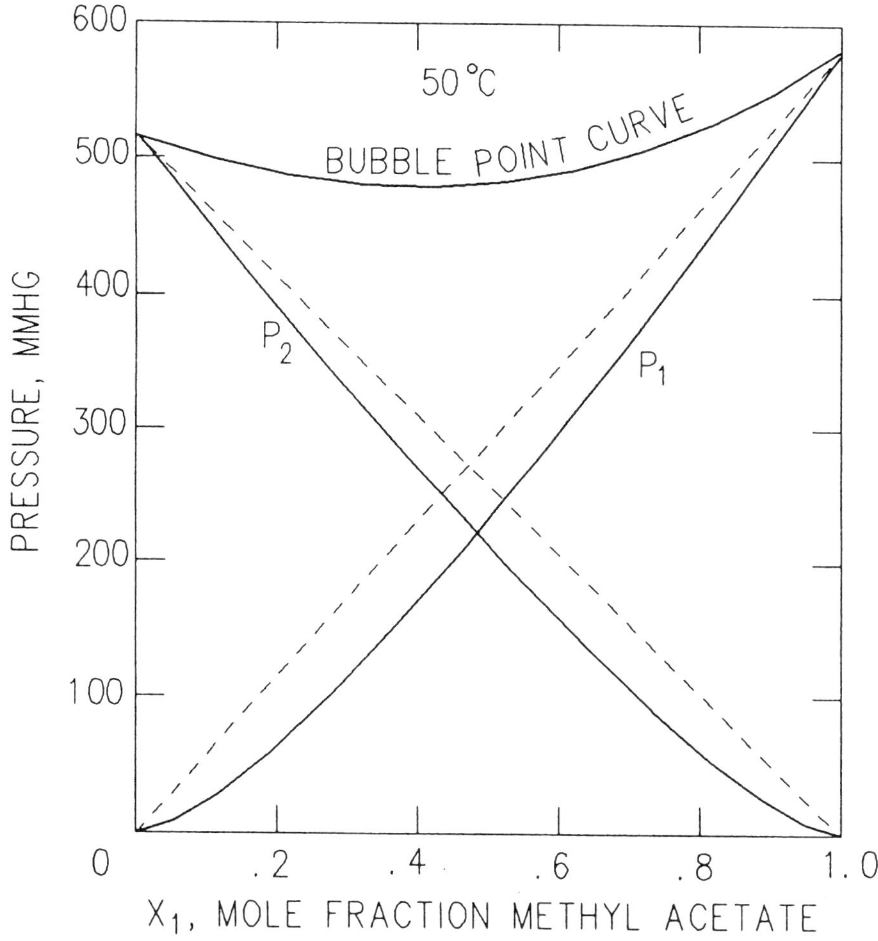

Figure 1.10 Pressure–composition diagram for methyl acetate–chloroform (Example 1.4).

liquids and a vapor. When the deviations from ideality are large enough, mixtures can form two immiscible liquids at equilibrium with each other. It is easy to see that an ideal solution cannot form two liquid phases at equilibrium. In order for this to occur, the condition $(p_i^0 X_i)^\alpha = (p_i^0 X_i)^\beta$, where α and β designate each liquid phase, must be satisfied. p_i^0 is the vapor pressure of component i at system temperature; therefore $(p_i^0)^\alpha = (p_i^0)^\beta$. It follows that $(X_i)^\alpha = (X_i)^\beta$; that is, the two liquid phases are identical.

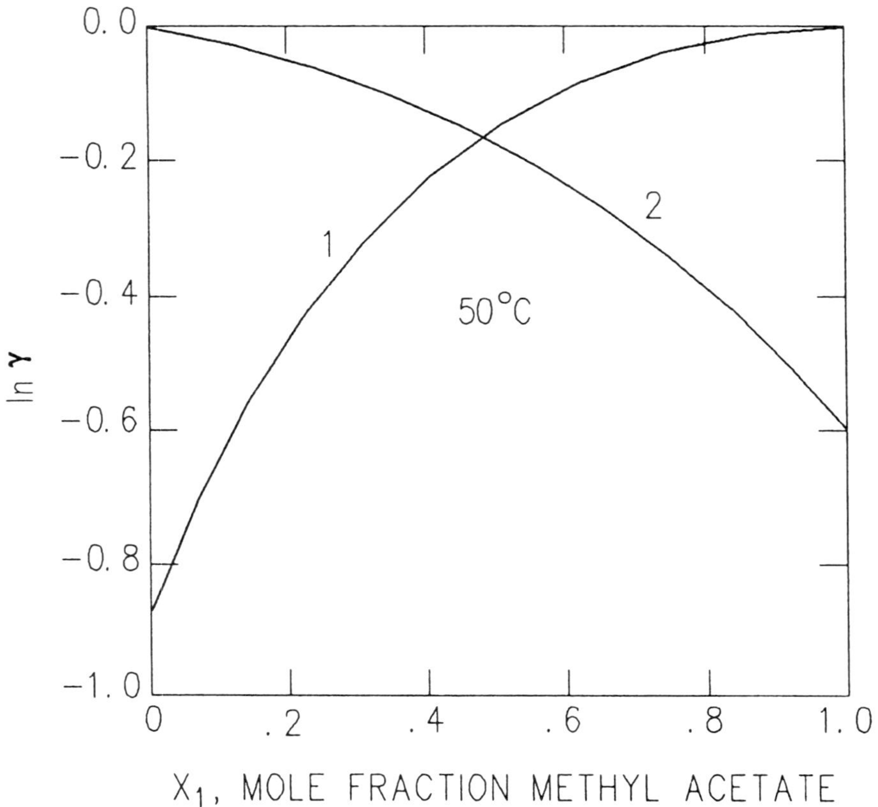

Figure 1.11 Activity coefficients vs. composition for methyl acetate–chloroform (Example 1.4).

Binary Systems

A nonideal binary mixture may exist as a single liquid phase at certain compositions, temperatures, and pressures, or as two liquid phases at other conditions. Also, depending on the conditions, a vapor phase may or may not exist at equilibrium with the liquid. When two immiscible liquid phases coexist at equilibrium, their compositions are different, but the component fugacities are equal in both phases.

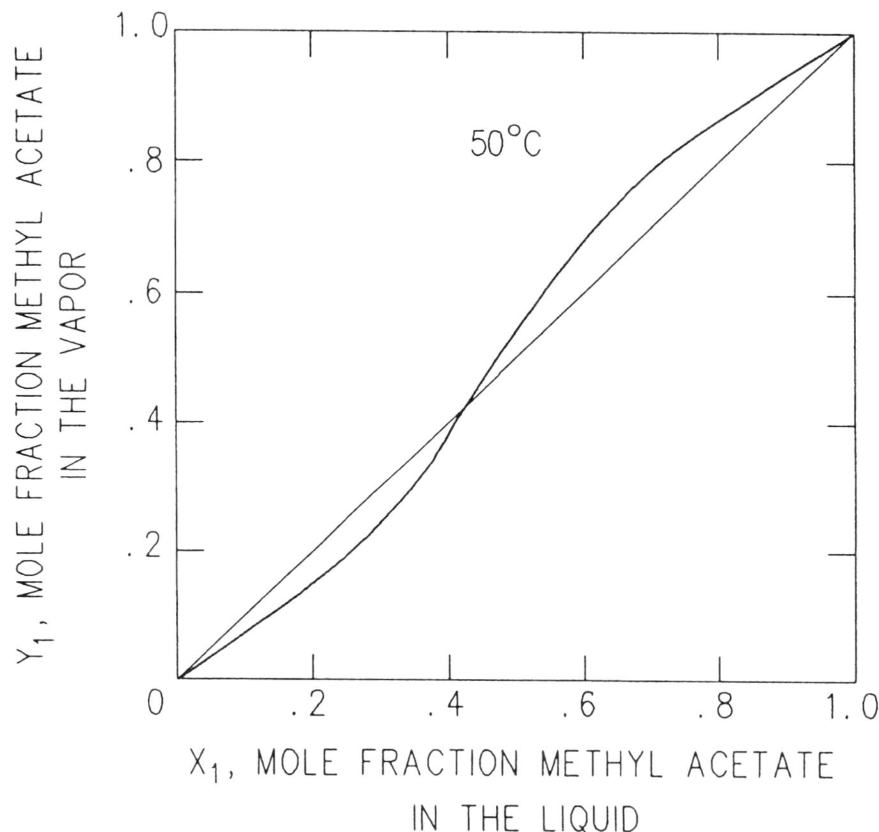

Figure 1.12 Y–X diagram for methyl acetate–chloroform (Example 1.4).

Equilibrium compositions of liquid phases at equilibrium are calculated by equating the component fugacities, similar to vapor–liquid equilibrium calculations, described in more detail in Chapter 2. The activity coefficients may be calculated by equations presented in Section 1.3.3, in particular the UNIQUAC and NRTL equations. The composition dependence of these equations is developed to the point where the same equation with the same constants can predict activity coefficients

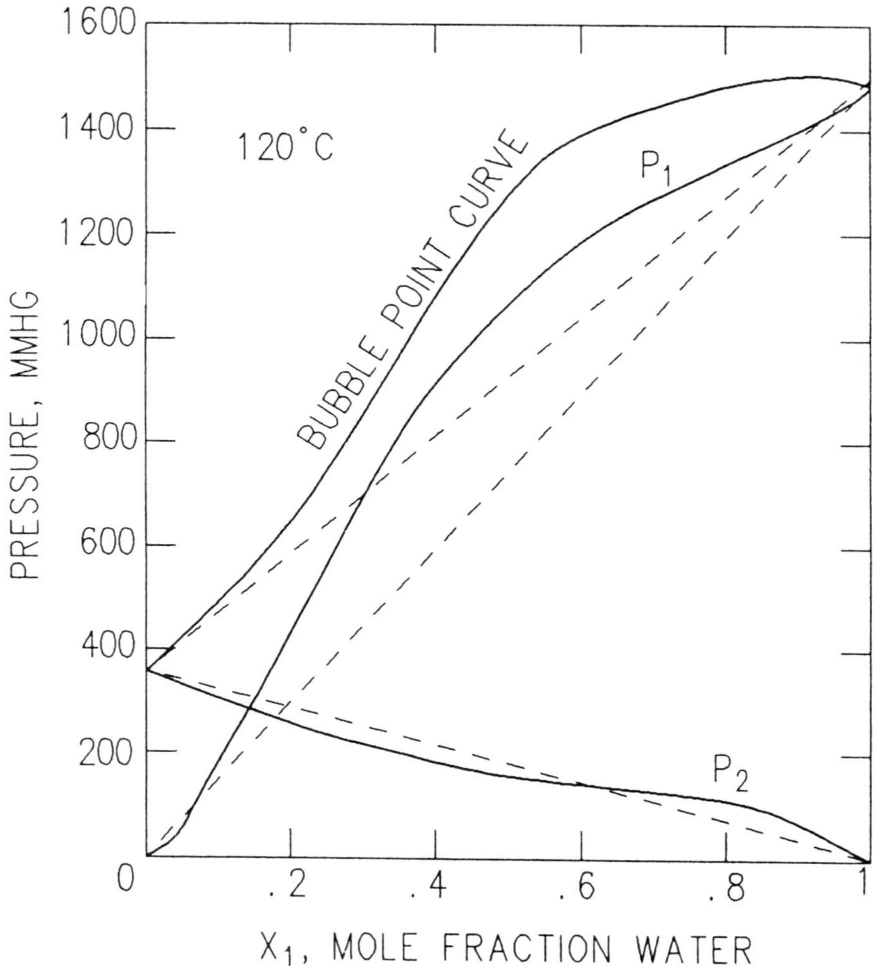

Figure 1.13 Pressure–composition diagram for water-3-hydroxy-2-butanone (Example 1.5).

over wide ranges of composition, thus allowing it to predict two immiscible liquid phases at equilibrium.

Example 1.6 N-Butanol–Water

Typical of nonideal binaries forming two liquid phases is the n-butanol–water system at 1 atmospheric pressure. A solution with approximately 2 mole % n-butanol

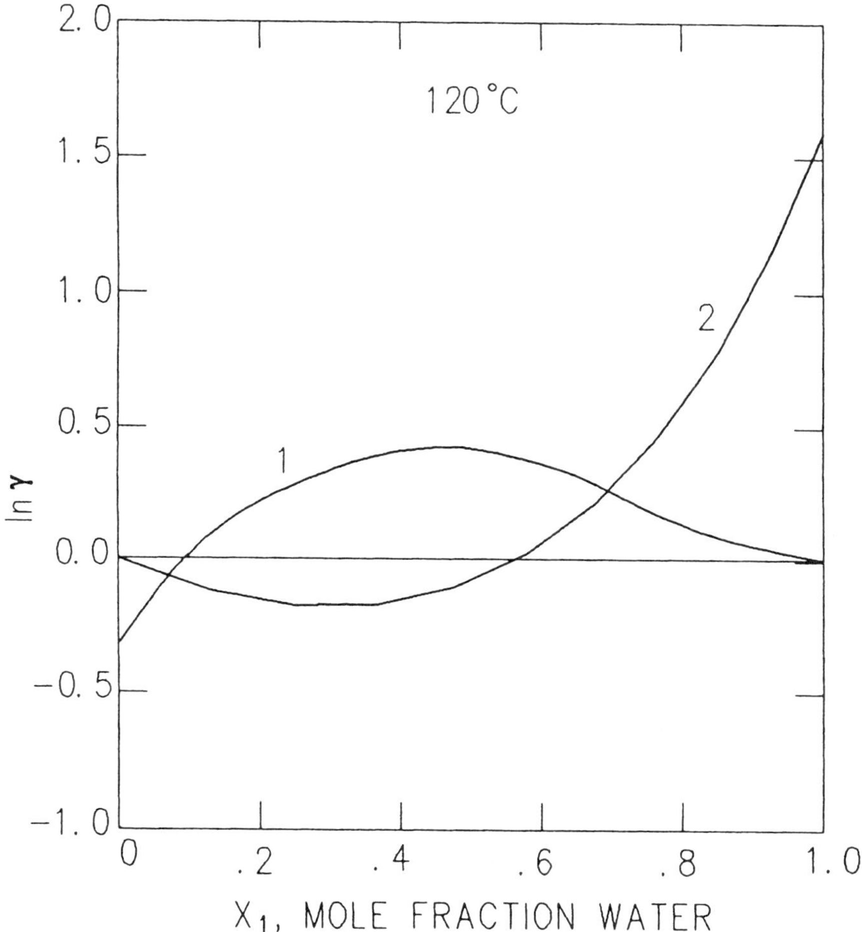

Figure 1.14 Activity coefficients vs. composition for water-3-hydroxy-2-butanone (Example 1.5).

in water exists at equilibrium with another liquid phase with approximately 38 mole % n-butanol in water. The fugacity of n-butanol in both phases is about 0.48. A phase diagram of this binary is illustrated in Figure 1.16. The curves, which closely match experimental data, are based on calculations using the NRTL equation for activity coefficients.

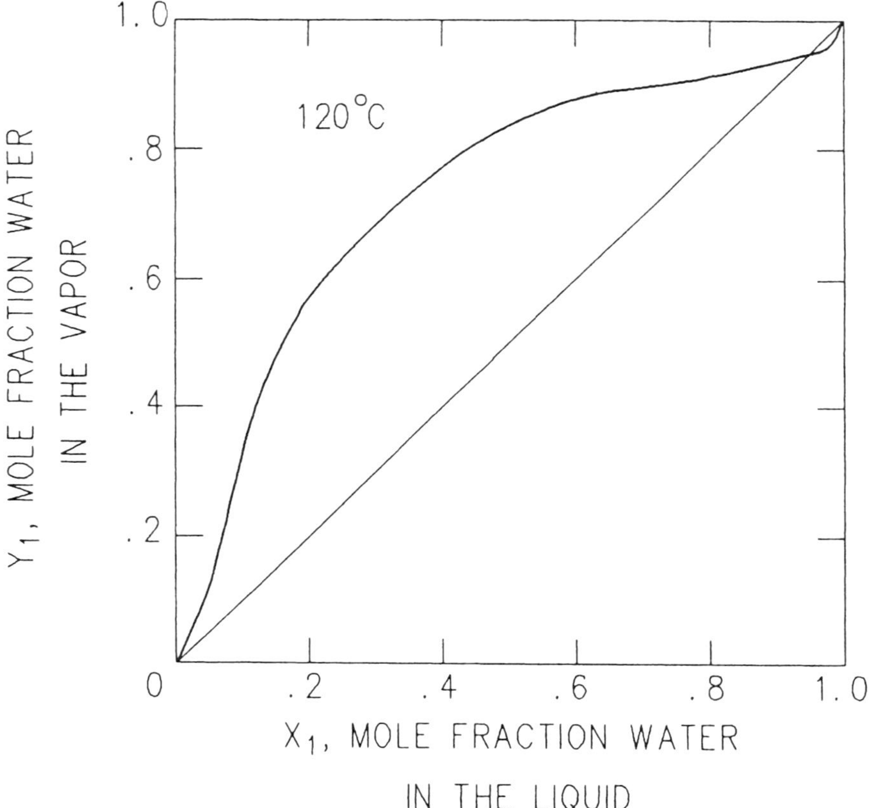

Figure 1.15 Y–X diagram for water-3-hydroxy-2-butanone (Example 1.5).

The liquids in the immiscible region boil at a constant temperature of about 200°F. As long as two liquid phases coexist at equilibrium, the boiling point and the equilibrium vapor composition are constant. This is a heterogeneous azeotrope: the mixture has a minimum boiling point and the vapor is at equilibrium with two liquid phases. The appearance of a second liquid phase is a phenomenon that is used to effect separation in certain processes, described in Chapter 10, by splitting the phases and processing each liquid phase separately.

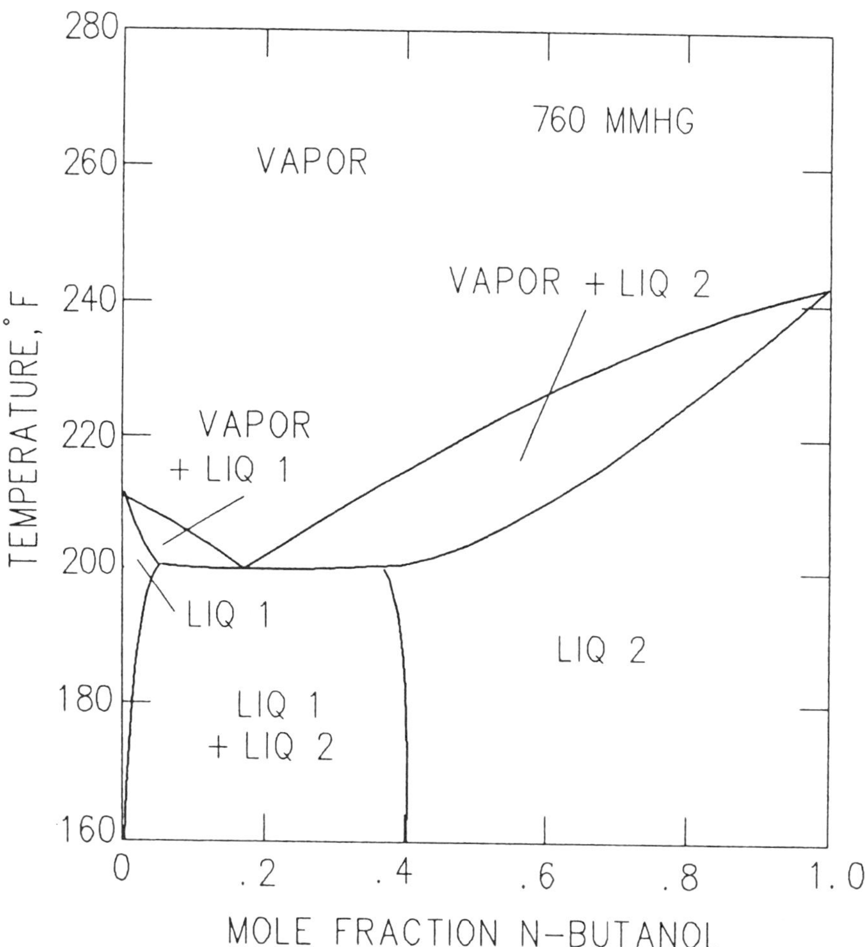

Figure 1.16 Temperature–composition for n-butanol–water (Example 1.6).

Ternary Systems

The phase compositions of an equilibrium binary are determined if the temperature and pressure are given. In the n-butanol–water binary, for instance, the liquid phase compositions are given by the upper curves of the liquid 1 and liquid 2 regions

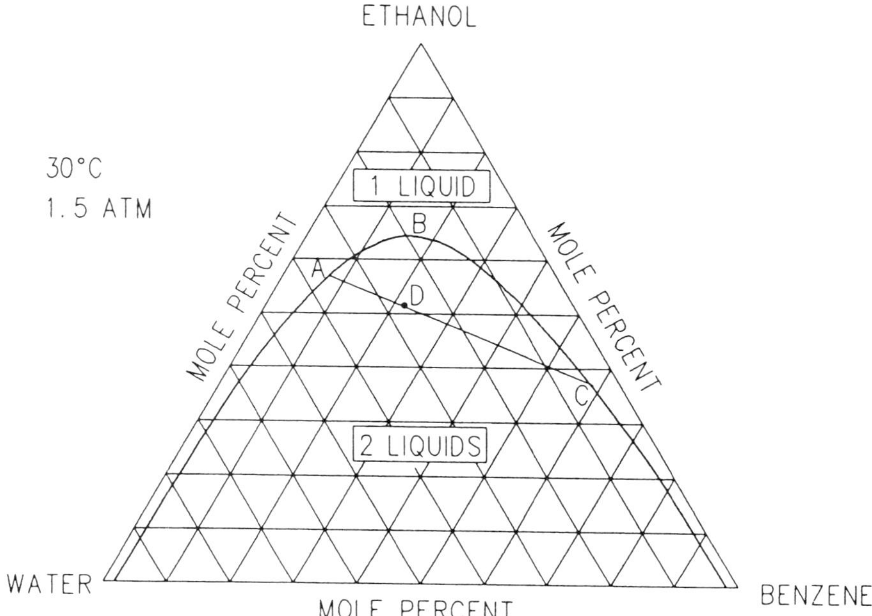

Figure 1.17 The ethanol–benzene–water system (Example 1.7).

in Figure 1.16. In ternaries or multicomponent systems, the overall composition, as well as the temperature and pressure, must be given in order to determine the phase compositions. The difference between the two cases is a result of the phase rule, discussed in Chapter 2. A convenient means for graphically describing ternary systems forming two liquid phases is the triangular diagram such as the one described in the following example.

Example 1.7 Ethanol–Benzene–Water

Figure 1.17 depicts a triangular diagram plotted for the ethanol–benzene–water ternary at a temperature of 30°C and a pressure of 1.5 atm. The plots are based on calculations using the NRTL activity coefficient equation. Each of the vertices of the triangle represents a pure component, and the bases of the triangle represent binaries of the other two components. The mole percent of a component at any point in the triangle is measured by its distance from the triangle base opposite the vertex of that component.

The ethanol–benzene–water diagram has a single-liquid phase region and a two-liquid phase region, separated by the equilibrium curve ABC (Figure 1.17). If the composition of a single-liquid phase corresponding to point A is slightly altered such that the equilibrium curve is crossed to the two-liquid phase region, a second liquid phase is formed with a composition corresponding to point C. The tieline AC is the locus of points representing equilibrium liquid phases with compositions corresponding to points A and C. Thus, point D on the tieline represents two liquid phases with compositions given by points A and C. The relative quantity of the phase with composition A to the phase with composition C equals the ratio of segments CD to DA, according to the *lever rule.*

Since ternary diagrams represent one temperature and one pressure, a set of diagrams would be required to describe the effects of temperature and pressure, as well as the effect of composition. A three-dimensional diagram could be constructed to describe the phase compositions–temperature relationship at a fixed pressure.

1.4 ENTHALPY

The calculation of multistage separation processes involves the solution of phase equilibrium relationships, mass balances, and energy balances. Energy balances require the computation of enthalpies of streams entering and leaving an equilibrium stage. The enthalpy is a function of state, defined in Section 1.1.2 as $H = U + PV$. It is a function of the stream composition, its temperature, and its pressure.

Energy balance calculations require evaluation of enthalpy changes rather than absolute values of the enthalpy. The variation of enthalpy with temperature and pressure is given by the equation

$$dH = \left(\frac{\partial H}{\partial T}\right)_P dT + \left(\frac{\partial H}{\partial P}\right)_T dP = C_P dT + \left(\frac{\partial H}{\partial P}\right)_T dP \tag{1.37}$$

where C_P is the heat capacity at constant pressure. The enthalpy change resulting from a transition from T_1 and P_1 to T_2 and P_2 may be represented by three steps: an isothermal process from T_1 and P_1 to T_1 and P_0, where P_0 is a low pressure such as atmospheric; an isobaric process from T_1 and P_0 to T_2 and P_0; and an isothermal process from T_2 and P_0 to T_2 and P_2.

In the low-pressure isobaric step, ideal gas behavior is assumed and ideal gas heat capacities are calculated from experimental pure-component data. The mixture low-pressure enthalpy change is calculated as

$$H_2^0 - H_1^0 = \sum_i X_i \left(H_{i2}^0 - H_{i1}^0\right)$$

$$= \sum_i X_i \int_{T_1}^{T_2} C_{Pi}^0 dT \tag{1.38}$$

where H_{i1}^0 and H_{i2}^0 are pure-component ideal gas enthalpies and C_{Pi}^0 is pure-component ideal gas heat capacity. Component ideal gas heat capacities are measured relative to some standard reference point, such as 0°C.

The isothermal steps represent the effect of pressure on the enthalpy. The enthalpy difference between the real state and ideal gas at the same temperature is the enthalpy departure and is generally calculated from PVT data, either experimental or predicted by an equation of state. From Equation 1.37, the enthalpy departure, or the variation of enthalpy from 0 pressure to system pressure at constant temperature, is given by

$$\Delta H = \int_0^P \left(\frac{\partial H}{\partial P} \right)_T dP$$

From the definition of enthalpy (Section 1.1.2), a differential enthalpy change is given by

$$dH = dU + PdV + VdP$$

Substituting the expression for dU from Equation 1.3 gives

$$dH = TdS + VdP$$

At constant temperature,

$$\left(\frac{\partial H}{\partial P} \right)_T = T \left(\frac{\partial S}{\partial P} \right)_T + V$$

Using the Maxwell relation,

$$\left(\frac{\partial S}{\partial P} \right)_T = -\left(\frac{\partial V}{\partial T} \right)_P$$

the expression for the enthalpy departure is written as

$$\Delta H = \int_0^P \left[V - T \left(\frac{\partial V}{\partial T} \right)_P \right] dP$$

$$= PV - RT - \int_\infty^V \left[P - T \left(\frac{\partial P}{\partial T} \right)_V \right] dV$$

(1.39)

The second form of the equation may be derived by starting with the differential enthalpy written as

$$dH = d(PV) + dU = d(PV) + TdS - PdV$$

Equation 1.39 is general and applies to mixtures as well as pure substances in either the vapor or liquid phase. To use the equation, however, accurate PVT data for the system are required. These data may either be experimental or predicted such as by an equation of state. If the PVT method is capable of representing mixtures and the effect of composition, then the prediction of the enthalpy of mixing is inherent in Equation 1.39. Enthalpy changes accompanying phase changes can also be calculated if PVT data are available for both phases or if the equation of state is applicable to both phases. Equations of state, such as the Soave or Peng–Robinson equations, have been used for calculating enthalpy departures for both vapor and liquid phases. These equations are explicit in the pressure and may be conveniently incorporated in the second form of Equation 1.39, as detailed in the original equation of state references. The first form of Equation 1.39 may be used with equations of state that are explicit in the volume.

For liquid solutions that cannot be adequately represented by equations of state or other PVT predictive methods, the enthalpy of a mixture may be calculated by adding the excess enthalpy term to the enthalpy of an ideal solution:

$$H = \sum_i \overline{H}_i X_i + H^{ex}$$

In this equation, $\overline{H}_i$ is the enthalpy of pure component i at system temperature and pressure, and H^{ex} is the excess enthalpy. Equation 1.39 together with the equations defining the activity coefficient and the fugacity provide the basis for deriving an expression for the excess enthalpy in terms of the derivatives of the activity coefficients with respect to temperature. The result is

$$H^{ex} = -RT^2 \sum_i X_i \left[\frac{\partial \ln \gamma_i}{\partial T} \right]_P$$

Evaluation of the excess enthalpy by means of this equation requires activity coefficient data as a function of temperature.

Example 1.8 Excess Enthalpy for Methanol–Water Mixture

The enthalpy of a mixture of 0.5 lbmole methanol and 0.5 lbmole water at 50°F and 25 psia is calculated as

$$H = 0.5H(MeOH) + 0.5H(Water) + H^{ex}$$

At 50°F and 25 psia, the enthalpies of methanol and water are given as 329 BTU/lbmole and 325 BTU/lbmole, respectively. The enthalpy of 1 lbmole mixture at the same conditions is

$$H = (0.5)(329) + (0.5)(325) + H^{ex} = 327 + H^{ex}$$

The activity coefficients of methanol and water in solution at 25 psia and two temperatures 5°F apart in the vicinity of 50°F are calculated by the van Laar equation. The derivatives are approximated by the ratio of finite differences. The results are

	γ(50°F)	γ(55°F)	$\ln\gamma$(50°F)	$\ln\gamma$(55°F)
Methanol	0.987	1.004	−0.0131	0.0040
Water	0.892	0.920	−0.1143	−0.0834

The summation term in the excess enthalpy expression is approximated as

$$0.5(0.0040 + 0.0131)/5 + 0.5(-0.0834 + 0.1143)/5 = 0.00480R^{-1}$$

The excess enthalpy is then

$$H^{ex} = -1.987(50 + 460)^2(0.0048) = -2481 \text{BTU/lbmole}$$

The mixture enthalpy is

$$H = 327 - 2481 = -2154 \text{BTU/lbmole}$$

Enthalpy Balances Involving Phase Change

Multistage separation calculations require heat balances involving transition between the phases. The enthalpy balance equations must, therefore, include heats of vaporization and condensation. Usually these heats are not calculated separately but are implied by using appropriate methods for calculating liquid enthalpies and vapor enthalpies. Consider, for instance, a system where a vapor stream and a liquid stream enter with enthalpies H_1 and h_1, respectively, and leave as a mixed phase stream with vapor enthalpy H_2 and liquid enthalpy h_2. These are total stream enthalpy rates with units such as BTU/hr. If no heat is added to or removed from the system, a heat balance is written as

$$(H_2 + h_2) - (H_1 + h_1) = 0$$

Any heat of condensation or vaporization that accompanies this step is included in this heat balance.

It is possible to cause phase transition without the addition or removal of heat and without any pressure change by bringing together streams with distinct compositions. The process of mixing and attaining equilibrium may involve the transition of certain components from one phase to the other. As this takes place, temperature changes are expected due to heat of condensation or vaporization. The following example illustrates these effects.

Example 1.9 Enthalpy Balances in the NC4–NC6 Binary

Normal butane and normal hexane are two chemically similar components, and a mixture of them at low pressures is expected to have characteristics approaching those of an ideal solution. Thus, for instance, the enthalpy of a liquid mixture of these two components may be approximated by the summation of the products of each pure component enthalpy at mixture conditions times its mole fraction. The excess enthalpy could be neglected.

The following are enthalpy data for a vapor stream 1 and a liquid stream 2 that mix, reach equilibrium adiabatically at the same pressure, and separate into a vapor stream 3 and a liquid stream 4. The calculations are based on the Soave equation of state.

	Stream 1 Vapor	Stream 2 Liquid	Stream 3 Vapor	Stream 4 Liquid
NC4, lbmole/hr	64.2	35.8	59.6	40.4
NC6, lbmole/hr	21.1	98.9	21.9	98.1
T, °F	150.0	150.0	153.3	153.3
P, psia	40.0	40.0	40.0	40.0
Enthalpy, MMBTU/hr	1.17	0.68	1.14	0.71

A temperature rise from 150.0°F to 153.3°F is observed as a result of a net mass transfer from the vapor to the liquid (condensation). Since the system is assumed to behave as an ideal solution, no excess enthalpy exists and no heat of mixing contributed to the temperature rise. In nonideal mixing involving phase change, both effects, heat of mixing and phase change, can contribute to the temperature change. The two effects could pull either together or in opposite directions.

To determine the heat associated with the net phase change, the above process could be carried out isothermally:

	Stream 1 Vapor	Stream 2 Liquid	Stream 3 Vapor	Stream 4 Liquid
NC4, lbmole/hr	64.2	35.8	53.6	46.4
NC6, lbmole/hr	21.1	98.9	17.6	102.4
T, °F	150.0	150.0	150.0	150.0
P, psia	40.0	40.0	40.0	40.0
Enthalpy, MMBTU/hr	1.17	0.68	0.98	0.74

Maintaining the temperature at 150.0°F enhances the phase transition by removing the net heat of condensation, calculated as

$$\Delta H = (0.98 + 0.74) - (1.17 + 0.68) = -0.13 \, \text{MMBTU/hr}$$

NOMENCLATURE

A	Helmholtz function		S	Entropy
C	Number of components		T	Temperature
C_P	Heat capacity at constant pressure		U	Internal energy
f_i	Fugacity of pure component i		V	Volume
G	Gibbs free energy		W	Work
H	Enthalpy		X	Mole fraction in the liquid
K	Vapor–liquid distribution coefficient		Y	Mole fraction in the vapor
k_H	Henry's law constant		z	Compressibility factor
n	Number of moles		γ	Activity coefficient
P	Pressure		η	Carnot efficiency
p	Partial pressure		μ	Chemical potential
Q	Amount of heat		ω	Acentric factor
			ϕ	Fugacity coefficient

SUBSCRIPTS

c	Critical conditions
i,j	Component designations
r	Reduced conditions

SUPERSCRIPTS

α,β	Phase designations
0	Ideal gas conditions or standard state conditions

REFERENCES

Abrams, D. S. and J. M. Prausnitz, *A.I.Ch.E. Journal*, 21, 116, 1975.

API Technical Data Book — Petroleum Refining, 5th Revision, American Petroleum Institute, Washington, D.C., 1978.

Benedict, M., G. B. Webb, and L. C. Rubin, *Chem. Eng. Progress*, 47(8), 419, 1951; 47(9), 449, 1951.

Black, C., *A.I.Ch.E. Journal*, 5(2), 249, 1959.

Carlson, H. D. and A. P. Colburn, *I & EC*, 34, 581, 1942.

Gmehling, J. and U. Onken, *Vapor–Liquid Equilibrium Data Collection*, Dechema Chemistry Data Series, Port Washington, NY, Scholium International, Inc., 1977.

Lee, B. I. and M. G. Kesler, *A.I.Ch.E. Journal*, 21, 510, 1975.

Peng, D. Y. and D. B. Robinson, *Ind. Eng. Chem. Fundam.*, 15, 59, 1976.

Plocker, U., H. Knapp, and J. M. Prausnitz, *Ind. Eng. Chem. Process Des. Dev.*, 17, 324, 1978.

Rackett, J., *Chem. Eng. Data*, 15, 514, 1970.

Renon, H. and J. M. Prausnitz, *A.I.Ch.E. Journal*, 14, 135, 1968.

Soave, G., *Chemical Engineering Science*, 22, 1197, 1972.

Starling, K. E., *Fluid Thermodynamic Properties for Light Petroleum Systems*, Houston, Gulf
    Publishing Company, 1973.
Walas, S. M., *Phase Equilibria in Chemical Engineering*, Butterworth Publishers, 1985.
Wilson, G. W., *J. Am. Chem. Soc.*, 86, 127, 1964.
Wohl, K., *Trans. Am. Inst. Chem. Engr.*, 42, 215, 1946.

PROBLEMS

1.1 An ideal gas enclosed in a cylinder is maintained at a constant temperature
T. The cylinder is partitioned by a frictionless movable disk into two sections, one
containing n_1 moles and the other n_2 moles. The starting pressure is P_1 in the first
section and P_2 in the second. P_2 is greater than P_1.

a. Show that the system reaches equilibrium when the two pressures are equal. Hint:
   Calculate the entropy change. Start with Equation 1.3.
b. Calculate the entropy change if T = 100°F, P_1 = 1 atm, P_2 = 2 atm, n_1 = 75 lbmole,
   and n_2 = 25 lbmole.

1.2 An alternative form of the virial equation of state, truncated after the third
virial coefficient is written as

$$z = 1 + B'/V + C'/V^2$$

Derive expressions for the virial coefficients in terms of the critical properties. What
is the value of the critical compressibility factor, z_C?

1.3 Equations of state may be applied to mixtures by determining the equation
constants for the mixture from constituent pure component properties and constants
using appropriate combining rules. The combining rules for the Redlich–Kwong
equation constants for a binary mixture are given as

$$a = a_1 Y_1^2 + 2 Y_1 Y_2 \sqrt{(a_1 a_2)} + a_2 Y_2^2$$

$$b = b_1 Y_1 + b_2 Y_2$$

where Y_1 and Y_2 are the component mole fractions, and a_1, b_1, a_2, b_2 are the pure
component Redlich–Kwong constants. Alternatively, mixture pseudocritical properties are calculated, and these are used to compute the mixture equation of state
constants. The pseudocritical properties may be calculated using Kay's rules:

$$P_{pc} = \sum_i Y_i P_{ci}$$

$$T_{pc} = \sum_i Y_i T_{ci}$$

Compare the compressibility factors obtained by these two methods for a 75% mole propane–25% mole carbon dioxide mixture at 300°K and a specific volume of 700 ml/gmole. The critical properties are given:

	T_c, °K	P_c, atm
1. Propane	369.7	41.9
2. Carbon dioxide	304.3	72.8

$R = 82.06$ ml-atm/gmole-°K

1.4 Derive equations to calculate component fugacity coefficients in a binary mixture using the virial equation of state truncated after the second virial coefficient. The mixture second virial coefficient is given as

$$B = Y_1 B_1 + Y_2 B_2 + Y_1 Y_2 \delta_{12}$$

where

$$\delta_{12} = 2 B_{12} - B_1 - B_2$$

B_1 and B_2 are the second virial coefficients of components 1 and 2, and B_{12} is the interaction second virial coefficient. Express the virial equation of state in the following alternative form, truncated after the second virial coefficient:

$$\frac{PV}{RT} = z = 1 + \frac{BP}{RT}$$

Hint: The component fugacity coefficient in a mixture is a partial property, and is calculated as the partial derivative of the corresponding total property with respect to the number of moles of the component in question in the mixture.

1.5 Vapor–liquid equilibrium distribution coefficients, or K-values, may be calculated with varying degrees of accuracy. For an equimolar propylene–isobutane mixture at 120°F, compare the K-values at 125, 200, and 275 psia using each of the following methods:

a. Raoult's law
b. The virial equation of state for the vapor, truncated after the second virial coefficient. (See Problem 1.4). The following second virial coefficients are given at 120°F:

$$B_1 \text{ (propylene)} = -4.809 \text{ cuft/lbmole}$$

$$B_2 \text{ (isobutane)} = -8.496 \text{ cuft/lbmole}$$

The interaction second virial coefficient is given as

$$B_{12} = \left(1 - c_{12}\right)\sqrt{B_1 B_2}$$

The parameter c_{12} may be assumed close to zero for these components. What assumptions may be made to simplify the liquid fugacity calculations?
Vapor pressures are obtained from the equation

$$\ln p_i^0(\text{psia}) = a - \frac{b}{c + T(°F)}$$

	a	b	c
1. Propylene	11.757	3253.6	412.6
2. Isobutane	11.592	3658.8	400.0

$$R = 10.73 \ \text{ft}^3\text{-psia/lbmole-}°R$$

1.6 Activity coefficient data for heptane and toluene in their binary solution is given at 1 atm as follows:

X(heptane)	0.0	0.1	0.2	0.3	0.4	0.5	0.6	0.7	0.8	0.9	1.0
γ(heptane)		1.288	1.233	1.183	1.143	1.096	1.067	1.032	1.016	1.000	0.995	1.000
γ(toluene)		1.000	1.007	1.023	1.040	1.067	1.084	1.132	1.189	1.247	1.256	1.230

Check if this data is thermodynamically consistent based on the Gibbs–Duhem equation.

1.7 Derive expressions to calculate the van Laar constants from activity coefficient data. Such data are often available from azeotropic or infinite dilution measurements.

1.8 Hydrofluoric acid and water form an azeotrope at 250°F and atmospheric pressure, with a hydrofluoric acid mole fraction of 0.35. Calculate the activity coefficients of hydrofluoric acid and water at the azeotrope, and the van Laar constants. Assume ideal gas behavior in the vapor phase. The vapor pressures of hydrofluoric acid and water are given as follows:

$$\ln p_1^0(\text{HF}) = 15.35 - \frac{8092.6}{571.94 + T}$$

$$\ln p_2^0(\text{H}_2\text{O}) = 14.72 - \frac{7254.4}{391.00 + T}$$

where p^0 is in psia and T is in °F.

1.9 Calculate the activity coefficients, bubble point pressure, and the vapor composition as a function of liquid composition for the hydrofluoric acid–water binary at 250°F. Use the van Laar equation with the constants calculated in Problem 1.8. Vapor pressure data may also be obtained from Problem 1.8. Assume ideal gas behavior in the vapor phase.

1.10 Derive equations to calculate the enthalpy departure using each of the following methods: (a) the ideal gas equation, (b) the virial equation of state truncated after the second virial coefficient, (c) the Soave–Redlich–Kwong equation of state.

CHAPTER 2

The Equilibrium Stage

Multistage separation processes operate on the principle that when binary or multicomponent mixtures form two or more phases at equilibrium with each other, each phase generally assumes a distinct composition. If, for example, a multicomponent vapor mixture is at equilibrium with a liquid phase, the vapor will, in general, be richer in the "lighter" components than the liquid while the liquid will be richer in the "heavier" components. Thus, if one desires to separate a given feed stream into lighter and heavier components, a step in the right direction would be to bring the stream to such temperature and pressure that would result in the formation of two phases — a vapor and a liquid, then to separate the two phases in a flash drum. As a closer look will indicate, however, the "separation power" of a single equilibrium stage is rather limited.

A stage in a separation process could, for instance, be a tray in a distillation column, a section in a packed absorption column, or a partial condenser or reboiler. An *equilibrium stage* is a contacting device — a vessel — in which two or more phases can mix thoroughly and interact with each other for a sufficient length of time until the number of moles of any component moving from one phase to another equals the number of moles of that component moving in the opposite direction, i.e., until no more net mass transfer takes place between the phases. The composition of each phase is then the *equilibrium composition*.

In the majority of multistage separation processes, the phases in question are vapor and liquid. In liquid–liquid extraction, the interacting phases are two immiscible liquids. For the most part, this book is focused on vapor–liquid separation, although Chapter 11 is devoted to liquid–liquid extraction.

The equilibrium stage may be thought of as the building block of multistage processes. The operation of a particular process depends on the number of stages and their configuration and on the way they are designed to interact with each other. This chapter deals with the equilibrium stage, its behavior, the relationships between the parameters that determine its operation, and methods for solving the equilibrium stage equations.

Although real-life processes may not actually reach equilibrium, the assumption of an equilibrium stage is essential for a rigorous solution of a problem and for providing a sound basis for column design and performance evaluation. The application of equilibrium stage calculation results to actual performance information is usually accomplished by utilizing tray efficiencies, a concept that will be discussed in Chapter 16.

2.1 PHASE BEHAVIOR

The thermodynamic conditions of a system (i.e., mixture) may be defined by a set of parameters, the independent variables. Once their values are fixed, other parameters, the dependent variables, assume certain values that may be either measured or otherwise determined. The selection of the particular set of independent variables is generally arbitrary, although these parameters must uniquely define the system. In the case of the phase behavior of a mixture in an equilibrium stage, the independent variables could be the overall composition of the mixture, its temperature, and pressure. Fixing these parameters determines whether the system is single- or multiple-phase, the relative amounts of the phases, the composition of each phase, its enthalpy, etc. Quantitative information about these dependent variables must rely on experimental measurements or may be predicted on the basis of thermodynamic principles plus data representing the constituents and the interactions among them.

2.1.1 Phase Diagrams

Graphical representation of phase behavior provides a useful tool for studying the relationships among the various interacting variables that define the system. It also helps understand many terms that are frequently used in describing multistage separation processes. The most common vapor–liquid equilibrium diagrams are the phase envelope, or P–T diagram, for binary or multicomponent systems and the T–Z and Y–X diagrams for binary systems.

The Phase Envelope

The phase envelope of a mixture is analogous to the vapor pressure curve of a pure component. The vapor pressure curve defines temperature and pressure conditions at which a pure component can exist as vapor and liquid at equilibrium. In this two-phase regime only one parameter, the temperature or the pressure, may be varied independently. This observation is consistent with the phase rule, which may be expressed as follows:

$$F = C - \phi + 2 \tag{2.1}$$

where F is the number of degrees of freedom, C is the number of components, and ϕ is the number of phases. For a single–component, two–phase system, there is 1 degree of freedom, the temperature or the pressure.

For a system consisting of C components, the phase rule indicates that, in the two-phase region, there are $F = C - 2 + 2 = C$ degrees of freedom. That is, it takes C independent variables to define the thermodynamic state of the system. The independent variables may be selected from a total of 2C *intensive variables* (i.e., variables that do not relate to the size of the system) that characterize the system: the temperature, pressure, C-1 vapor-component mole fractions, and C-1 liquid-component mole fractions. The number of degrees of freedom is the number of intensive variables minus the number of equations that relate them to each other. These are the C vapor–liquid equilibrium relations, $Y_i = K_i X_i$, $i = 1, \ldots, C$. The equilibrium coefficients, K_i are themselves functions of the temperature, pressure, and vapor and liquid compositions. The number of degrees of freedom is, thus, $2C - C = C$, which is the same as that determined by the phase rule.

If the overall composition of a multicomponent mixture is set, two degrees of freedom remain, the temperature and the pressure, and another variable becomes calculable by material balance, the vapor-to-liquid ratio. The phase representation of the system is bivariant on a P–T diagram rather than univariant as in the case of a pure-component vapor pressure curve.

A schematic of a pressure–temperature diagram for a fixed composition mixture is shown in Figure 2.1. At a temperature, T_1, and pressure, P_1, represented by point

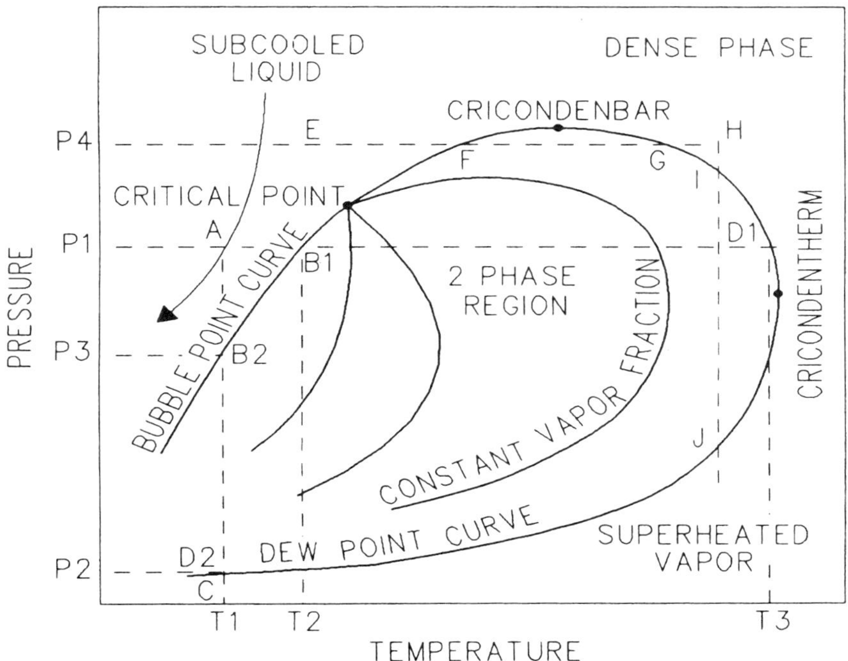

Figure 2.1   Schematic of a phase envelope.

A, the mixture is a subcooled liquid, i.e., no change of phase (vaporization) results from small changes in temperature and/or pressure. If the temperature is gradually increased at constant pressure, the vapor pressures of the constituents rise, causing the composite vapor pressure of the mixture to go up. The bubble point, B1, is reached at temperature T_2 when the mixture vapor pressure equals the system pressure, P_1, and the first vapor bubble appears. Further increase in the temperature generates an increasing amount of vapor at equilibrium with the remaining liquid. The composition of the liquid changes continuously with temperature so that, at any given temperature in the two-phase region, the composite vapor pressure of the liquid equals the system pressure. The dew point, D1, is reached at temperature T_3 when the last liquid drop evaporates. With additional isobaric temperature rise, the vapor becomes superheated. Within the two-phase region, the vapor and liquid are saturated, i.e., the vapor is at its dew point and the liquid is at its bubble point.

The process may be traced back from a superheated vapor to a subcooled liquid, and it does not have to be isobaric. For example, a superheated vapor at point C is compressed isothermally at temperature T_1 until condensation starts at pressure P_2, corresponding to the dew point (point D2). Further isothermal compression results in more condensation until the last vapor bubble has condensed at pressure P_3, which defines bubble point B2. If the pressure is raised isothermally above the bubble point, the liquid becomes subcooled.

Bubble points and dew points may be generated as described above for a given mixture over ranges of temperature and pressure. The locus of bubble points is the bubble point curve and the locus of dew points is the dew point curve. The two curves together define the phase envelope. In addition to the bubble point curve (saturated total liquid) and the dew point curve (saturated total vapor), other curves may be drawn representing constant vapor mole fraction. All these curves meet at one point, the critical point, where the vapor and liquid phases lose their distinctive characteristics and merge into a single, dense phase.

The critical point of a mixture is dependent on the critical points of the constituents; however, no explicit relationship is available relating the former to the latter. Mixture pseudocritical temperatures and pressures are empirically defined averages of one kind or another of the individual components' critical temperatures and pressures and are used for correlating thermodynamic data. The true critical point, as defined above and represented in Figure 2.1, is best determined as the limit of the bubble point and dew point curves. Moving along the bubble point curve, bubble point pressures are determined at incremental temperatures. The lowest temperature at which no phase transition can be detected by raising or lowering the pressure is the critical temperature. Similarly, the lowest pressure at which no phase transition can be detected by raising or lowering the temperature is the critical pressure. Approaching the critical point along the dew point curve is, in general, more involved since the dew point curve may include a region of *retrograde condensation*, which is discussed next.

Unlike a pure component, the critical point of a mixture is not necessarily the highest temperature or pressure at which two phases can coexist. This fact may be observed by studying the phase envelope shown in Figure 2.1 and in particular the dew point curve, whose shape is characteristic of a variety of mixtures. Starting at point E in the single-phase region where the pressure, P_4, is higher than the critical and the temperature is lower than the critical, the temperature is raised isobarically to point F on the phase envelope. As the mixture starts separating into two phases, it is the vapor and not the liquid that makes up most of the mixture. Thus, raising the temperature causes condensation, hence the term retrograde condensation. Point F is on the retrograde segment of the dew point curve. As the temperature is further raised isobarically, the amount of liquid increases to a maximum then starts decreasing until another dew point G is reached at the same pressure, P_4. Similar behavior occurs if the pressure is lowered isothermally from certain parts in the single-phase region, such as point H, to where retrograde condensation starts at point I. The amount of liquid then goes through a maximum, followed by complete vaporization at point J, another dew point at the same temperature. Condensation that takes place with temperature rise or with pressure drop is known as retrograde condensation, and the temperature/pressure conditions where this occurs define the retrograde region. In such systems, vapor and liquid can coexist at temperatures and pressures above the critical point. The highest temperature and the highest pressure where two-phases can coexist are called, respectively, the *cricondentherm* and *cricondenbar*, also shown in Figure 2.1.

In areas surrounding the phase envelope, the system exists as a single phase. Below the dew point curve and at higher temperatures, it is a superheated vapor, while areas above the bubble point curve and to the left of it represent a subcooled liquid.

In areas above the phase envelope between subcooled liquid and superheated vapor, the mixture is a dense or supercritical fluid. This is a transitional area where the fluid may be classified either as liquid or vapor.

The T–Z Diagram

In this diagram, applicable mainly to binary systems, the temperature, pressure, and overall composition are the independent variables, with the pressure held constant. The diagram, shown schematically in Figure 2.2, consists of an upper curve representing dew points and a lower curve representing bubble points. The Z coordinate represents overall mole fraction of component 1, usually chosen as the more volatile component. A vertical line at $Z = 0$ corresponds to pure component 2, and point A represents its boiling point at the fixed system pressure. Similarly, pure component 1 is represented by a vertical line at $Z = 1$, and its boiling point by point B. Points above the dew point curve are in the vapor phase, and those below the bubble point curve are in the liquid phase, while the area between the two curves corresponds to the mixed phase.

The curves on a T–Z diagram are isobaric. If the pressure effect is to be included, a three-dimensional plot may be visualized, where the third coordinate would represent the pressure or, more simply, a series of curves similar to Figure 2.2 may be drawn, each representing a different pressure.

If a mixture with composition Z_1 is brought to the pressure corresponding to a given T–Z diagram and the temperature is fixed at T_0 in the two-phase region, the mixture separates into a liquid phase with composition X_1 and a vapor phase with

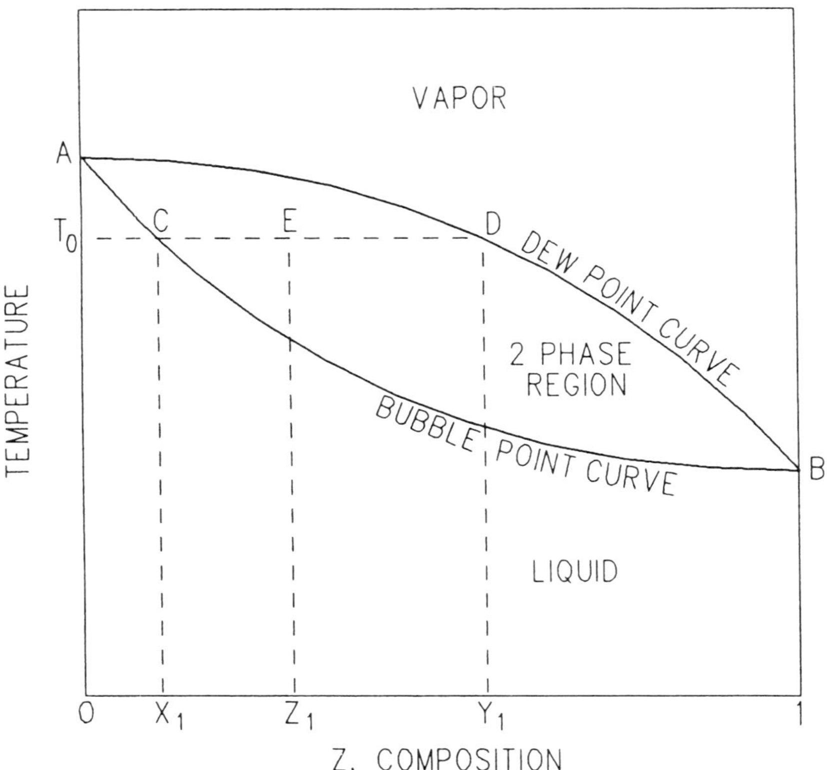

Figure 2.2   Schematic of T–Z diagram.

composition Y_1. Equilibrium liquid and vapor compositions are determined by the intersection of a horizontal line through T_0 with the bubble point and dew point curves, respectively (points C and D in Figure 2.2).

The relative amounts of vapor and liquid may also be evaluated from the T–Z diagram. A material balance on component 1 is written as

$$FZ_1 = FVY_1 + F(1 - V)X_1 \tag{2.2}$$

where F is the total amount of vapor and liquid and V is the vapor mole fraction. Rearranging the equation, the vapor-to-liquid ratio is calculated as

$$\frac{V}{1-V} = \frac{Z_1 - X_1}{Y_1 - Z_1} = \frac{CE}{ED} \tag{2.3}$$

where CD and ED are the line segments on the bubble point side and dew point side, respectively. This relationship is known as the *lever–arm rule*, which is a graphical representation of the material balance.

The Y–X Diagram

As in the case of the T–Z diagram, this phase representation is applicable only to binary systems. Isobaric binary vapor–liquid equilibrium data are plotted as the mole fraction of the lower boiling component in the vapor, Y_1, versus the mole fraction of the same component in the liquid X_1, as shown in Figure 2.3. A 45-degree diagonal representing $Y_1 = X_1$ is drawn for reference. In general, the vapor is consistently richer in the lower boiling component than the liquid, i.e., $Y_1 > X_1$ over the entire composition range. Therefore, in general, the equilibrium curve lies entirely above the diagonal, as typified by curve A in Figure 2.3.

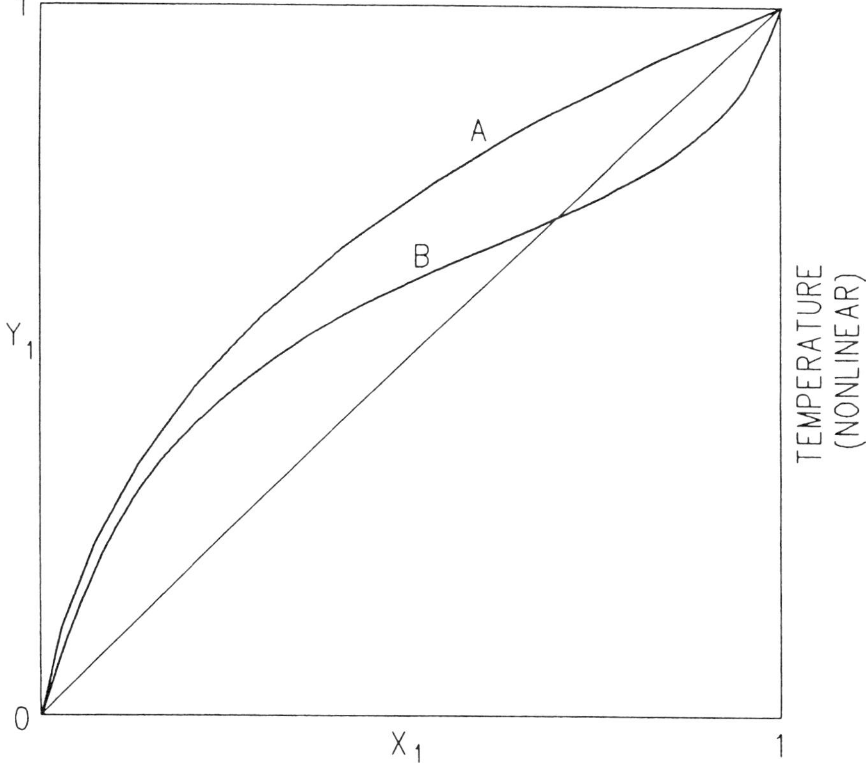

Figure 2.3 Schematic of Y–X diagram.

Systems that form azeotropes, however, exhibit unique characteristics as discussed in Section 1.3.4. In a typical azeotrope-forming binary, the concentration of the low boiling component in the vapor, Y_1, is greater than that in the liquid, X_1, at low concentrations of the light component. This behavior is similar to the general case. At the azeotropic point, the vapor and liquid compositions are identical, $Y_1 = X_1$, at which point the equilibrium curve crosses the diagonal. At higher concentrations of the low boiling component, the liquid becomes richer in that component,

$Y_1 < X_1$, and the equilibrium curve falls below the diagonal. Curve B in Figure 2.3 represents this type of azeotropic system.

Points plotted on Y–X diagrams normally represent isobaric vapor–liquid equilibrium data. Data from a T–Z diagram may be easily replotted on a Y–X diagram. The temperature obviously changes along the equilibrium curve on a Y–X diagram. In order to show its variation on the same diagram, a nonlinear temperature coordinate may be included, parallel to the Y coordinate, as shown in Figure 2.3.

2.1.2   Distribution Coefficients

It takes two independent variables to define uniquely a two-phase system having a fixed overall composition. Quantitative determination of the other variables requires thermodynamic data such as enthalpy and component equilibrium distribution coefficients, or K-values. Distribution coefficients are defined as

$$K_i = \frac{Y_i}{X_i} \tag{2.4}$$

where Y_i and X_i are mole fractions of component i in the vapor and liquid, respectively. The thermodynamic principles underlying the distribution coefficients and enthalpy, as well as methods for predicting them, are discussed in Chapter 1. Of consequence at this juncture is the fact that these properties are, in general, functions of temperature, pressure, and composition. Thus, if a mixture has fixed overall composition, its thermodynamic properties, whether measured or predicted, are uniquely determined by two independent variables. This statement applies to multiphase as well as single-phase systems. In fact, the two independent variables, plus pertinent thermodynamic data, determine whether the system exists as single phase or mixed phase.

2.1.3   Flash Operations

The main variables associated with phase relationships include the overall composition, Z_i; temperature; pressure; liquid composition, X_i; vapor composition, Y_i; vapor mole fraction, V; and heat duty, Q. A process in which Z_i and two other independent variables are set and equilibrium separation of the phases is allowed to take place is called a *flash operation*. A general flash operation is shown in Figure 2.4. A feed stream initially at conditions T_1 and P_1 is controlled so that its final conditions satisfy two specifications. The feed is of fixed rate and composition, F and Z_i. A heat duty, Q, may be added to or removed from the system as required. The feed is flashed to generate a vapor product with flow rate FV and a liquid product with flow rate $F(1 - V)$, where V is the vapor mole fraction at flash conditions T_2 and P_2. In general, V may be equal to 0 or 1 or any value in between. The total enthalpy of the products is H_2. The type of flash operation depends on which two independent variables are specified. Following are different possible cases.

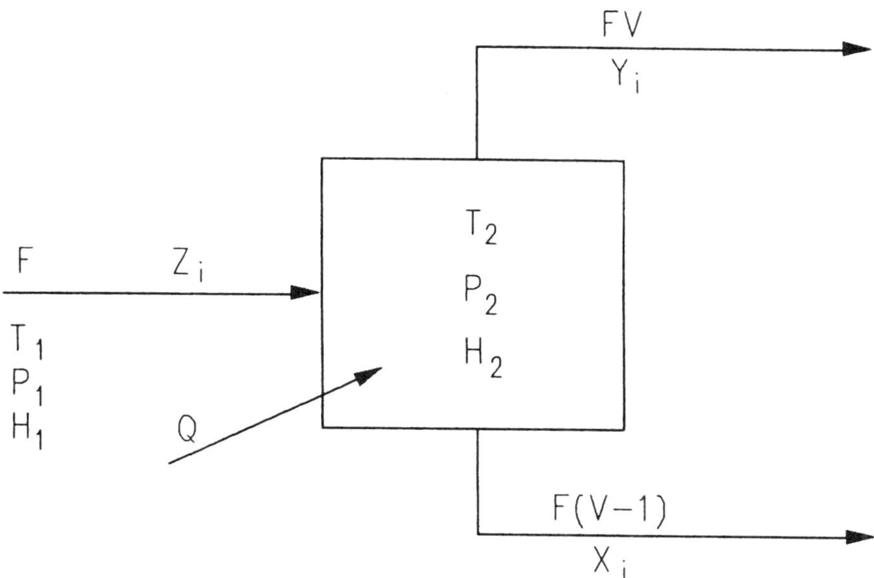

Figure 2.4 General flash operation.

Isothermal Flash

The specified variables are the final temperature and pressure, T_2 and P_2. The dependent variables are the vapor fraction, V, the liquid and vapor compositions, X_i and Y_i; and the total enthalpy of the two phases, H_2. The term isothermal should not be interpreted to imply that the transition from initial conditions to final conditions is at constant temperature; T_1 is, in general, different from T_2. It simply means that within the flash drum the temperature, as well as the pressure, is fixed. The duty required to bring about the final conditions is equal to the enthalpy change, $Q = H_2 - H_1$, where H_1 is the enthalpy at T_1 and P_1. Isothermal flash conditions may be represented by a point (T_2, P_2) on the phase envelope diagram. It is clearly possible that this point may fall either within the phase envelope or outside it, in which case the system would be all vapor or all liquid (or dense phase). A flash drum operating at such conditions would have a single product and no phase separation would take place. In a single-phase situation, the dependent variables are the vapor fraction, V (0 or 1, depending on whether the phase is liquid or vapor), and the enthalpy, H_2. The liquid or vapor composition is, of course, identical to the feed or overall composition, Z_i. Note that any set of temperature and pressure specifications is feasible.

Adiabatic Flash

In this operation the specified variables are the temperature or pressure and the duty, Q. Specifying the duty is equivalent to specifying the final enthalpy, H_2. The dependent variables are the pressure or temperature, the vapor fraction, and the vapor and liquid compositions. In a truly adiabatic process $Q = 0$ (or $H_2 = H_1$), but the term *adiabatic flash* is generally applied to a process where the duty is specified. The problem is to determine a temperature (or pressure if the temperature is specified) at which the total enthalpy of the products satisfies the duty specification. Once T_2 and P_2 are known, the problem is handled as an isothermal flash. Again, in this case, the solution could result in a single phase or a mixed phase, and any set of temperature (or pressure) and duty specifications is feasible.

Bubble Point

If one independent variable is the temperature or pressure and the only product is a saturated liquid (i.e., the other independent variable is the vapor fraction set at the 0 limit), then the operation is a *bubble point flash*. The dependent variables are the pressure or temperature, the duty, and liquid and vapor compositions. Since the liquid is the only product, its composition is identical to the feed composition. The vapor composition refers to the composition of a vapor at equilibrium with the liquid. Although its rate is 0, its composition is determinable and is equal to the composition of the first vapor bubble resulting from infinitesimal vaporization. Bubble points are not feasible at all temperatures and pressures. As discussed in Section 2.1.1, Figure 2.1 indicates that no bubble points can exist at temperatures or pressures above the critical point.

Dew Point

In a *dew point flash*, the fraction vapor is set at the unity limit, at a specified temperature or pressure. The dependent variables are the pressure or temperature, the duty, and the liquid and vapor compositions. The only product is a saturated vapor with composition equal to that of the feed. The equilibrium liquid composition corresponds to the first liquid drop resulting from infinitesimal condensation. As in the bubble point flash, dew points can exist only within certain ranges of temperature and pressure. Referring to Figure 2.1, no dew points can exist at temperatures above the cricondentherm or at pressures above the cricondenbar. At temperatures or pressures where retrograde condensation can occur, there can be two dew points: a normal dew point and a retrograde dew point.

General-Type Flash

In general, one may wish to set any two specifications on flash products or conditions. For example, one may specify component concentrations in one of the products, or a product rate, plus the temperature or pressure. A flash may be operated to satisfy separation specifications, such as in multistage separation processes. The next section discusses the performance of a single equilibrium stage as a separation process for different types of systems.

2.2 PERFORMANCE OF THE EQUILIBRIUM STAGE

Consider an equilibrium stage with a feed stream at fixed rate, composition, and thermal conditions (temperature and pressure). Let the two independent variables defining the system be the pressure and duty, making it an adiabatic flash. If now it is desired to specify another variable such as the temperature or the vapor or liquid rate or composition, then at least one of the independent variables — the pressure or duty — must be allowed to vary in order for the new specification to be satisfied. Certain parameters are inherently or by necessity fixed for a given system and may not be varied. In an equilibrium stage, for instance, the fact that it is a single stage is a fixed parameter. The feed rate, composition, and thermal conditions are also fixed parameters. Other parameters, such as the duty, temperature, or pressure, and, in the case of multiple feeds, some of the feed rates, may be varied to satisfy process requirements. The distinction between fixed and variable parameters reflects actual operating practices. Thus, in order to control the vapor rate in a flash drum, one would normally vary the duty (or temperature) rather than the feed rate, which is usually controlled by upstream process considerations.

This section examines the effect of temperature, pressure, and, in certain situations, the feed rate on the performance of an equilibrium stage. Several examples are presented, typifying various separation processes. The different situations are mainly characterized by the nature of the feed makeup.

Although the obvious purpose of separation processes is to raise or lower the concentration of certain components to certain levels, another factor must be considered and that is the recovery of those components. Thus the amount of a product as well as its purity are of concern to the process engineer. The concentration, commonly expressed as mole fraction of the components of interest, X_i or Y_i, is a measure of quality, while the total product rate or component recovery is a measure of quantity. The recovery of a component or group of components is defined as the fraction or percentage of these components in the feed that are recovered in a given product. If, for instance, the mole fraction of component i in the feed to a flash drum is Z_i and its mole fraction in the vapor product is Y_i, the amount of i in the feed is FZ_i and in the vapor product is VFY_i, where F is the molar feed rate and V is the vapor mole fraction. The recovery of component i in the vapor product is, therefore,

$$\text{Recovery} = \frac{VFY_i}{FZ_i}$$

$$= \frac{VY_i}{Z_i} \tag{2.5}$$

or

$$\text{Percent recovery} = 100\,\frac{VY_i}{Z_i} \tag{2.6}$$

The performance of an equilibrium stage may be examined in terms of components' concentration and recovery in a given product and the dependence of these quantities on the independent variables and the phase equilibrium characteristics of the system.

It should be noted that a single state is inefficient and inflexible as a separation process. The independent variables, namely the temperature and pressure, can vary only over limited ranges, bounded by the phase envelope of the mixture. In a single stage, a component concentration can go through only a single-step change. Thus, the concentration of component i can change from X_i in the liquid to $Y_i = K_i X_i$ in the vapor. By contrast, in a multiple-stage operation, a number of steps or equilibrium stages make up the process. This allows the composition of both phases to change progressively from stage to stage, resulting in more efficient separation.

In certain systems, phase separation is enhanced by introducing an additional feed to the separation device such as in absorption or stripping and in azeotropic and extractive distillation. Although these processes are usually multistage, their characteristics are discussed in this section using the single-stage model.

2.2.1 Single-Feed Systems

In the simplest form of phase separation, the stream to be separated is flashed into a vapor product and a liquid product, each having a different composition. Following are examples of single-stage separations of certain mixtures. The distribution coefficients are predicted by the Soave–Redlich–Kwong equation of state (Soave, 1972).

Example 2.1 Binary Mixtures — Wide Boiling

Two miscible components having an appreciable difference in their boiling points form a mixture with widely separated dew points and bubble points. The mixture has a wide boiling range, and the distribution coefficients of the two components differ considerably. An example of this type of mixture is the ethane–n-hexane system, with 50%–50% mole composition. The equilibrium temperature, distribution coefficients, compositions, and recovery of ethane in the vapor are calculated at a fixed pressure of 100 psia and over a vapor mole fraction ranging from 0 to 1. Selected results are given in Table 2.1. The temperature, mole fraction ethane in the vapor, and ethane recovery in the vapor are plotted vs. mole fraction vapor in Figure 2.5.

Table 2.1   Single-Stage Performance

50% Mole Ethane (1)–50% Mole n-Hexane (2) at 100 psia						
T, F	V	K_1	K_2	X_1	Y_1	VY_1/Z_1
−1.19	0	2.00	4.67E-3	0.500	0.998	0
7.82	0.1	2.24	6.19E-3	0.445	0.997	0.199
21.50	0.2	2.64	9.31E-3	0.376	0.994	0.398
141.18	0.5	7.22	1.38E-1	0.122	0.878	0.878
213.53	0.8	10.07	4.15E-1	0.061	0.610	0.976
232.06	1.0	10.65	5.25E-1	0.047	0.500	1.000

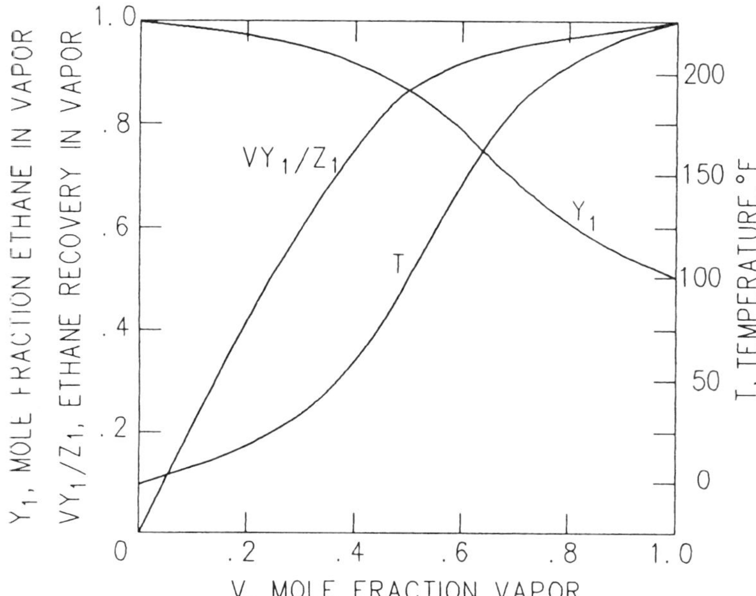

Figure 2.5   Single-stage performance. 50% Mole ethane (1)–50% mole n-hexane (2) at 100 psia.

Because of the large difference between the boiling points of the constituents, the temperature spans a wide range between the bubble point and the dew point, about 233°F. As a result, the separation is relatively easy, in the sense that a large recovery is attainable for a given product purity. If, for instance, a target purity is set at 80% ethane in the vapor, a recovery of about 93% can be achieved, as shown in Figure 2.5.

Example 2.2 Binary Mixtures — Close Boiling

The constituents of this type of mixture have relatively close boiling points. The mixture dew points and bubble points are not far apart, and the distribution coefficients are not much different for each component. An example of such a system is an equimolar ethane–propane mixture. Data representing the separation of this mixture at 100 psia in an equilibrium stage are given in Table 2.2 and Figure 2.6. Because of the relatively small difference in the boiling points and distribution coefficients, the two components do not separate readily as shown by the low ethane recovery in the vapor of 16.2%, obtainable at an ethane purity of 80%.

Example 2.3 Multicomponent Mixtures

In general, mixtures occurring in chemical processes are multicomponent rather than binary. It is often desired to concentrate one of the constituents, a *key component*, in a product in order to increase its value. To accomplish the separation, an

Table 2.2   Single-Stage Performance

50% Mole Ethane (1)–50% Mole Propane at 100 psia						
T, F	V	K_1	K_2	X_1	Y_1	VY_1/Z_1
−13.00	0	1.66	0.342	0.500	0.830	0
−9.82	0.1	1.73	0.363	0.466	0.806	0.161
−6.30	0.2	1.81	0.387	0.430	0.780	0.312
5.45	0.5	2.10	0.476	0.323	0.677	0.677
16.19	0.8	2.38	0.570	0.238	0.566	0.906
21.77	1.0	2.53	0.623	0.198	0.500	1.000

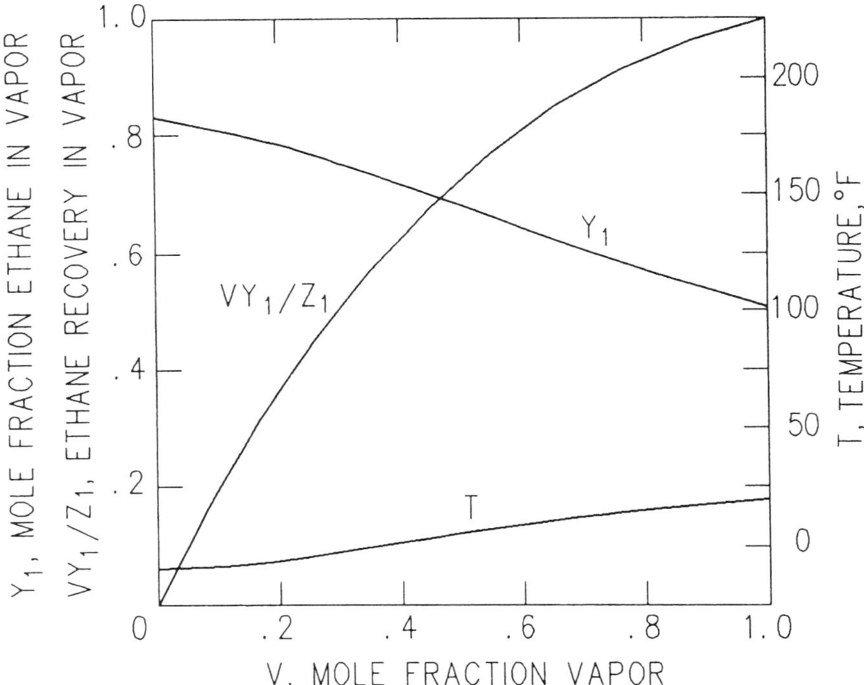

Figure 2.6   Single-stage performance. 50% Mole ethane (1)–50% mole propane (2) at 100 psia.

elaborate processing scheme may be required. In a multicomponent situation, components both lighter and heavier than the key component may be present. If the key component is recovered in the vapor product of a single equilibrium stage, undesired lighter components are more highly concentrated in the product than in the feed. Similarly, if the key component is recovered in the liquid product, heavier components are more highly concentrated in the product than in the feed.

An example of the performance of a single stage with a multicomponent feed at 100 psia is presented in Table 2.3 and Figure 2.7. The composition of the feed, made up of nitrogen and light hydrocarbons through hexane, is also given in Table

Table 2.3 Single-Stage Performance

Multicomponent Mixture at 100 psia

Component	Mole Percent
Nitrogen	1
Methane	1
Ethane	50
Propane	15
n-Butane	13
n-Pentane	10
n-Hexane	10

Ethane

T, F	V	X	Y	VY/Z
−95.02	0.0	0.500	0.191	0
−13.84	0.1	0.468	0.784	0.157
−1.65	0.2	0.419	0.825	0.330
41.42	0.5	0.236	0.764	0.764
111.99	0.8	0.102	0.600	0.960
156.20	1.0	0.065	0.500	1.000

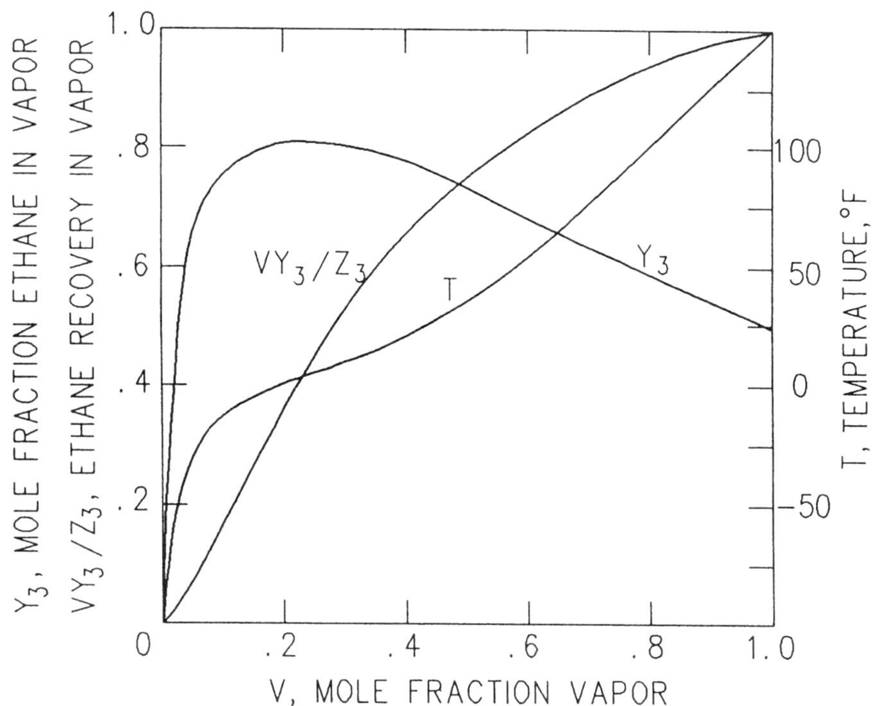

Figure 2.7 Single-stage performance. Multicomponent mixture at 100 psia.

2.3. Consider ethane as the key component to be recovered in the vapor. Since it is not the lightest component, initial vaporization does not produce the highest ethane concentration. In fact, the initial vapor product is poorer in ethane than the feed. As more of the mixture is vaporized, the ethane concentration in the vapor increases, reaches a maximum, then starts declining back to the feed concentration at the limit when all the feed is vaporized. Ethane recovery at maximum ethane concentration of 82.5% is about 33%. The equilibrium temperature varies as the liquid and vapor compositions change. A steep temperature rise is noticed at very low vapor fractions, where the bubble point is highly sensitive to the nitrogen concentration in the liquid.

2.2.2 Single-Stage Absorption/Stripping

The phase separation of certain mixtures may take place at economically impractical temperatures and pressures. An example of this type of separation might be the recovery of light hydrocarbon gases dispersed in air. In these situations, a second stream may be added to the mixture to shift the vapor–liquid equilibrium region. Such a process is equivalent to a single-stage absorption or stripping device.

Example 2.4 Single-Stage Absorber: Nitrogen–Ethane

Nitrogen is to be separated out of a mixture containing 50% mole nitrogen and 50% mole ethane. Table 2.4 and Figure 2.8 represent the binary phase characteristics of this system at 100 psia. One way of determining the level of separation is to measure the amount of nitrogen separated at a given nitrogen purity. A nitrogen recovery of about 95% is expected at a purity of 98%. In order for this to happen, however, the mixture must be brought to the two-phase region, at approximately −260°F, requiring cryogenic distillation.

Table 2.4 Single-Stage Performance

50% Mole Nitrogen (1)–50% Mole Ethane (2) at 100 psia

			Nitrogen		
T, F	V	X	Y	K	VY/Z
−289.62	0.0	0.500	0.999	2.000	0
−178.14	0.5	0.027	0.973	36.037	0.973
−95.22	0.8	0.013	0.622	47.846	0.995
−82.95	1.0	0.010	0.500	50.000	1.000

As an alternative process, a liquid such as hexane may be added to the vapor mixture to form a two-phase ternary. Because of the large difference between the distribution coefficients of nitrogen and ethane, most of the nitrogen goes in the vapor phase, while most of the ethane goes in the liquid. Separating out the nitrogen in this manner is, in effect, the equivalent of a single-stage absorption process.

Table 2.5 shows the nitrogen composition and recovery in the vapor product of an equilibrium stage at different hexane-to-vapor feed ratios. The temperature and

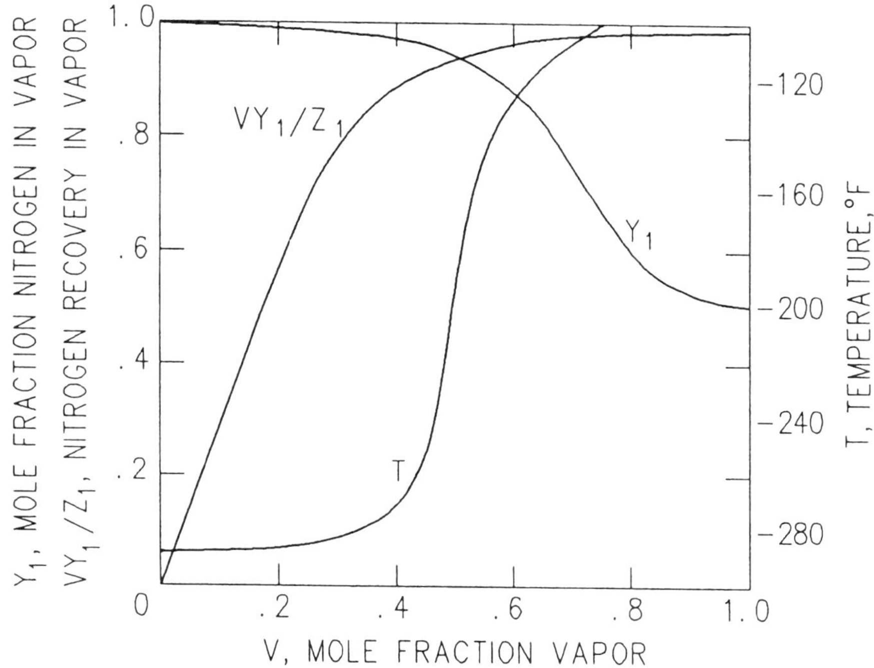

Figure 2.8 Single-stage performance. 50% Mole nitrogen (1)–50% mole ethane (2) at 100 psia.

pressure are held at −100°F and 100 psia. A nitrogen purity of 98%, comparable to that obtained in the binary at −260°F, is attainable at −100°F by introducing hexane at a molar ratio of 7.5 to the vapor feed. The recovery of nitrogen in the vapor product is about 84%, somewhat lower than in the binary separation.

Table 2.5 Single-Stage Performance with Multiple Feeds

F_L/F_V	Nitrogen			
	X	Y	K	Recovery
1.0	0.01	0.898	91.6	0.971
2.5	0.01	0.947	97.0	0.941
5.0	0.01	0.971	99.3	0.892
7.5	0.01	0.980	100.1	0.842
10.0	0.01	0.984	100.5	0.792

F_V — Vapor Feed, 50% Mole Nitrogen–50% Mole Ethane

F_L — Liquid Feed, n-Hexane

Stage Temperature −100°F

Stage Pressure 100 Psia

2.2.3 Close Boilers and Azeotropes

Other systems that require the addition of a separating agent to enhance or effect their separation include azeotropes and mixtures of close boiling components.

The separation of close boiling components by ordinary fractionation may require too many stages or could be practically impossible because of the proximity of the distribution coefficients (K-values). The compositions of equilibrium vapor and liquid phases in such systems are almost identical. In general, close boiling components are likely to have different chemical structures, and would, therefore, interact differently with a third component. In extractive distillation, a solvent that is less volatile than the feed components is added to the mixture for the purpose of preferentially depressing the volatility of one of the feed components. In azeotropic distillation, the added component, or *entrainer*, forms an azeotrope with one of the components to be separated. The azeotrope may have either a higher or lower boiling temperature than the other component and may, therefore, leave the separation device with either the bottoms or overhead.

Azeotropes, discussed in Chapter 1, form constant boiling mixtures and are, therefore, impossible to separate unless the azeotrope is broken by the addition of an external agent, using either extractive or azeotropic distillation as in the case of close boiling mixtures.

The following examples demonstrate the above principles applied to a single stage.

Example 2.5 Single-Stage Azeotropic Separation

Benzene and cyclohexane are difficult to separate because of their close boiling points (176°F and 177°F). They also form an azeotrope at a composition of about 60% mole benzene. The two components may be separated by adding ethanol, which forms azeotropes with each component. At 15 psia, the cyclohexane–ethanol azeotrope boils at 153°F and the benzene–ethanol azeotrope at 156°F. The two azeotropes are separated, followed by removal of ethanol by water wash. The effect of adding ethanol to the benzene–cyclohexane mixture is shown in Figure 2.9 for a single-stage separation. Adding increasing amounts of ethanol (the entrainer) results in higher quantities of the benzene and cyclohexane azeotroping with the ethanol, thereby improving the separation. At higher ethanol concentrations, the ratio of cyclohexane to benzene in the vapor phase increases. This ratio approaches a limit after all the cyclohexane and benzene have azeotroped. In this example the stage pressure is held at 15 psia. At all ethanol feed rates, 50% of the combined benzene–cyclohexane feed is vaporized. The liquid activity coefficients were calculated with the NRTL equation.

Example 2.6 Single-Stage Extractive Separation

Isobutane and 1-butene are close boilers and, although they do not form an azeotrope, are difficult to separate by conventional distillation. Using a single stage, a feed stream containing 40% mole isobutane and 60% mole 1-butene is flashed at 75 psia so that 50% of this stream is vaporized. With no solvent added, the vapor

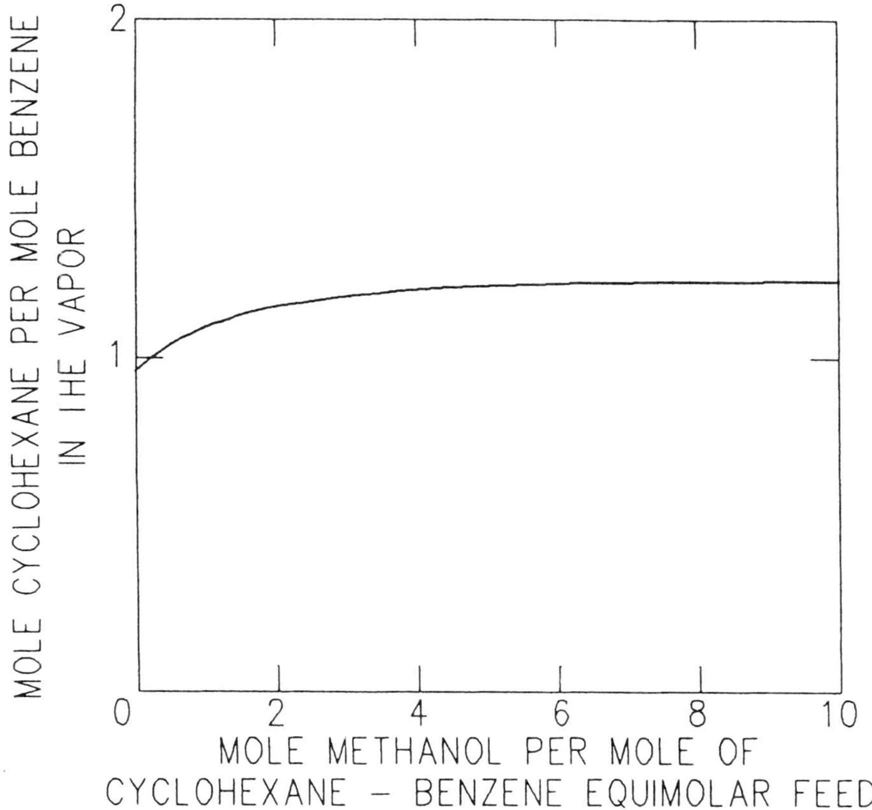

Figure 2.9   Single-stage azeotropic separation. Benzene–cyclohexane–ethanol at 15 psia.

and liquid product compositions are about the same. The ratio of 1-butene to isobutane in the liquid product is about 1.6 (Figure 2.10) compared to 1.5 in the feed. The two components are chemically different, 1-butene being unsaturated and isobutane being saturated. Furfural, which is less volatile than both components, is added as a solvent to alter the 1-butene–isobutane relative volatilities, that is, to depress the 1-butene K-value relative to that of isobutane. This is equivalent to a statement that 1-butene is more soluble than isobutane in furfural. As a result, the ratio of 1-butene to isobutane in the liquid product is enhanced as more solvent is added while maintaining the amount of C4s vaporized at 50% mole of the feed (Figure 2.10). Furfural can be recovered from the hydrocarbons in downstream processing with relative ease because of its substantially lower volatility.

The calculations for this example are based on the Margules liquid activity coefficient equation with the following parameters:

$$\text{Isobutane - furfural} \quad A_{12} = 1.14 \quad A_{21} = 1.31$$
$$\text{1 - butene - furfural} \quad A_{12} = 0.84 \quad A_{21} = 1.03$$

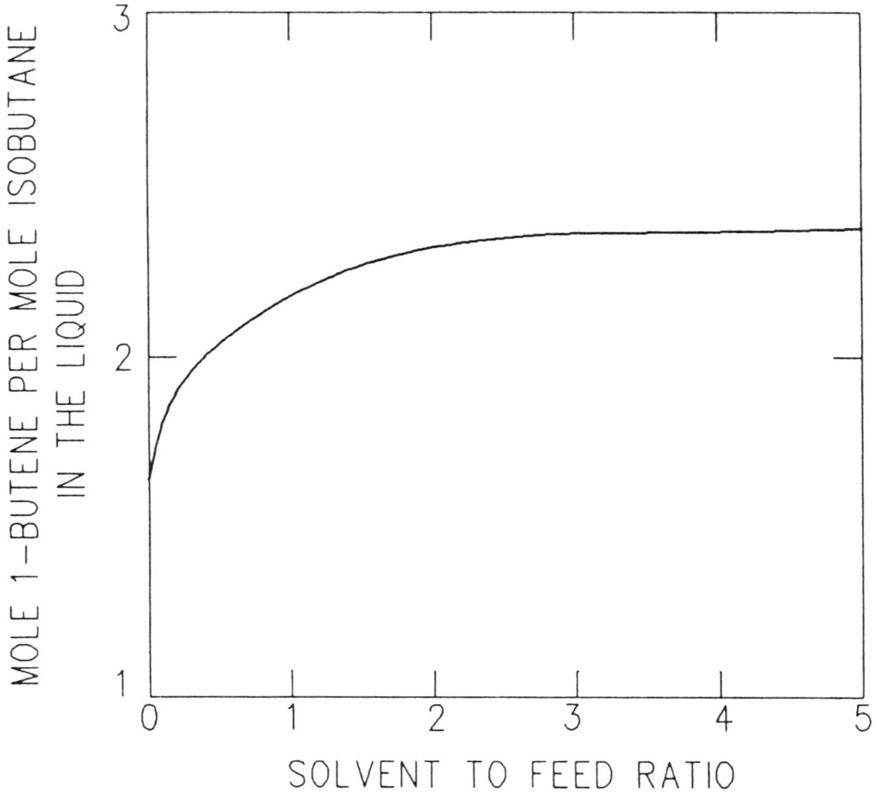

Figure 2.10 Single-stage extractive separation. 1-butene–isobutane–furfural at 75 psia.

2.3 SOLUTION METHODS

In order to define the quantitative relationships among the various parameters in a unit operation, a mathematical model is employed, in which the physical relationships are expressed as mathematical equations. Thus, the equilibrium stage may be simulated by a model for which the mathematical solution represents physical performance. When physical relations are translated into analytical expressions, certain assumptions must be made and the accuracy of the simulation model depends on the validity of these assumptions. For an equilibrium stage model, it is assumed that the stage is essentially at equilibrium. Additionally, it is assumed that the thermodynamic models used for predicting thermophysical properties, namely the distribution coefficients and enthalpy, are accurate. To the extent that these assumptions are met, the performance of the equilibrium stage may be accurately predicted.

The physical relations that must be satisfied in an equilibrium stage are the material balance, the enthalpy balance, and the phase equilibrium relations. Referring to Figure 2.4, material balance equations may be written for each component i as

$$FZ_i = FVY_i + F(1 - V)X_i$$

or

$$Z_i = VY_i + (1 - V)X_i \tag{2.7}$$

for $i = 1, \ldots, C\text{-}1$, where C is the number of components. The problem definition is such that the feed conditions are set, i.e., the feed composition, Z_i, is known. Expression 2.7 represents C-1 independent equations, each corresponding to one of C-1 components. The last component mole fraction in each phase is determined from the relations

$$\sum_i X_i = \sum_i Y_i = 1$$

An enthalpy balance on the equilibrium stage may be stated as

$$H_F + Q = h(T, P, X_i) + H(T, P, Y_i) \tag{2.8}$$

where H_F is the total feed enthalpy, Q is heat duty added to (if positive) or removed from (if negative) the system, h is the total liquid enthalpy, and H is the total vapor enthalpy. H_F is known since the feed conditions are defined, and Q is a variable parameter. H and h are functions of the stage temperature and pressure and the compositions of the respective phase. Subscripts 2 are dropped for outlet conditions.

The last set of equations governing the equilibrium stage consists of the phase equilibrium relations, which may be expressed in terms of the distribution coefficients as

$$\frac{Y_i}{X_i} = K_i(T, P, X_i, Y_i) \tag{2.9}$$

$$i = 1, \ldots, C$$

These are C equations, corresponding to the C components. The distribution coefficients are functions of the temperature, pressure, and liquid and vapor compositions. Equation 2.9 is based on the thermodynamic condition that at phase equilibrium the fugacities of each component in the liquid phase and the vapor phase are equal. Referring to Section 1.3, the fugacities may be expressed in terms of fugacity coefficients as follows:

$$f_i^V = PY_i \Phi_i^V$$
$$\tag{2.10}$$
$$f_i^L = PX_i \Phi_i^L$$

At equilibrium,

$$f_i^V = f_i^L$$

or

$$PY_i \Phi_i^V = PX_i \Phi_i^L$$

for i = 1, …, C. These may be rearranged to give

$$K_i = Y_i / X_i = \Phi_i^L / \Phi_i^V \tag{2.11}$$

The fugacity coefficients may be derived from equations of state or activity coefficients (Chapter 1). In any case, the fugacities are functions of temperature, pressure, and composition. An alternative to Equation 2.9 for expressing phase equilibrium is, therefore,

$$f_i^L\left(T, P, X_i\right) = f_i^V\left(T, P, Y_i\right) \tag{2.12}$$

for i = 1, …, C. Equations 2.7, 2.8, and 2.12 bring the total number of independent equations to 2C. These equations involve 2C + 2 variables, namely the temperature, T; the pressure, P; C-1 component liquid mole fractions, X_i; C-1 component vapor mole fractions, Y_i; the vapor mole fraction, V; and the duty, Q. 2C equations determine 2C variables, leaving two independent variables that must be fixed in order to define the equilibrium stage. These are the two degrees of freedom that were discussed with the phase rule in Section 2.1. The above is a mathematical statement of the same principle.

The set of 2C simultaneous equations is nonlinear and fairly complex since it involves calculating fugacities and enthalpies, themselves nonlinear functions of the temperature, pressure, and composition. Because of this complexity, the equations cannot be solved directly and some iterative method must be used. In general, the computational methods depend on which two variables are selected as the independent variables. Although in principle any two variables may be fixed, the problem complexity may vary from case to case. It is found, for instance, that a solution is more readily reached if P and T rather than P and Q are the independent variables. Since most of these calculations are carried out on computers, the solution methods should be designed for speed of convergence and reliability. Several methods have been proposed for handling the different types of flash calculations which appear to have received the most acceptance. Some of these are discussed here.

2.3.1 The Isothermal Flash Method

This method starts off by fixing the temperature and pressure and iterating around the vapor fraction to calculate the equilibrium phase separation and compositions. This first step is an isothermal flash calculation. If T and P are in fact the independent variables, the solution obtained in the first step is the desired solution. If T or P is fixed and one more variable is specified, then another, outer iterative loop is required, which iterates around P or T (whichever is not fixed) until the other specified variable is satisfied.

Basic Algorithm

The component material balance and distribution coefficient relations, Equations 2.7 and 2.9, are combined to express X_i and Y_i in terms of feed composition and vapor fraction:

$$X_i = \frac{Z_i}{1 + V(K_i - 1)} \tag{2.13a}$$

$$Y_i = \frac{Z_i K_i}{1 + V(K_i - 1)} \tag{2.13b}$$

Since

$$\sum X_i = \sum Y_i = 1$$

either of the above equations may be summed up over all the components $i = 1, \ldots,$ C to form one equation in V for a given set of K_i. Alternatively, the equation

$$\sum (X_i - Y_i) = 0$$

may be solved for V. This equation represents a smoother function of V than $\sum X_i = 1$ or $\sum Y_i = 1$ and is, therefore, more amenable to solution. The function takes the form

$$f(V) = \sum_{i=1}^{C} \frac{Z_i(K_i - 1)}{1 + V(K_i - 1)} = 0 \tag{2.14}$$

If K_i were a function of temperature and pressure only and if a set of K_i values were available at a specified T and P, a solution of Equation 2.14 would satisfy both the material balance and equilibrium relations. In general, however, K_i is also a function of X_i and Y_i, themselves unknown until Equations 2.13a and 2.13b are solved. This entails the need for an iterative scheme, whereby a set of distribution coefficients is assumed and Equation 2.14 is solved. The resulting compositions are checked to determine if the equilibrium condition, Equation 1.19, is met for all the components. The distribution coefficients are updated based on calculated compositions, and the process is repeated until the equilibrium relations are met. The algorithm may be summarized in the following steps:

1. For a given feed composition, temperature, and pressure, assume initial estimates for the distribution coefficients, K_i. These estimates may be based on Raoult's law (Section 1.3).
2. Based on current K_i values, solve Equation 2.14 for V. The solution to this equation is itself iterative since it is implicit in V and nonlinear. A method such as Newton–Raphson may be used.
3. Compute the liquid and vapor compositions using Equations 2.13a and 2.13b.
4. Compute the liquid and vapor fugacities of each component in the liquid and vapor phase, f_i^L and f_i^V (Section 1.3).
5. If $f_i^L = f_i^V$ for all components, then the thermodynamic conditions for equilibrium are satisfied and the solution is reached at the current values of vapor fraction and vapor and liquid compositions. If the liquid and vapor fugacities are not equal within a set tolerance for any component, the distribution coefficients are updated on the basis of Equation 2.11, rewritten as

$$K_i = \frac{Y_i}{X_i} \frac{f_i^L}{f_i^V} \tag{2.15}$$

The iterative computation is restarted at step 2. Once a converged solution is reached, the duty associated with the isothermal flash may be calculated using Equation 2.8.

If a single equation of state is used in step 4 to calculate liquid and vapor densities and fugacities, situations can arise where the distinction between liquid and vapor compressibilities may not be obvious. Using the wrong compressibility may result in nonconvergence or convergence to a false solution. A temperature perturbation technique may be used in these situations to identify the liquid and vapor compressibilities (Khoury, 1978). The method involves searching for multiple compressibility roots at temperatures around the system temperature, then selecting the smallest root as a "pseudo-vapor" compressibility and the largest root as a "pseudo-liquid" compressibility. The pseudo compressibilities are used in intermediate iterations. In subsequent iterations the temperature is reset to system temperature and pseudo compressibilities are replaced by true compressibilities.

Extension to General Flash Calculations

The isothermal flash algorithm described above may be incorporated into an iterative scheme for solving other types of flash calculations. The isothermal flash routine becomes a module in an outer computational loop in which either the temperature or pressure is varied to satisfy a given specification.

For example, consider an adiabatic flash malculation at fixed pressure. A feed with a given composition and initial thermal conditions is flashed at a fixed pressure, and, in the process, a fixed heat duty, Q, is added or removed. In the special case of an adiabatic flash, $Q = 0$. The equilibrium temperature and phase compositions at the flash pressure must be determined. The temperature is determined from a heat balance, by solving Equation 2.8. The iterative solution consists of the following steps:

1. Assume a flash temperature, T.
2. Carry out an isothermal flash calculation at the given pressure, P, and the current temperature, T.
3. If Equation 2.8 is satisfied within some given tolerance, a solution to the problem has been reached. If the error is greater than the tolerance, proceed to the next step.
4. Use an interpolative–extrapolative technique to predict the next temperature approximation and repeat the calculations beginning at step 2. In the first two iterations, the temperature is estimated independently to start the calculations.

2.3.2 Phase Boundary Calculations

Bubble points or dew points can be determined by flash calculations with V in Equations 2.7 or 2.13 set to either 0 or 1. Although the general flash calculation method described in the previous section may be applied to bubble points and dew points, phase boundary calculations are handled more efficiently by other methods. At the bubble point, the liquid composition equals the feed composition since all the feed remains in the liquid phase. The composition of a vapor bubble at equilibrium with the liquid is given by Equation 2.13b, which reduces to

$$Y_i = K_i Z_i = K_i X_i$$

At the bubble point, the following condition must, therefore, be satisfied:

$$\sum Y_i = \sum K_i X_i = 1 \tag{2.16}$$

Similarly, at the dew point, the vapor composition equals the feed composition, and the composition of a liquid drop at equilibrium with the vapor is given by Equation 2.13a as

$$X_i = Z_i / K_i = Y_i / K_i$$

resulting in the dew point condition:

$$\sum X_i = \sum (Y_i / K_i) = 1 \tag{2.17}$$

A bubble point or dew point calculation consists of determining the pressure at a given temperature, or the temperature at a given pressure, which would result in K_is that satisfy Equation 2.16 or 2.17. In addition, the phase equilibrium condition, Equation 2.12, must be satisfied.

Bubble Point–Dew Point Calculations for Composition-Independent K-Values

For ideal solutions, the distribution coefficients may be assumed to be functions of temperature and pressure only, and their dependence on composition may be neglected. With these assumptions the distribution coefficients are given by Raoult's law (Section 1.3):

$$K_i = \frac{p_i^0}{P} \tag{2.18}$$

where p_i^0 is the vapor pressure of component i, and P is the system pressure. Equation 2.18, together with a suitable expression for the vapor pressure, may be used to compute approximate values of the bubble point or dew point pressure or temperature. These may be used as initial estimates in the iterative computations of true bubble points and dew points. Vapor pressures may be calculated from the Antoine equation:

$$p_i^0 = \exp\left(A_i - \frac{B_i}{T + C_i} \right) \tag{2.19}$$

The constants A_i, B_i, and C_i are characteristic to each component. Combining Equations 2.18 and 2.19 with 2.16 gives an equation for the bubble point; combining with Equation 2.17 gives an equation for the dew point. Thus, at the bubble point,

$$\frac{1}{P} \sum_i \frac{Y_i}{\exp\left(A_i - \dfrac{B_i}{T + C_i} \right)} = 1 \tag{2.20}$$

and at the dew point,

$$P \sum_i \frac{Y_i}{\exp\left(A_i - \dfrac{B_i}{T + C_i} \right)} = 1 \tag{2.21}$$

The summations in Equations 2.20 and 2.21 are carried over all components in the system. If the temperature is fixed, the bubble point and dew point pressures may be calculated directly from Equations 2.20 and 2.21. If, however, the pressure is fixed and the bubble point or dew point temperature is required, Equations 2.20 or 2.21 must be solved by an iterative technique such as the Newton–Raphson method because the equations are implicit in the temperature.

Heuristic Method

When the distribution coefficients are composition dependent, the above method must be modified to account for the effect of composition. A search for the unknown bubble point or dew point temperature or pressure is started on the basis of some composition-independent relationship between K-values and the temperature and pressure, such as an expression combining Equations 2.18 and 2.19. Component fugacities are then calculated for the vapor phase and the liquid phase, and the K-values are updated using Equation 2.15. The calculations are repeated until Equations 2.16 or 2.17, as well as Equation 2.12, are satisfied. The iterative scheme for the bubble point pressure calculation may proceed with the following steps:

1. Assume a bubble point pressure based, for example, on the composition-independent K-value method.
2. Using current K-values, calculate $Y_i = K_i X_i$ for $i = 1, \ldots, C$.
3. Normalize the vapor composition by replacing Y_i with $Y_i/\Sigma Y_i$.
4. Calculate vapor and liquid fugacities, f_i^V and f_i^L, for $i = 1, \ldots, C$.
5. Update the K-values using Equation 2.15.
6. Update the pressure using an approximate relationship such as Equation 2.20.

Repeat the computations at step 2. A solution is reached when $\Sigma Y_i = 1$ and $f_i^L = f_i^V$ for all components.

Simultaneous Method

In this method the equations describing the system are all solved simultaneously using, for instance, the Newton–Raphson technique. Applied to phase boundary conditions, the basic relationships in Equations 2.7, 2.8, and 2.12 take a special form. Since V is either 0 at the bubble point or 1 at the dew point, Equation 2.7 reduces to the trivial relationship $Z_i = X_i$ or $Z_i = Y_i$. Equation 2.8 need not be involved in the simultaneous solution of the system equations. Once the phase boundary conditions are determined, this equation may be solved independently to calculate the heat duty required to bring the feed to its bubble point or dew point. Thus, the only equations to be solved for bubble or dew point calculations are Equations 2.12.

At the bubble point the X_is are known ($X_i = Z_i$) and the temperature or pressure is given. The unknowns are C-1 vapor compositions and the pressure or temperature. At the dew point the Y_is are known ($Y_i = Z_i$) as well as the temperature or pressure, and the unknowns are C-1 liquid compositions plus pressure or temperature. In either case, C equations ($i = 1$ to C) must be solved for C unknowns.

Applying the method to bubble point pressure prediction, for example, where the liquid composition and temperature are known, Equation 2.12 may be written in the form (Peng et al., 1977)

$$F_i(P, Y_1, Y_2, \ldots, Y_{C-1}) = 0 \tag{2.22}$$

for $i = 1, 2, \ldots, C$, the number of components. The problem is to find a set of parameters $P, Y_1, Y_2, \ldots, Y_{C-1}$ that satisfy the set of Equations 2.22. The solution is found iteratively, beginning with parameter estimates $P^0, Y_1^0, Y_2^0, \ldots, Y_{C-1}^0$, where the superscript refers to iteration number. The estimates may be obtained using the approximate bubble point method (Equation 2.20). A correction vector

$$\Delta Y_1^k, \Delta Y_2^k, \ldots, \Delta Y_{C-1}^k$$

is calculated from the coefficient matrix

$$\frac{\partial F_i}{\partial P} \frac{\partial F_i}{\partial Y_i} \frac{\partial F_i}{\partial Y_2} \ldots \frac{\partial F_i}{\partial Y_{C-1}}$$

for $i = 1, \ldots, C$. New parameter values are then calculated as

$$P^{k+1} = P^k + \Delta P^k$$

$$Y_1^{k+1} = Y_1^k + \Delta Y_1^k$$

$$\vdots$$

$$Y_{C-1}^{k+1} = Y_{C-1}^k + \Delta Y_{C-1}^k$$

The iterations are repeated until Equations 2.22 are satisfied within a specified tolerance.

NOMENCLATURE

A, B, C	Antoine vapor pressure constants	p^0	Vapor pressure
		Q	Heat duty
C	Number of components	T	Temperature
F	Degrees of freedom	V	Vapor fraction
F	Feed flow rate	X	Mole fraction in liquid
f	Fugacity	Y	Mole fraction in vapor
h	Liquid enthalpy	Z	Mole fraction in feed
H	Vapor enthalpy	Φ	Fugacity coefficient
K	Distribution coefficient	ϕ	Number of phases
P	Pressure		

SUBSCRIPTS

i (or 1, 2, ...)   Component designation

SUPERSCRIPTS

k   Iteration number designation
L   Liquid phase
V   Vapor phase

REFERENCES

Khoury, F. M., *Hydrocarbon Processing*, December 1978, pp. 155–157.
Peng, D. Y. and D. B. Robinson, A rigorous method for vapor–liquid equilibrium calculations, AIChE 70th Annual Meeting, New York, 1977.
Soave, G., *Chemical Engineering Science*, 27, 1197, 1972.

PROBLEMS

2.1 A hydrocarbon stream is separated into vapor and liquid streams in a flash drum maintained at 80°F and 150 psia. The feed composition and component K-values (assumed composition-independent) are given below. Calculate the vapor rate as a fraction of the feed, and the vapor and liquid compositions.

i	Component	Z_i	K_i
1	Ethane	0.40	2.90
2	Propane	0.35	0.95
3	n-Butane	0.15	0.33
4	n-Pentane	0.10	0.11

2.2 A stream containing ethane, butane, and pentane is fed to a temperature-controlled vessel where phase separation takes place between the vapor and the liquid. The vessel pressure is maintained at 100 psia. The K-values, assumed composition-independent, are given by the equation

$$\ln K_i = A_i - B_i/(T + C_i)$$

where T is in degrees Rankine. The feed composition and the equation constants are given below:

		Z_i	A_i	B_i	C_i
1	Ethane	0.27	5.75	2155	−30.3
2	n-Butane	0.34	6.21	3386	−60.1
3	n-Pentane	0.39	6.52	3973	−72.9

Determine the bubble point and dew point temperatures at 100 psia, and the vapor fraction and temperature that results in a maximum n-butane mole fraction in the vapor.

2.3 The starting conditions of a given hydrocarbon stream are 100°F and 100 psia. The stream is heated in an exchanger where 10,000 Btu/lbmole are transferred to the stream. Calculate the temperature, vapor fraction, and vapor and liquid compositions at the exchanger outlet. Assume the pressure remains constant at 100 psia. Stream composition:

		z_i
1	Ethane	0.25
2	n-Butane	0.35
3	n-Pentane	0.40

Use the K-values given in Problem 2.2. The enthalpies may be calculated on the assumption of ideal solution (zero excess enthalpy). The following data are given for the component enthalpies:

$$H(\text{Btu/lbmole}) = a + bT \ (°F)$$

		Liquid		Vapor	
		a	b	a	b
1	Ethane	2160	22.0	7200	13.5
2	n-Butane	2668	32.5	12590	20.0
3	n-Pentane	2880	40.0	15150	28.5

2.4 Derive equations to calculate the K-values and composition of a vapor at equilibrium with a binary liquid solution having a given composition and a given relative volatility α_{12}, or ratio of K-values. Calculate the composition of the vapor at equilibrium with a liquid having 0.40 mole fraction n-hexane and 0.60 mole fraction n-heptane. The relative volatility of n-hexane to n-heptane is 3.40.

2.5 A vapor mixture of n-hexane, n-heptane, and water is condensed at 1 atm. The hydrocarbon K-values are given by the equation

$$\ln K_i = A_i - B_i/T$$

where T is in degrees Rankine. The mixture composition and the values of the constants are as follows:

		z_i	A_i	B_i
1	n-Hexane	0.62	11.08	6821
2	n-Heptane	0.05	11.29	7545
3	Water	0.33	—	—

Assuming water and the hydrocarbons are immiscible,

a. Calculate the water and hydrocarbon dew point temperatures and equilibrium compositions.

b. Outline an algorithm for calculating the temperature and phase compositions of a hydrocarbon–water mixture when the fraction of the hydrocarbons condensed is β. Assume composition-independent K-values.

2.6 A stream of hydrofluoric acid in water at 250°F and 30 psia contains 12% mole HF. It is proposed to concentrate the HF in solution by partial vaporization in a single stage, by means of temperature and pressure control. Calculate the resulting liquid composition and the fraction vaporized at 250°F and 20 psia. Can this process be used to concentrate the liquid for any starting composition? Use the van Laar equation for liquid activity coefficients and assume ideal gas behavior in the vapor phase. The vapor pressures of HF and water at 250°F are 245.74 psia and 30.04 psia, respectively, and the van Laar constants are (see Problems 1.8 and 1.9)

$$A_{HF} = -6.0983, \ A_{H20} = -6.9658$$

2.7 In a single equilibrium stage absorber at 25°C and 1 atm, water is used to reduce the amount of acetone in 1000 kg-mole/hr air from 2% mole to 0.5% mole. Assuming the amounts of air and water leaving the absorber to be the same as the inlets, calculate the amount of water required. The following information may be used:

Acetone vapor pressure at 25°C = 229 mmHg
Activity coefficient of acetone at low concentrations (infinite dilution) at 25°C = 6.5
Water vapor pressure at 25°C = 24 mmHg
Assume air is insoluble in water–acetone. Assume ideal gas behavior in the vapor phase.

Using the calculated water rate, calculate the compositions of the outlet streams when the assumption of constant liquid and gas rates is dropped.

CHAPTER 3

Fundamentals of Multistage Separation

In the discussion in Chapter 2 on the performance of single stages, it was noted that, for a feed stream of given composition, the range of possible product compositions obtainable with an equilibrium stage at a given pressure is bounded by the bubble point and the dew point compositions at that pressure. In order to extend this range to higher or lower component concentrations, multiple stages are used.

Multistage processes may be considered made of a number of equilibrium stages stacked together to perform various separation functions. Using these building blocks, configurations may be designed to perform functions such as distillation, absorption, stripping, and rectification. This chapter discusses the conceptual construction of several multistage processes from equilibrium stages.

Multiple stages enhance separation in two ways. One is that temperature variation from stage to stage affects the distribution coefficients so that at lower temperatures the lighter components' concentration is driven up and vice versa at higher temperatures. The other way is the progressive enrichment or depletion of different volatility components from stage to stage resulting from changes in the compositions and flows of fluids in the process. The two factors are generally at work together to varying degrees in most multistage separation processes. A purely temperature effect process may be seen in binary distillation, while a purely compositional effect would be a constant temperature absorption process. The two effects are analyzed separately in detail in the following sections.

The separation characteristics of a process are determined by the number of stages that make it up, their configuration, and the operating conditions. In general, as the process becomes more complex, the number of variables required to define it increases, making it more versatile and enabling it to handle wider separation ranges. The specification-variable relationships associated with a number of fundamental configurations are discussed next and illustrated with examples.

3.1 CASCADED STAGES

Figure 3.1 illustrates a process in which a feed stream, F, is flashed in stage 2 at temperature T_2 and pressure P_2 to produce a saturated vapor stream, V_2 at its dew point and a saturated liquid stream, L_2 at its bubble point. The vapor is sent to stage 1 where it is partially condensed at temperature T_1, and the liquid is sent to stage 3 where it is partially vaporized (or reboiled) at temperature T_3. The resulting vapor and liquid streams are V_1, L_1, V_3, and L_3 as indicated.

The pressures in the three stages are assumed equal. In order for condensation to take place in stage 1, T_1 must be below the dew point of $V_2 (T_1 < T_2)$. Also, for vaporization to occur in stage 3, T_3 must be above the bubble point of $L_2 (T_3 > T_2)$.

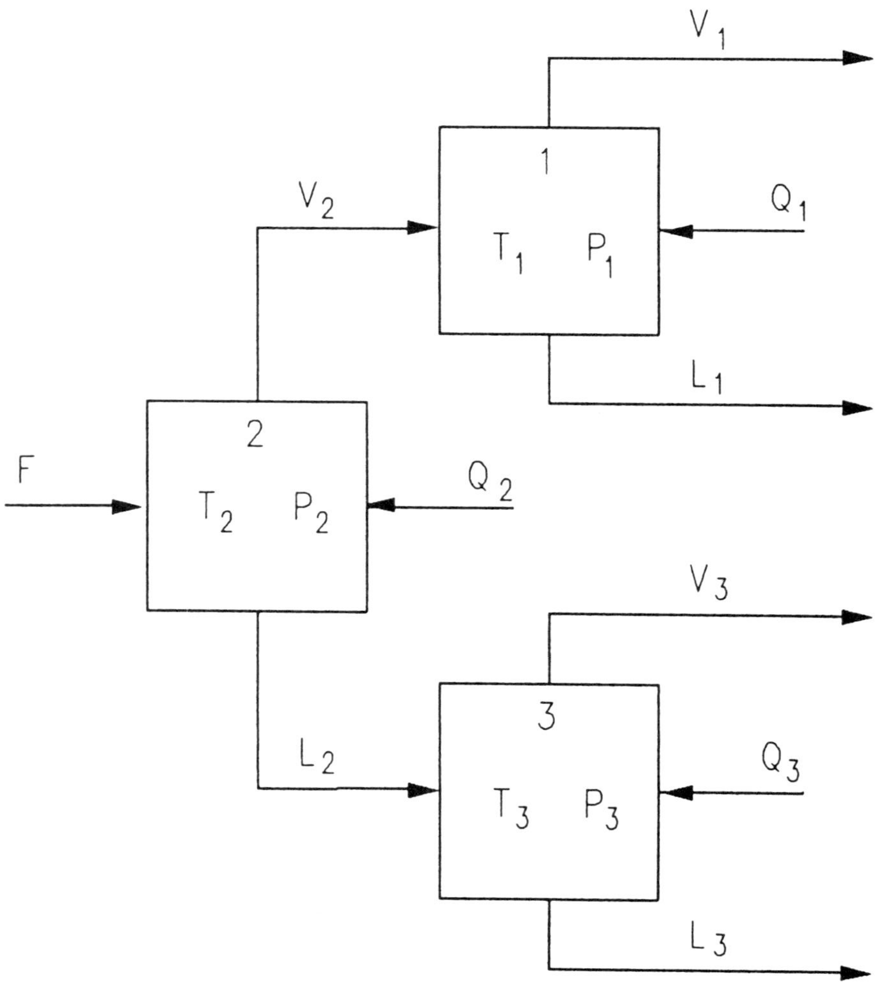

Figure 3.1   Cascaded stages.

The system of cascaded equilibrium stages therefore results in a wider temperature spread between the separated products' dew point and bubble point than in a single stage. Stream V_1 is richer in lighter components than V_2, and stream L_3 is richer in heavier components than L_2, the net result being sharper separation.

3.1.1 Graphical Representation

The operation of cascaded stages may be described for a binary on a temperature–concentration diagram, Figure 3.2. Starting with a feed stream to stage 2 with mole fraction component 1 (the lighter component) equal to Z_{21}, the highest possible concentration of 1 at that stage would be Y_{21}, which is the concentration of the first vapor bubble. (The first subscript refers to the stage and the second subscript to the component.) The lowest possible concentration would be X_{21}, which corresponds to the first condensing liquid drop when the dew point curve is approached from higher temperatures.

If the feed to stage 2 is raised to some temperature T_2, intermediate between the bubble point and dew point, the vapor would have a concentration Z_{11} and the liquid

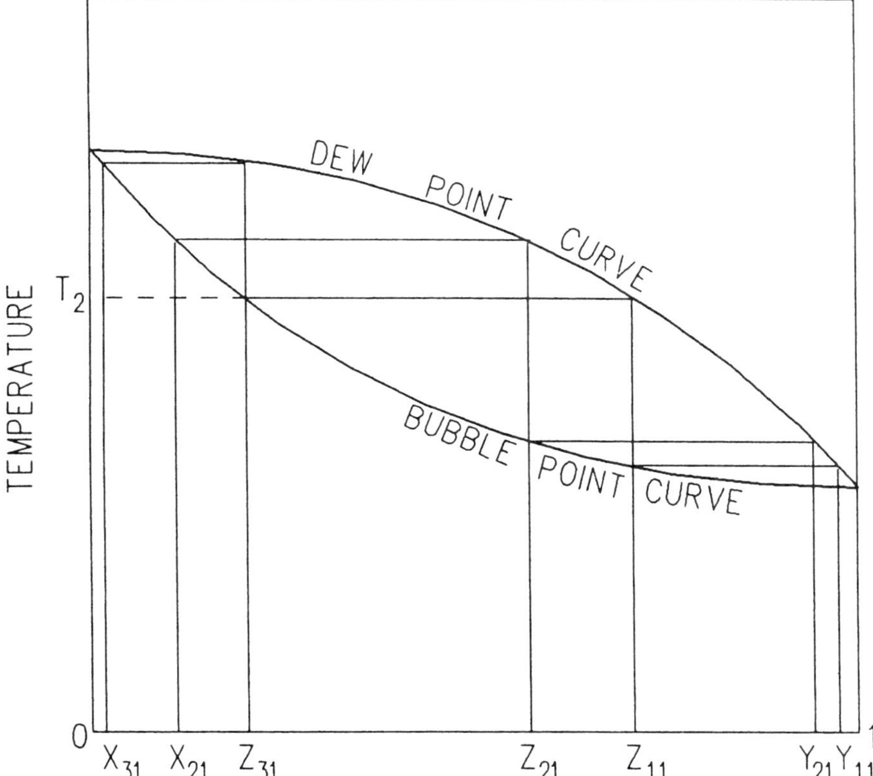

Figure 3.2 Graphical representation of cascaded stages.

Z_{31}. If the vapor with concentration Z_{11} is fed to stage 1, then the maximum possible concentration of 1 at stage 1 would be Y_{11}. Similarly, if the liquid with concentration Z_{31} is fed to stage 3, then the minimum possible concentration of 1 at stage 3 would be X_{31}. Clearly, Y_{11} is greater than Y_{21}, and X_{31} is smaller than X_{21}. Thus, a wider spread of product concentrations is obtainable from cascaded stages than from a single stage.

3.1.2 Equilibrium Relationships

The performance of cascaded stages may also be described analytically on the basis of material balances and phase equilibrium relations. Consider, for simplicity, a binary mixture of components 1 and 2 and again assume component 1 to be the lower boiling, or more volatile, component. The equilibrium equations for stage j are

$$\frac{Y_{j1}}{X_{j1}} = K_{j1} \tag{3.1}$$

$$\frac{Y_{j2}}{X_{j2}} = \frac{1 - Y_{j1}}{1 - X_{j1}} = K_{j2} \tag{3.2}$$

Equation 3.2 makes use of the fact that $X_{j1} + X_{j2} = 1$ and $Y_{j1} + Y_{j2} = 1$. Assuming the distribution coefficients K_{j1} and K_{j2} to be known quantities or estimated by some data predicting method, Equations 3.1 and 3.2 may be solved for the unknowns X_{j1} and Y_{j1}:

$$X_{j1} = \frac{1 - K_{j2}}{K_{j1} - K_{j2}} \tag{3.3}$$

and

$$Y_{j1} = \frac{K_{j1}\left(1 - K_{j2}\right)}{K_{j1} - K_{j2}} \tag{3.4}$$

The relative volatility, defined as the ratio of distribution coefficients,

$$\alpha_{j12} = \frac{K_{j1}}{K_{j2}}$$

is now combined with Equations 3.3 and 3.4 to give

$$X_{j1} = \frac{\alpha_{j12}}{\alpha_{j12} - 1}\frac{1}{K_{j1}} - \frac{1}{\alpha_{j12} - 1} \tag{3.5}$$

$$Y_{j1} = \frac{\alpha_{j12}}{\alpha_{j12} - 1} - \frac{K_{j1}}{\alpha_{j12} - 1} \tag{3.6}$$

For systems that are not highly nonideal, in which the distribution coefficients are not strong functions of the composition, it may be assumed that the relative volatility is fairly constant over reasonably wide temperature ranges.

In the cascaded flash, since $T_3 > T_2 > T_1$ and because distribution coefficients generally go up with temperature, it follows from Equations 3.5 and 3.6 that Y_{11} is greater than Y_{21} and X_{32} is greater than X_{22}. To arrive at this, recall that component 1 is the more volatile component; hence, $K_{j1} > K_{j2}$ and $\alpha_{j12} > 1$, which implies that the ratio

$$\alpha_{j12}/(\alpha_{j12} - 1)$$

is always greater than 1 and

$$1/(\alpha_{j12} - 1)$$

is always positive. If the relative volatility, and hence the two ratios above, are assumed to have the same values on stages 1 and 2, Equation 3.6 indicates that as K_{j1} decreases, Y_{j1} increases, and, since T_1 is less than T_2, K_{11} is less than K_{21}, and Y_{11} is greater than Y_{21}. Similar reasoning may be applied to Equation 3.5 to conclude that X_{32} is greater than X_{22}. This provides an analytical verification that improved separation is obtainable with cascaded flashes.

3.1.3 Parameter Relationships

The cascaded flash is also more versatile than the single flash in that more parameters may be varied to manipulate its performance. Since stage 2 has a feed of fixed rate, composition, and thermal conditions, two parameters are required to define its operation, as discussed in Chapter 2. These may be chosen as the pressure and duty. If we assume the pressure is fixed because of practical process considerations, 1 degree of freedom remains.

In a three-stage cascaded flash system, each stage could have an associated heat source or sink. If these duties are fixed and the pressure on each stage is also set, the operation of the system is defined. This implies that all the other variables, such as stage temperatures and product rates and compositions, are determinate. The three duties may now be varied independently to allow three of the other variables to meet certain specifications. By contrast, in a single stage only one specification is possible at fixed pressure. The three specifications apply to the system as a whole, and it is not mandatory that each specification be associated with a particular stage.

Equations 3.1 and 3.2, or their alternative forms, Equations 3.3 and 3.4, and 3.5 and 3.6 represent the phase equilibrium relationships on stage j for a binary mixture. In addition to these, component material balances for each component on stage j generate the following equations:

$$F_j Z_{j1} = V_j Y_{j1} + L_j X_{j1} \tag{3.7}$$

$$F_j Z_{j2} = V_j Y_{j2} + L_j X_{j2} \tag{3.8}$$

Here F_j, V_j, and L_j are the feed, vapor product, and liquid product molar rates for stage j, and Z_{ji}, Y_{ji}, and X_{ji} are the feed, vapor, and liquid mole fractions of component i at stage j. Summation of Equations 3.7 and 3.8 gives total material balance,

$$F_j = V_j + L_j$$

which may be substituted back in Equation 3.7 or 3.8 to eliminate V_j or L_j. The result for the vapor rate is

$$V_j = \frac{F_j \left(Z_{j1} - X_{j1} \right)}{Y_{j1} - X_{j1}} \tag{3.9}$$

An energy balance equation may also be written for each stage. All in all, there are five independent equations for each stage with seven unknowns: X_{j1}, Y_{j1}, V_j, L_j, T_j, P_j, and O_j, where O_j is the heat duty on stage j. Two variables per stage must be given in order to define a unique solution. The following example illustrates the solution of a three-stage cascaded flash with a binary mixture for different sets of performance specifications.

Example 3.1 Propane–Butane Cascaded Flash Separation

A mixture of 50 moles/hr propane and 50 moles/hr butane at 100°F and 100 psia is to be separated into propane-rich and butane-rich products using a three-stage cascaded flash arrangement similar to Figure 3.1. The pressure on each stage is fixed at 100 psia. Calculate the product rates and compositions and the stage temperatures that correspond to each of the following operating conditions:

A. The stage temperatures are specified:

    $T_1 = 90°F$
    $T_2 = 100°F$
    $T_3 = 110°F$

B. The following are specified:

    Mole fraction propane in the vapor product from stage 1, $Y_{11} = 0.80$
    Flow rate of the vapor product from stage 2, $V_2 = 60$ mole/hr
    Flow rate of butane in the liquid product from stage 3, $L_3 X_{32} = 25$ mole/hr

C. Same as B, with one difference: specify the flow rate of propane in the vapor product from stage 1, $V_1 Y_{11} = 10$ mole/hr, instead of the vapor product from stage 2.

Distribution coefficients of propane and butane are given at 100 psia and at five different temperatures in Table 3.1.

Table 3.1   Distribution Coefficients of Propane and Butane at 100 psia

T, F	80	90	100	110	120
K (C3)	1.370	1.540	1.710	1.900	2.090
K (C4)	0.428	0.496	0.571	0.654	0.744
α	3.20	3.10	2.99	2.91	2.81

Solution. The three stages involve six independent variables. Since the three-stage pressures are fixed, three other specifications are made in each of the above cases. In principle, the solution involves setting up 15 equations with 21 variables, six of which are given. The simplicity of this problem makes it more practical to group a few equations at a time and eliminate some of the variables rather than solve the whole system of equations simultaneously. This method is not particularly systematic and cannot be generalized to more complex problems but is a useful exercise for studying the relationships between variables. The specifications for cases A and B are such that the stages may be solved sequentially, one at a time. In case C, two stages must be solved simultaneously. The heat duties for each stage are not calculated, although the temperatures are. Enthalpy data would be required to carry out heat balances for duty calculations.

A. Since the stage temperatures are specified in this case, the corresponding distribution coefficients are used directly in Equations 3.3 and 3.4 to calculate the component mole fractions in the liquid and vapor on each stage. The results are given in Table 3.2.

Table 3.2   Calculated Mole Fractions of Propane and Butane at 100 psia (Example 3.1)

Stage	1	2	3
T, F	90	100	110
K (C3)	1.540	1.710	1.900
K (C4)	0.496	0.571	0.654
X (C3)	0.483	0.377	0.278
X (C4)	0.517	0.623	0.722
Y (C3)	0.744	0.644	0.528
Y (C4)	0.256	0.356	0.472

Next, Equation 3.9 is used to calculate the vapor rate leaving stage 2:

$$V_2 = 100 \ (0.50 - 0.377)/(0.644 - 0.377) = 46.07 \ \text{mole/hr}$$

The liquid rate is calculated from a material balance on stage 2:

$$L_2 = 100 - 46.07 = 53.93 \ \text{mole/hr}$$

The feed to stage 1 is V_2 and the products from this stage are calculated as above:

$$V_1 = 46.07 - (0.644 - 0.483)/(0.744 - 0.483) = 28.42 \text{ mole/hr}$$

$$L_1 = 46.07 - 28.42 = 17.65 \text{ mole/hr}$$

Similarly, L_2 is the feed to stage 3 and the products are

$$V_3 = 53.93 \ (0.377 - 0.278)/(0.528 - 0.278) = 21.36 \text{ mole/hr}$$

$$L_3 = 53.93 - 21.36 = 32.57 \text{ mole/hr}$$

B. Stage 2 is solved first. Since the temperature is not known, the functional dependence of the distribution coefficients on the temperature must be determined. The
data in Table 3.1 show an approximately linear relationship between K-values and
temperature. A good representation of the propane distribution coefficients within
the indicated temperature range may be expressed by the equation

$$K(C_3) = 1.36 + 0.01805[T(F) - 80] \tag{3.10}$$

The butane distribution coefficients could also be expressed by a similar equation.
The computations may be simplified, however, by assuming constant relative volatility, which has an average value of 3.00, based on data in Table 3.1.

The phase equilibrium and material balance relations are applied to stage 2
using Equations 3.6 and 3.7:

$$Y_{21} = 3.00/(3.00 - 1) - K_{21}/(3.00 - 1)$$

$$(100)(0.50) = 60Y_{21} + 40X_{21}$$

Substituting $X_{21} = Y_{21}/K_{21}$ in the second equation and simplifying, the above two
equations are rewritten as

$$Y_{21} = 1.5 - 0.5K_{21} \tag{3.11}$$

$$Y_{21}(6 + 4/K_{21}) = 5 \tag{3.12}$$

Equations 3.11 and 3.12 are combined by eliminating Y_{21}, resulting in a quadratic
equation in K_{21}:

$$3K_{21}^2 - 2K_{21} - 6 = 0$$

The positive root of this equation is $K_{21} = 1.786$. The temperature is now computed
directly from Equation 3.10, $T_1 = 103.60°F$ and the butane distribution coefficient
is calculated from the relative volatility,

$$K_{22} = K_{21}/3 = 1.786/3.00 = 0.595$$

The distribution coefficient of propane on stage 1 is calculated from the equivalent of Equation 3.11 written for stage 1, with $Y_{11} = 0.80$:

$$0.80 = 1.5 - 0.5K_{11}$$

$$K_{11} = 1.400$$

and for butane

$$K_{12} = 1.400/3 = 0.467$$

Stage 1 temperature, T_1, is calculated from Equation 3.10:

$$T_1 = 80 + (K_{11} - 1.36)/0.01805 = 82.22°F$$

The calculations for stage 3 follow. The feed to stage 3 is the liquid from stage 2:

$$L_2 = 40 \text{ mole/hr}$$

From Equation 3.11,

$$Y_{21} = 1.5 - (0.5)(1.786) = 0.607$$

$$X_{21} = Y_{21}/K_{21} = 0.607/1.786 = 0.340$$

$$X_{22} = 1 - X_{21} = 0.660$$

A butane balance is made on stage 3:

$$L_2X_{22} = V_3Y_{32} + L_3X_{32}$$

The problem specifies that $L_3X_{32} = 25$ mole/hr, hence,

$$V_3Y_{32} = (40)(0.660) - 25 = 1.4 \text{ mole/hr}$$

From a total material balance on stage 3,

$$V_3 + L_3 = 40$$

Substituting for V_3, L_3, and $K_{32} = Y_{32}/X_{32}$,

$$1.4/Y_{32} + 25K_{32}/Y_{32} = 40$$

or

$$40Y_{32} = 25K_{32} + 1.4$$

Using the equivalent of Equation 3.6 for component 2 and assuming a constant relative volatility of 3.00, the following is obtained:

$$Y_{32} = \frac{1/3.00}{1/3.00 - 1} - \frac{K_{32}}{1/3.00 - 1}$$

or

$$Y_{32} = -0.5 + 1.5\, K_{32}$$

This equation is combined with the previous one to eliminate Y_{32}, yielding $K_{32} = 0.611$. The propane distribution coefficient is

$$K_{31} = (3.00)(0.611) = 1.833$$

Stage 3 temperature is calculated from Equation 3.10:

$$T_3 = 80 + (K_{31} - 1.36)/0.01805 = 106.2°F$$

C. Two specifications on stage 1 must be satisfied by varying conditions on stages 1 and 2. Therefore, stages 1 and 2 should be solved simultaneously. The following equations apply to stage 1:
The material balance:

$$V_2 = V_1 + L_1$$

Propane material balance:

$$V_2 Y_{21} = V_1 Y_{11} + L_1 X_{11}$$

Phase equilibrium relationship for propane:

$$K_{11} = Y_{11}/X_{11}$$

The following equations apply to stage 2:
Total material balance:

$$F_2 = V_2 + L_2$$

Propane material balance:

$$F_2 Z_{21} = V_2 Y_{21} + L_2 X_{21}$$

Phase equilibrium relationship for propane:

$$K_{21} = Y_{21}/X_{21}$$

The following variables are given or specified:

$$F_2 = 100.00 \text{ mole/hr}$$

$$Z_{21} = 0.50$$

$$Y_{11} = 0.80$$

$$V_1 Y_{11} = 10.00 \text{ mole/hr}$$

$$V_1 = 10/0.80 = 12.50 \text{ mole/hr}$$

These values are substituted in the equations for stage 1, which are combined to give the following:

$$V_2 Y_{21} = 10.00 + (V_2 - 12.50)(0.80)/K_{11}$$

Stage 2 equations result in

$$50 = V_2 Y_{21} + (100.00 - V_2)Y_{21}/K_{21}$$

In addition to these, Equation 3.6 is applied to stages 1 and 2, assuming the relative volatility has a constant value of 3.00:

$$Y_{11} = 1.5 - 0.5 \, K_{11} = 0.80$$

$$Y_{21} = 1.5 - 0.5 \, K_{21}$$

The last four equations are to be solved for Y_{21}, V_2, K_{11}, and K_{21}. K_{11} is solved directly:

$$K_{11} = (1.5 - 0.8)/0.5 = 1.40$$

K_{21} is eliminated, leaving two equations in V_2 and Y_{21}. An iterative solution gives $V_2 = 50.736$ mole/hr and $Y_{21} = 0.632$. The other variables are readily solved using the above system of equations. Stage 3 is solved independently as in Case B.

3.2 DISTILLATION BASICS

When the vapor product from a feed stage is condensed in another stage at a temperature lower than the feed stage temperature, a two-stage cascaded operation is formed. Vapor from the lower temperature stage is richer in lighter components and leaner in heavier components than vapor from the feed stage, as discussed in the previous section. The temperature at each stage is now held constant and the liquid from the upper stage is recycled to the feed stage (Figure 3.3). If the feed stream is a binary mixture, the compositions of streams V_1 and L_2 are essentially unchanged from the cascaded arrangement since the compositions in such systems

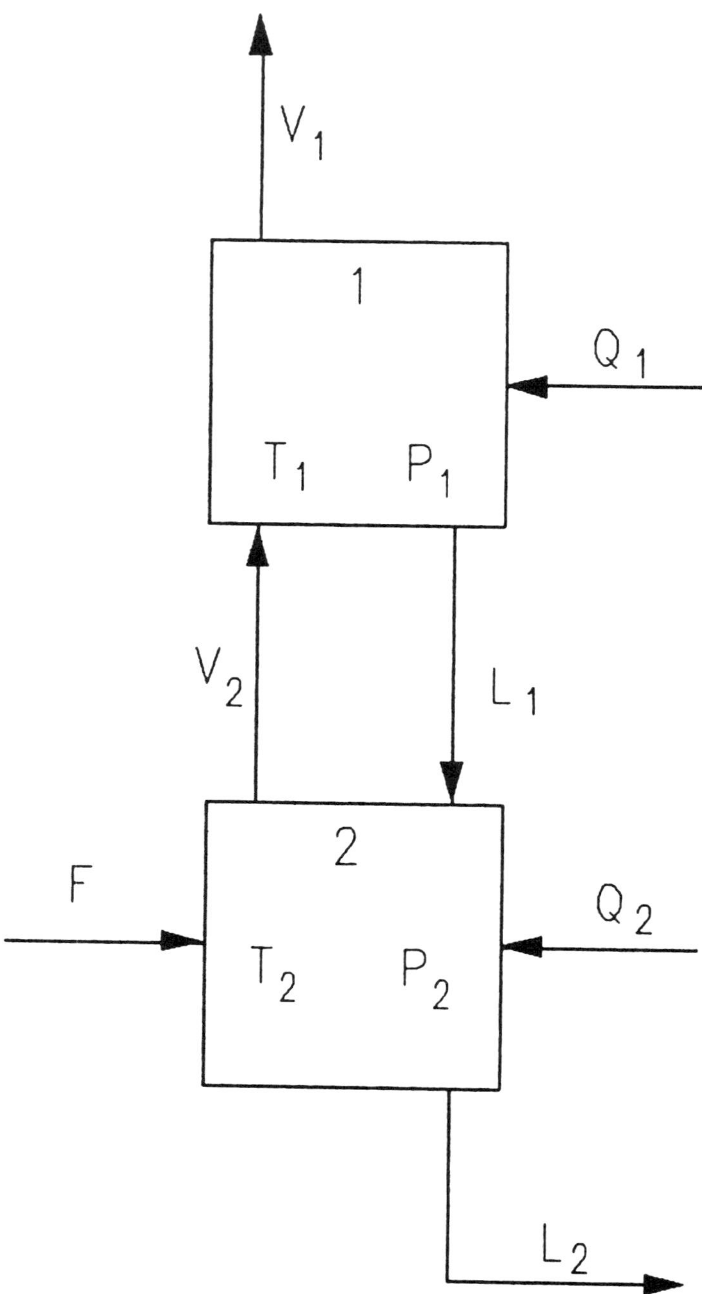

Figure 3.3   Two-stage distillation column.

are primarily functions of the temperature, as demonstrated in Section 3.2.2. The rates of products V_1 and L_2, however, are now higher since L_1 is recycled instead of being taken as an additional product.

This arrangement represents a basic multistage separation process. Figure 3.3 illustrates a two-stage distillation column where heavier components of the vapor rising from the bottom stage are condensed and refluxed back to the lower stage, resulting in an overhead vapor product that has a higher concentration of lighter components. Liquid flowing down to the lower stage is reboiled and lighter components are vaporized and sent back to the upper stage so that a bottoms liquid stream is produced with a higher concentration of heavier components. In order to effect condensation at the upper stage and vaporization at the bottom stage, heat duties are applied to each stage: Q_1 negative (heat removed) and Q_2 positive (heat added).

3.2.1 Temperature Effect on Separation

As pointed out earlier in this chapter, it is helpful in understanding multistage separation to distinguish between two effects: temperature and composition. (Other factors, such as pressure, number of stages, column configuration, etc., are assumed fixed in this context.) Of the two effects, the more predominant in distillation is temperature. In the case of binary distillation where the composition effect on distribution coefficients may be neglected, temperature is essentially the only factor. Under these conditions the equations derived in Section 3.1.2 are applicable.

The added separation power of a two-stage distillation column (Figure 3.3) over a single stage results from the fact that the two stages can be maintained at two different temperatures. For a binary system the compositions are given by Equations 3.3 and 3.4 or 3.5 and 3.6. These equations indicate that separation, expressed in terms of mole fractions of the products, is a function of K-values. In this simple model the K-values are a function of temperature only. Hence, the temperature effect is the only active one in binary distillation.

If more stages are added, the temperature spread becomes wider and therefore separation becomes sharper. Also, as discussed later on, increased vapor rate from stage 2 to 1 and liquid rate from 1 to 2 results in a larger temperature difference between the stages and, hence, better separation.

3.2.2 Mathematical Representation

A mathematical formulation of the column of Figure 3.3 follows for a binary mixture. Material balances for components 1 and 2 are written for stages 1 and 2, with the first subscript referring to the stage and the second subscript to the component.

$$FZ_{21} + L_1X_{11} = V_2Y_{21} + L_2X_{21} \tag{3.13}$$

$$F(1 - Z_{21}) + L_1(1 - X_{11}) = V_2(1 - Y_{21}) + L_2(1 - X_{21}) \tag{3.14}$$

$$V_2 Y_{21} = V_1 Y_{11} + L_1 X_{11} \tag{3.15}$$

$$V_2(1 - Y_{21}) = V_1(1 - Y_{11}) + L_1(1 - X_{11}) \tag{3.16}$$

Additionally, Equations 3.1 and 3.2 derived for cascaded stages in Section 3.2.2 are rewritten for the present two-stage column:

$$\frac{Y_{11}}{X_{11}} = K_{11} \tag{3.17}$$

$$\frac{Y_{21}}{X_{21}} = K_{21} \tag{3.18}$$

$$\frac{1 - Y_{11}}{1 - X_{11}} = K_{12} \tag{3.19}$$

$$\frac{1 - Y_{21}}{1 - X_{21}} = K_{22} \tag{3.20}$$

Two more equations may be derived to represent enthalpy balances on each stage. The enthalpy balance equations provide functional relationships between stage temperatures T_1 and T_2 and duties Q_1 and Q_2.

All in all, 10 equations exist with 12 unknowns: X_{11}, X_{21}, Y_{11}, Y_{21}, L_1, L_2, V_1, V_2, T_1, T_2, Q_1, and Q_2. Hence, in order to solve the column, two of the variables must be specified. The distribution coefficients may be supplied as additional data or may be calculated from appropriate correlations. The column equations may be manipulated in different ways, depending on which of the variables are specified and which are to be solved. For instance, a chain of eliminations and substitutions may be carried out to develop an expression for L_1/V_1, the ratio of reflux to vapor from the top tray, or *reflux ratio*:

$$\frac{L_1}{V_1} = \frac{Y_{21} - Y_{11}}{X_{11} - Y_{21}} \tag{3.21}$$

The application of these equations to specific cases is demonstrated in Example 3.2. The number and complexity of the equations for multistage, multicomponent columns are much higher, which rules out elimination and substitution methods. General systematic techniques for solving systems of column equations are discussed in Chapter 13.

3.2.3 Parameter Relationships

In the two-stage distillation column (Figure 3.3), the operating pressure and the feed flow rate, composition, and thermal conditions are assumed fixed. Each stage with an associated heat source or sink has 1 degree of freedom (Chapter 2). The column as a whole then has 2 degrees of freedom. The duties Q_1 and Q_2 may be considered the independent variables. Their magnitudes control the column operation and, if they are allowed to vary, two other performance variables may be specified. For instance, the liquid rate from the bottom stage and the reflux rate (liquid from the top stage flowing down the column) may be specified. This completely defines the column operation; hence, all the other variables, including the two duties, may be calculated.

If the process is operated at a specified liquid rate from stage 2 (L_2), the vapor rate from stage 1 (V_1) is fixed by virtue of overall material balance. A second independent variable, the liquid from stage 1, or the reflux rate L_1, is now specified. For a given bottoms rate L_2 or overhead rate V_1, the effect of reflux rate on stage temperatures and separation is examined next.

As L_1 increases, the fraction of feed vaporized on stage 2 increases because of the following. The vapor fraction is

$$V_2/(F + L_1)$$

A material balance on stage 2 gives

$$V_2 = F + L_1 - L_2$$

which is substituted in the vapor fraction expression, resulting in

$$(F + L_1 - L_2)/(F + L_1)$$

or

$$1 - L_2/(F + L_1)$$

It may be readily verified that, as L_1 increases, the vapor fraction increases since F and L_2 are fixed. A higher vapor fraction means higher temperature. Therefore, an increased reflux rate implies an elevated temperature at the vaporization state. The reverse is true for the condensation stage. As the reflux rate increases, so does the vaporization rate, V_2, in order to maintain a constant bottoms rate, L_2. The fraction vapor on stage 1 is V_1/V_2, and, as V_2 increases, this fraction decreases since V_1 is constant. Thus a lower temperature is obtained on the condensation stage as a result of higher reflux. The duties Q_1 and Q_2 also go up with increasing reflux rate. More cooling is required for condensation and more heating for vaporization.

Another result of higher reflux rate or ratio is sharper separation: V_1 becomes richer in lighter components and L_2 becomes richer in heavier components. At higher reflux, the temperature gradient from one stage to the other becomes steeper, resulting in higher K-values in the lower stage and lower K-values in the upper stage. The effect of that is higher concentration of the light components in the upper stage and the opposite on the lower stage, as may be inferred from Equations 3.5 and 3.6 for a binary system.

The reflux ratio may vary between the limits of 0 and infinity. At 0 reflux ratio, the operation of a multistage process reduces to that of a single stage. This may be verified by setting L_1/V_1 to 0 in Equation 3.21, which implies that $Y_{21} = Y_{11}$. Thus at 0 reflux, the separation is lowest, being the same as in a single stage.

As the reflux ratio is increased, the separation becomes sharper and, as the limit of $L_1/V_1 = \infty$ is approached, the maximum separation that can be achieved with a given number of stages is reached. The column is then said to operate at total reflux. In practical terms, total reflux may be approached by operating the column so that the reflux rate is very large compared to the feed and product rates. From Equation 3.21, at total reflux, $X_{11} = Y_{21}$; that is, the liquid composition from stage 1 equals the vapor composition from stage 2.

The effect of reflux ratio on the performance of a two-stage distillation column is demonstrated numerically for a binary system in the following example.

Example 3.2 Propane–Butane Two-Stage Distillation

The propane–butane equimolar mixture of Example 3.1 at 100°F and 100 psia is fed at a rate of 100 moles/hr to the bottom stage of a two-stage distillation column as in Figure 3.3. The liquid rate, L_2, is specified at 50 mole/hr, and it is desired to determine the effect of reflux ratio, L_1/V_1, on the products composition and stage temperatures.

Solution. At the limit of 0 reflux ratio, Equation 3.21 reduces to

$$Y_{21} = Y_{11}$$

From a propane material balance on the entire column,

$$FZ_{21} = V_1 Y_{11} + L_2 X_{21}$$

$$(100)(0.50) = 50Y_{11} + 50X_{21}$$

$$1 = Y_{11} + X_{21}$$

Vapor–liquid equilibria are expressed in the form of Equation 3.6 where an average relative volatility of 3.00 is assumed, as in Example 3.1. The following equations are obtained:

$$Y_{11} = 3X_{11}/(1+2X_{11})$$

$$Y_{21} = 3X_{21}/(1+2X_{21})$$

The above four equations are solved for X_{11}, X_{21}, Y_{11}, and Y_{21}. By eliminating all these variables except X_{11}, the following quadratic equation is obtained:

$$2X_{11}^2 + 2X_{11} - 1 = 0$$

for which one positive root exists, $X_{11} = 0.366$. The other variables are solved by substitution, giving $Y_{11} = 0.634$, $X_{21} = 0.366$, and $Y_{21} = 0.634$. The two stages have identical compositions at 0 reflux ratio. The K-value of propane is calculated from its definition

$$K_{11} = K_{21} = Y_{11}/X_{11} = Y_{21}/X_{21} = 0.634/0.366 = 1.732$$

The temperature is also the same on both stages and is calculated using Equation 3.10:

$$T_1 = T_2 = 80 + (K_{11} - 1.36)/0.01805 = 100.6°F$$

As the reflux ratio approaches infinity (total reflux), Equation 3.21 requires that

$$X_{11} = Y_{21}$$

The other equations are the same as in the case of 0 reflux ratio. By use of simple eliminations and substitutions, the four equations are solved simultaneously, giving the following results:

$$X_{11} = 0.5$$

$$Y_{11} = 0.75$$

$$X_{21} = 0.25$$

$$Y_{21} = 0.50$$

The K-values are calculated as before:

$$K_{11} = Y_{11}/X_{11} = 0.75/0.50 = 1.5$$

$$K_{21} = Y_{21}/X_{21} = 0.50/0.25 = 2.0$$

and the stage temperatures from Equation 3.10:

$$T_1 = 80 + (1.5 - 1.36)/0.01805 = 87.8°F$$

$$T_2 = 80 + (2.0 - 1.36)/0.01805 = 115.5°F$$

The computations for reflux ratios between 0 and infinity are more complicated. Systematic solution methods of the set of generally nonlinear equations are employed. For the present two-stage binary system, the equations are solved for a reflux ratio $L_1/V_1 = 1$ by elimination.

Since $L_2 = 50$ mole/hr, it follows by overall material balance that $V_1 = 50$ mole/hr. And since $L_1/V_1 = 1$, $L_1 = 50$ mole/hr. Also,

$$V_2 = F + L_1 - L_2 = 100 + 50 - 50 = 100 \text{ mole/hr}$$

Substituting the value of L_1/V_1 in Equation 3.21 gives

$$(Y_{21} - Y_{11})/(X_{11} - Y_{21}) = 1$$

or

$$2Y_{21} - Y_{11} - X_{11} = 0$$

The other equations are the same as for $L_1/V_1 = 0$ or infinity. Eliminating Y_{11} and Y_{21} results in the following two equations in X_{11} and X_{21}:

$$\frac{6X_{21}}{1 + 2X_{21}} - \frac{3X_{11}}{1 + 2X_{11}} = 0$$

$$X_{21} = \frac{1 - X_{11}}{1 + 2X_{11}}$$

By eliminating X_{21} between these two equations, the following quadratic equation in X_{11} is obtained:

$$3X_{11}^2 + X_{11} - 1 = 0$$

The positive root is calculated to be $X_{11} = 0.434$. Working back through the other equations, the remaining compositions are calculated as $X_{21} = 0.303$, $Y_{11} = 0.697$, and $Y_{21} = 0.566$. The K-values and stage temperatures are calculated as in the first two cases:

$$K_{11} = Y_{11}/X_{11} = 0.697/0.434 = 1.606$$

$$K_{21} = Y_{21}/X_{21} = 0.566/0.303 = 1.868$$

$$T_1 = 80 + (1.606 - 1.36)/0.01805 = 93.6°F$$

$$T_2 = 80 + (1.868 - 1.36)/0.01805 = 108.1°F$$

The above results for reflux ratios of 0, 1, and infinity are given in Table 3.3. Also tabulated are data points at other values of reflux ratio, calculated by computer simulation using the Soave–Redlich–Kwong (SRK) equation of state for predicting the K-values. The concentrations are also plotted in Figure 3.4, showing the effect of reflux ratio on separation. In this example, since $V_1 = L_2$, an overall material balance indicates that at any point $Y_{11} + X_{21} = 1$.

Table 3.3 Performance of Two-Stage Distillation Column (Example 3.2)

Reflux Ratio, L_1/V_1	T_1, F	T_2, F	Y_{11}	X_{21}
0*	100.60	100.60	0.634	0.366
0.2**	98.97	103.05	0.655	0.345
1.0*	93.60	108.10	0.697	0.303
20**	88.89	111.49	0.742	0.258
∞*	87.80	115.50	0.750	0.250

* As computed in Example 3.2.
**Simulation results, using the Soave–Redlich–Kwong equation of state for predicting K-values.

Feed Rate, F = 100 mole/hr: 50 mole/hr propane
                              50 mole/hr butane

Overhead rate, V_1 = 50 mole/hr

Bottoms rate, L_2 = 50 mole/hr

Pressure,   P = 100 psia

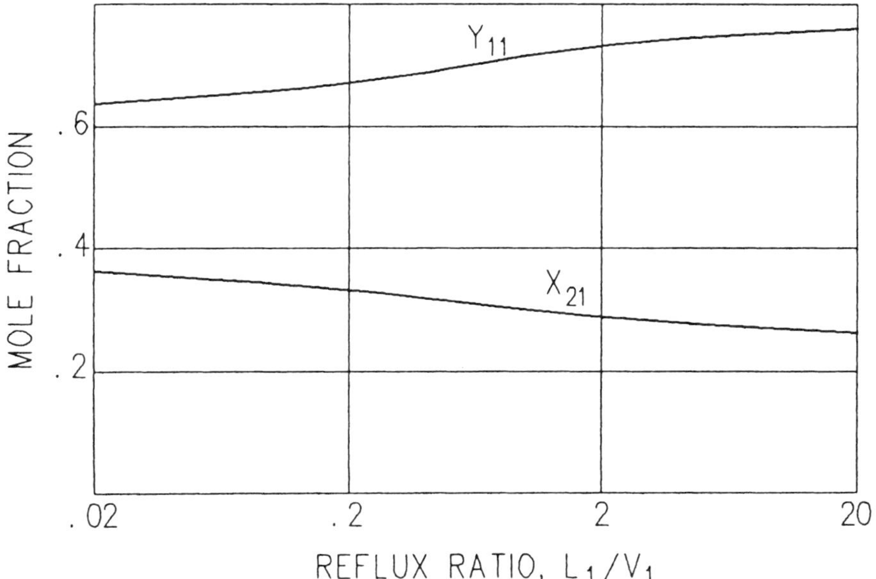

Figure 3.4   Performance of two-stage distillation column (Example 3.2).

3.3 ABSORPTION/STRIPPING BASICS

It was noted in Section 2.2 that the phase distribution of a mixture may be altered by introducing an additional feed to the equilibrium stage. The second feed influences the separation in different ways: by changing the values of the distribution coefficients, by shifting the azeotropic compositions, or simply by changing the vapor and liquid compositions as a result of the change in overall composition. A purely absorptive or stripping process is one in which the equilibrium phase distribution is controlled by manipulating the overall composition only and not by other factors such as temperature. In "ideal" absorption or stripping, the distribution coefficients are assumed composition-independent.

Absorption and stripping are similar in principle. The basics are discussed in this section. In some instances each process is discussed separately. In most cases the same principles apply to both, and, to avoid redundancy, both processes are simply referred to as "absorbers."

In a binary mixture equilibrium stage, the vapor and liquid compositions are a function only of distribution coefficients and hence only of temperature if the pressure is held fixed (Equations 3.3 and 3.4). Thus, in a binary system, the phase compositions are independent of feed composition. If a third component is added to the mixture, the equilibrium vapor and liquid compositions are influenced by the amount and identity of the third component. This phenomenon is the basis of absorption and stripping processes. The following is a simple mathematical analysis of the effect of a third component on phase distribution.

3.3.1 Ternary Systems

A binary mixture of components 1 and 2 is fed to an equilibrium stage held at constant temperature and pressure. The sum of mole fractions in the vapor and liquid is unity:

$$Y_1^b + Y_2^b = 1 \tag{3.22}$$

$$Y_1^b/K_1 + Y_2^b/K_2 = 1 \tag{3.23}$$

Superscript b designates binary compositions, and K_1 and K_2 are the distribution coefficients. Component 1 is assumed more volatile than 2, i.e., $K_1 > K_2$. Solving Equations 3.22 and 3.23 gives the following results, already arrived at in Equation 3.4:

$$Y_1^b = \frac{K_1(1 - K_2)}{K_1 - K_2} \tag{3.24}$$

$$Y_2^b = \frac{K_2(K_2 - 1)}{K_1 - K_2}$$ (3.25)

A third component is now introduced to the equilibrium stage at a given rate, keeping the stage temperature unchanged. The sums of the mole fractions in the liquid and vapor are written with superscript t designating ternary composition.

$$Y_1^t + Y_2^t + Y_3^t = 1$$ (3.26)

$$Y_1^t/K_1 + Y_2^t/K_2 + Y_3^t/K_3 = 1$$ (3.27)

The distribution coefficients are assumed independent of composition; hence, K_1 and K_2 appearing in Equation 3.27 are equal to those in Equation 3.23. Equations 3.26 and 3.27 are solved for Y_1^t and Y_2^t in terms of K_1, K_2, K_3, and Y_3^t:

$$Y_1^t = \frac{K_1(1 - K_2)}{K_1 - K_2} + \frac{K_1}{K_3}\frac{K_2 - K_3}{K_1 - K_2}Y_3^t$$ (3.28)

$$Y_2^t = \frac{K_2(K_1 - 1)}{K_1 - K_2} - \frac{K_2}{K_3}\frac{K_1 - K_3}{K_1 - K_2}Y_3^t$$ (3.29)

The effect of the third component on phase separation is investigated by comparing Y_1^t and Y_2^t with Y_1^b and Y_2^b. From the above equations,

$$Y_1^t - Y_1^b = \left(\frac{K_1}{K_3}\frac{K_2 - K_3}{K_1 - K_2}\right)Y_3^t$$ (3.30)

$$Y_2^t - Y_2^b = -\left(\frac{K_2}{K_3}\frac{K_1 - K_3}{K_1 - K_2}\right)Y_3^t$$ (3.31)

In a typical absorption process, a hydrocarbon stream of relatively low volatility (lean oil) is made to interact with a hydrocarbon gas mixture. The lean oil absorbs heavier gas components preferentially, recovering some of them and resulting in an effluent gas that is leaner in heavier components and richer in lighter components than the feed gas.

In the ternary system at hand, let component 3 represent the lean oil. It is less volatile than components 1 and 2; that is, $K_3 < K_2 < K_1$. Because of these inequalities, the coefficients of Y_3^t in parentheses in Equations 3.30 and 3.31 are always positive. This implies that $Y_1^t > Y_1^b$ and $Y_2^t < Y_2^b$. Hence, the addition of a heavier component 3 to a gas mixture of light components results in enrichment of lighter component 1 and depletion of heavier component 2 in the gas. Moreover, since Y_3^1 increases with the amount of component 3 in the feed, the degree of enrichment or depletion is in direct relationship to the rate of heavy component added.

In a stripping process the object is to lower the concentration of light components in a liquid mixture by interaction with a lighter gas. The stripping gas modifies the phase equilibrium by driving more of the lighter components in the liquid to the vapor phase than heavier components. A ternary model may be used to analyze stripping action in a manner analogous to absorption as discussed above. The reader may go through the exercise of deriving the mathematical relations for a ternary system that represents gas stripping.

3.3.2 Multistage Absorption

Improved separation is achieved by employing multistage absorbers instead of a single stage. A schematic of a two-stage absorber is shown in Figure 3.5. Vapor stream, V_2, which interacts with external liquid feed, L_0, in stage 1 had been enriched in lighter components and depleted in heavier components in stage 2. Separation by absorption is thus enhanced by increasing the number of stages. Better separation is also obtained by increasing the liquid feed rate (the external reflux).

Unlike distillation, an absorption or stripping process does not require the different stages to be at different temperatures. The "driving force" in absorption or stripping is a composition profile along the column induced by an external feed composition that is considerably different from the feed composition. In distillation the composition gradient is effected by a temperature profile. In absorption or stripping, the liquid and vapor streams flowing in the column originate from sources external to the column, whereas in distillation internal reflux and boilup vapor are generated by condensing part of the overhead vapor and reboiling part of the liquid bottoms. Because of this, the difference between the vapor and liquid compositions is usually greater in an absorption or stripping stage than in a distillation stage. Under characteristically absorptive conditions, the transfer of a light component from one phase to the other is accompanied by a relatively small transfer of the heavy component in the opposite direction.

Generally, multistage processes are neither purely absorptive nor purely distillative but a combination of both effects to varying degrees. Predominantly distillative processes are exemplified by binary distillation, while predominantly absorptive processes are exemplified by isothermal absorption. In isothermal absorption or stripping, heat must generally be removed or added on each stage to balance the heat associated with absorption or stripping.

3.3.3 Operating Parameters and Mathematical Formulation

The parameters required to define the operation of an absorber are discussed in relationship to a single stage. A single stage with a feed of fixed rate, composition, and thermal conditions has 2 degrees of freedom (Chapter 2). It is completely defined (0 degrees of freedom) if its pressure and duty are fixed. An absorber with a given

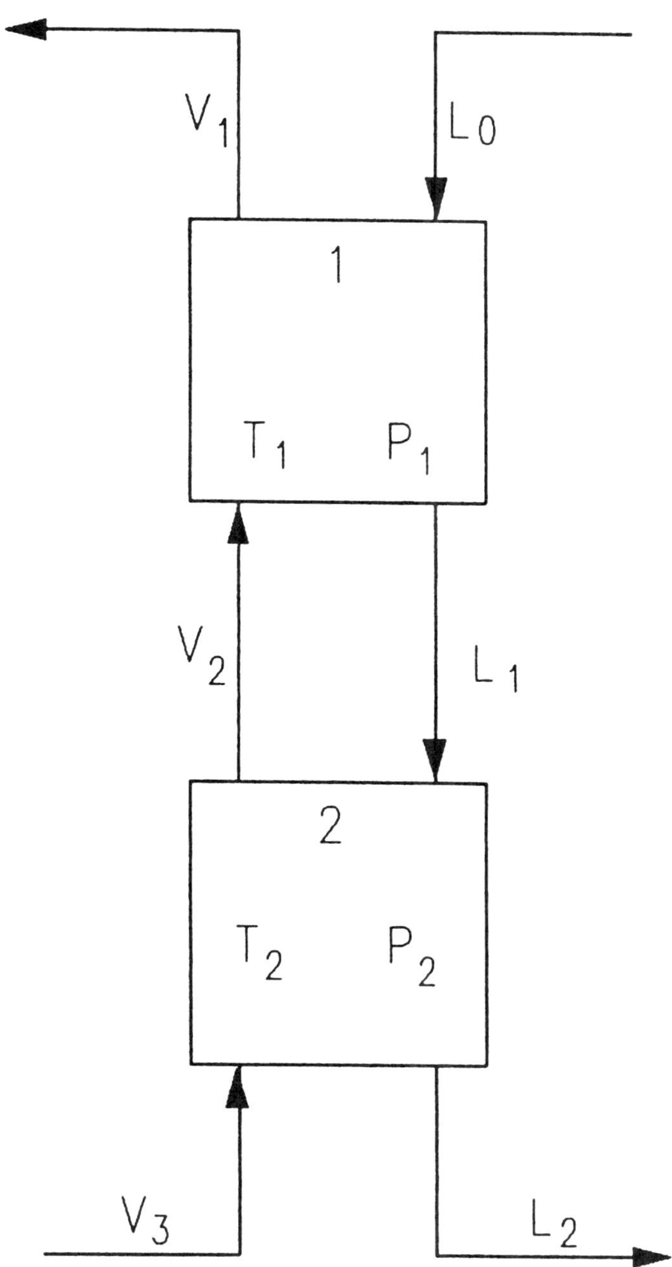

Figure 3.5   Two-stage absorber.

number of stages at a fixed pressure profile, with fixed external feeds, and having no heaters or coolers (duties are held at 0 by operating the column adiabatically) also has 0 degrees of freedom.

Degrees of freedom are determined in a similar manner for an absorber and a single stage. Each stage in an absorber receives feed streams that are either external or internal, coming from another stage. At steady state, both external and internal feed streams are of fixed rate, composition, and thermal conditions. Hence, the performance of an adiabatic absorber operating at fixed pressure is defined if the external feeds are defined.

The operation of an absorber may be controlled to meet performance specifications by varying one of the feed rates or by applying variable heating or cooling duties at certain stages. Consider an absorber with fixed lean oil and gas feeds. The liquid and vapor product rates, compositions, temperatures, and other properties are determined. If it is desired to meet a specified purity of certain components in one of the products such as the vapor, the lean oil rate may be used as a control variable. Its rate must be determined in order to meet the vapor purity specification.

Mathematical solution of absorbers, like other multistage separation processes, involves setting up material and energy balance equations and vapor–liquid equilibrium relations that describe the entire column. The resulting set of simultaneous equations is then solved by some suitable technique.

Referring to the model in Figure 3.5, a material balance for component i on stage j may be written as

$$L_{ji} + V_{ji} = L_{j-1,i} + V_{j+1,i} \tag{3.32}$$

$$i = 1, \ldots, C$$

$$j = 1, \ldots, N$$

Here L_{ji} and V_{ji} are equilibrium liquid and vapor molar rates of component i leaving stage j. C is the number of components and N is the number of stages. By extending this nomenclature, L_{0i} represents external liquid rate of component i entering stage 1 and $V_{N+1,i}$ stands for external vapor rate of component i entering stage N.

An enthalpy balance on stage j yields the following equation:

$$L_j h_j + V_j H_j = L_{j-1} h_{j-1} + V_{j+1} H_{j+1} \tag{3.33}$$

$$j = 1, \ldots, N$$

L_j and V_j are the total liquid and vapor molar rates leaving stage j, and h_j and H_j are the liquid and vapor enthalpies per mole on stage j.

Phase equilibrium relationships are expressed in terms of V_{ji}, V_j, L_{ji}, and L_j. Since

$$Y_{ji} = V_{ji}/V_j$$

$$X_{ji} = L_{ji}/L_j$$

the distribution coefficients are written as

$$K_{ji} = \frac{V_{ji}/V_j}{L_{ji}/L_j}$$

$$i = 1, \ldots, C$$

$$j = 1, \ldots, N$$

(3.34)

Each of equation sets 3.32 and 3.34 encompasses NC equations, and equation system 3.33 includes N equations. Altogether there are N(2C + 1) simultaneous equations. The unknowns are vapor and liquid molar rates of each component on each stage (V_{ji} and L_{ji}, j = 1, ..., N and i = 1, ..., C) and temperatures of each stage (T_j, j = 1, ..., N). The stage pressures are assumed at known, fixed values. The total number of unknowns is thus 2NC + N, identical to the number of equations with no degrees of freedom. Methods for solving the set of equations are discussed in Chapter 13.

NOMENCLATURE

C	Number of components	P	Pressure
F	Feed stream designation or molar flow rate	T	Temperature
H	Vapor stream enthalpy per mole	V	Vapor stream designation or molar flow rate
h	Liquid stream enthalpy per mole	X	Component mole fraction in the liquid
K	Distribution coefficient		
L	Liquid stream designation or molar flow rate	Y	Component mole fraction in the vapor
N	Number of equilibrium stages in a column	Z	Component mole fraction in the feed
		α	Relative volatility

SUBSCRIPTS

i   Component designation
j   Stage designation

SUPERSCRIPTS

b   Binary system
t   Ternary system

PROBLEMS

3.1 In the three-stage process described in Example 3.1, it is required to produce 20 mole/hr propane in the top stage vapor, 8 mole/hr propane in the top stage liquid, and 12 mole/hr propane in the bottom stage vapor. Calculate the temperature and heat duty on each stage. The vapor and liquid enthalpies of propane and butane are given by the following equations:

$$\text{Propane liquid enthalpy}: \quad h(\text{Btu/lbmole}) = 8932 + 27.1T(°F)$$

$$\text{Propane vapor enthalpy}: \quad H(\text{Btu/lbmole}) = 16701 + 9.1T(°F)$$

$$\text{Butane liquid enthalpy}: \quad h(\text{Btu/lbmole}) \quad = 2668 + 32.5T(°F)$$

$$\text{Butane vapor enthalpy}: \quad H(\text{Btu/lbmole}) \quad = 12590 + 20.0T(°F)$$

Assume additive enthalpies for the mixtures.

3.2 A cascaded three-stage process similar to Figure 3.1 is used for a crude separation of a three-component mixture. Determine the number of variables required to define the process.

Upstream control systems require continuous composition data for stream F. It is impractical to install an analyzer on F, but an existing analyzer is already installed on stream V_1. In addition, the following data are measured on a continuous basis: flow rate, temperature, and pressure of stream F, flow rates of L_1, L_3, and V_3, and the stage pressures P_1, P_2, and P_3. Outline a computational procedure that could be used to determine, at any given moment, the composition of stream F, as well as that of products L_1, L_3, and V_3, assuming steady-state conditions exist.

3.3 Alternative strategies are considered as backup to determine the composition of feed stream F in Problem 3.2 (Figure 3.1). Which of the following alternatives are feasible and what computational procedure could be used to determine the feed composition?

    a. Replace the flow meters on L_1 and V_3 with an analyzer on L_1.
    b. Replace the flow meters on L_1 and V_3 with temperature indicators on stages 1 and 3.
    c. Measure the following: feed flow, temperature, and pressure; temperature and pressure on each stage; and flow rates L_1 and L_3.

3.4 A two-stage distillation column similar to the column in Example 3.2 (Figure 3.3) is to be used for separating a mixture of 60 mole/hr propane and 40 mole/hr butane. The feed stream, at 100°F and 100 psia, is sent to the bottom stage, the reboiler. It is desired to recover 65% of the propane in the distillate with a purity of 75%.

a. What is the required reflux ratio and what are the two stage temperatures?
b. What reflux ratio is required to achieve a propane recovery of 90% at the same propane purity of 75%?
c. What is the maximum theoretical propane recovery in the distillate consistent with a 75% purity, and what are the stage temperatures?

3.5 A flash drum maintained at constant temperature and pressure separates a binary feed into vapor and liquid products for the purpose of lowering the concentration of the lighter component (2) in the liquid. A stripping stream consisting of a more volatile component (3) is added to the flash drum (a single-stage stripper) to enhance the removal of component 2 from the liquid.

a. Derive a relationship between the mole fraction of component 2 in the liquid product and the mole fraction of component 3 in the vapor product. The K-values of the components are K_1, K_2, and K_3.
b. Calculate the rate of stripping stream per 1000 mole feed stream required to lower component 2 mole fraction in the liquid to 0.35. The feed composition and K-values (assumed composition-independent) are given as follows:

| | Mole Fractions | | |
Component	Feed, F	Stripping Stream, S	K-value
1	0.54	0.00	0.72
2	0.46	0.00	1.36
3	0.00	1.00	3.50

3.6 A liquid absorbent A consisting of component 3 is used to selectively absorb component 2 from a binary gas mixture consisting of components 1 and 2 by contacting it in a single equilibrium stage. Determine the minimum rate of stream A required to maintain liquid flow in the absorber. Assume composition-independent K-values.

Calculate the minimum absorbent rate and the equilibrium compositions using the following data:

$$F = 100 \text{ mole / hr} \qquad K_1 = 18.0$$
$$X_{F1} = 0.95 \qquad K_2 = 1.5$$
$$X_{F2} = 0.05 \qquad K_3 = 0.1$$

Using the same data, calculate the minimum absorbent rate and the equilibrium compositions for a two-stage absorber.

3.7 Compute the products of a two-stage absorber with the feeds given below, using successive flash calculations. In each iteration solve both stages using stream data from the previous iteration.

Component, i	K_i	Gas Feed, Y_{3i}	Absorbent, X_{oi}
1	18.0	0.95	0.00
2	1.5	0.05	0.00
3	0.1	0.00	1.00
Mole/hr		100.00	50.00

CHAPTER 4

Material Balances in Multicomponent Separation

The separation of a multicomponent mixture into products with different compositions in a multistage process is governed by phase equilibrium relations and heat and material balances. It is not uncommon in simulation studies to require certain column product rates, compositions, or component recoveries to satisfy given specifications with no concern for conditions within the column. Such would be the case when downstream processing of the products is of primary interest. In these instances, one would be concerned only with overall component balances around the column but not necessarily with heat balances or equilibrium relations. Separation would thus be arbitrarily defined and the problem would be to calculate product rates and compositions.

The purpose of this chapter is to set up a mathematical model for component balances and to solve it for different sets of separation specifications.

4.1 MATHEMATICAL MODEL

The separation column is assumed to have one feed and two products: overhead (or distillate) and bottoms. The feed has a fixed flow rate and composition. Temperatures and pressures of the feed and products and of streams within the column may be fixed but are of no concern in the present context. Moreover, the column configuration, number of stages, and internal vapor and liquid flows are immaterial since the focus is on the net performance of a "black box" column.

Let the feed stream, F, contain C components with known mole fractions, Z_1, Z_2, ..., Z_C. The overhead and bottoms flow rates are D and B, respectively. As with the equilibrium stage discussed in Chapter 2, the products of the column may be completely defined in terms of the overhead mole fractions, Y_1, Y_2, ..., Y_C; the bottoms mole fractions, X_1, X_2, ..., X_C; and the distillate fraction, $V = D/F$.

These $2C + 1$ parameters may be considered the primary variables. They are interrelated by C material balance equations:

$$Z_i = VY_i + (1 - V)X_i$$
$$i = 1, \ldots, C$$

$$(4.1)$$

and one summation equation:

$$\sum Y_i = 1$$

$$(4.2)$$

Note that the summation $\Sigma X_i = 1$ does not constitute an additional independent equation since it may be obtained by summing up Equation 4.1 over all the components and combining with Equation 4.2.

There are, therefore, $C + 1$ equations with $2C + 1$ variables. In order to solve the equations, C constraints are required, making the total number of equations equal to the number of variables. The constraints define the component separation taking place in the column. In general, these constraints may be written as a function of the primary variables in the form

$$g_j\left(Y_1, \ldots, Y_c, X_1, \ldots, X_c, V\right) = 0$$
$$j = 1, \ldots, C$$

$$(4.3)$$

In rigorous treatment of columns, most of the constraints result from phase equilibrium and heat balance equations while the rest represent column performance specifications such as separation. In the present model, concerned mainly with material balances, all the constraints are imposed so as to achieve certain product rates, component purity, or recovery requirements. Equation sets 4.1, 4.2, and 4.3 include $2C + 1$ independent equations to be solved for the $2C + 1$ variables.

4.2 TYPES OF COLUMN SPECIFICATIONS

The g functions (Equation set 4.3) representing column constraints may be defined by specifying the values of some of the primary variables or those of other appropriate derived variables. The constraints could also be defined by any arbitrary functions designed to meet special separation specifications. Combinations of different types of constraints are also possible.

4.2.1 Primary Variable Specifications

The separation may be defined by specifying C of the primary variables ($X_1, \ldots,$ $X_C; Y_1, \ldots, Y_C; V$). The g functions representing these constraints would take the form

$$g_i = Y_i - \left(Y_i\right)_s = 0$$

$$g_i = X_i - \left(X_i\right)_s = 0$$

$$g = V - V_s$$

$$i = 1, \ldots, C$$

where $(X_i)_s$, $(Y_i)_s$, and V_s are the specified values.

Note that specifying only X_i, $i = 1, \ldots, C$ or only Y_i, $i = 1, \ldots, C$ does not constitute a complete definition of the separation because the mole fractions are interrelated by the summation equations

$$\sum_i Y_i = 1$$

$$\sum_i X_i = 1$$

Hence, providing all the X_i's or all the Y_i's does not result in C independent specifications. In order to define the separation uniquely, the C specified variables could be a mix of X_i's, Y_i's, and V.

If, for a given component i, both X_i and Y_i are given, then V is determined by Equation 4.1. Consequently, V may not be given as an additional constraint. Also, there could be only one component for which both X_i and Y_i are specified. Otherwise, the column would be overspecified since each pair of X_i and Y_i determines V. Because of these considerations, the allowable combinations of primary variable specifications are

$$Y_i \; i = 1, \ldots, C - 1 \text{ and } V$$

or

$$X_i \; i = 1, \ldots, C - 1 \text{ and } V$$

or

$$Y_i \; i = 1, \ldots, j \text{ and } X_i \; i = j + 1, \ldots, C$$

or

$$Y_i \; i = 1, \ldots, j \text{ and } X_i \; i = j, \ldots, C - 1$$

The obvious restrictions on specification values are

$$0 < V < 1$$

$$\sum Y_i \leq 1 \quad i = 1 \text{ through } j$$

$$\sum X_i \leq 1 \quad i = 1 \text{ through } j$$

where j is the number of components for which mole fractions in either product are specified.

Another restriction applies to situations where mole fractions of a given component are specified both in the overhead and distillate. If both X_i and Y_i are specified, then either of the following inequalities should hold:

$$Y_i > Z_i > X_i$$

or

$$Y_i < Z_i < X_i$$

That is, if a component gets concentrated in the overhead, it must get diluted in the bottoms, or vice versa. Violation of these inequalities results in values of the distillate fraction outside the permissible range of 0 to 1, as may be inferred from Equation 4.1.

The specifications must pass another feasibility check when the distillate fraction and product component mole fractions are specified. Equation 4.1 is rearranged as

$$X_i = \frac{Z_i - VY_i}{1 - V}$$

In order for X_i to be positive, the specified values of Y_i and V must be such that

$$Y_i \leq Z_i/V$$

When the constraints are given as primary specifications, Equation set 4.1 is easily decoupled and each equation is solved independently.

Example 4.1 Mole Fraction Specifications

A 100 mole/hr mixture of four components is to be separated into an overhead and bottoms products. The feed composition is as follows:

$$Z_1 = 0.1$$

$$Z_2 = 0.2$$

$$Z_3 = 0.3$$

$$Z_4 = 0.4$$

Calculate the product rates and compositions for the following sets of specifications:

A. $Y_1 = 0.21$, $Y_2 = 0.35$, $Y_3 = 0.20$, $V = 0.40$
B. $Y_1 = 0.23$, $Y_2 = 0.34$, $X_3 = 0.38$, $X_4 = 0.46$
C. $Y_1 = 0.20$, $Y_2 = 0.30$, $X_2 = 0.15$, $X_3 = 0.42$

Solution. Equations 4.1 and 4.2 together with each set of specifications are solved to determine product rates and compositions.

A. The specifications are substituted in the first three of Equation set 4.1:

$$0.10 = (0.40)(0.21) + (1 - 0.40)X_1$$

$$0.20 = (0.40)(0.35) + (1 - 0.40)X_2$$

$$0.30 = (0.40)(0.20) + (1 - 0.40)X_3$$

Each equation is solved separately, and the summation equations

$$\sum_i Y_i = 1$$

$$\sum_i X_i = 1$$

are solved for the fourth component mole fraction. The results are summarized in Table 4.1. The distillate rate is calculated from the definition of V:

$$D = FV = (100)(0.40) = 40 \text{ mole/hr.}$$

The bottoms rate, by material balance,

$$B = 100 - 40 = 60 \text{ mole/hr}$$

Table 4.1 Results for Example 4.1A

i	Z_i	Y_i	X_i
1	0.10	0.21	0.0267
2	0.20	0.35	0.1000
3	0.30	0.20	0.3667
4	0.40	0.24	0.5067

B. Equations 4.1 and 4.2 and the specifications are listed as nine equations with nine unknowns:

$$Y_1 = 0.23$$

$$Y_2 = 0.34$$

$$X_3 = 0.38$$

$$X_4 = 0.46$$

$$VY_1 + (1-V)X_1 = 0.1$$

$$VY_2 + (1-V)X_2 = 0.2$$

$$VY_3 + (1-V)X_3 = 0.3$$

$$VY_4 + (1-V)X_4 = 0.4$$

$$\sum_i Y_i = 1$$

The first eight equations are combined:

$$0.23V + (1-V)X_1 = 0.1$$

$$0.34V + (1-V)X_2 = 0.2$$

$$VY_3 + (1-V)(0.38) = 0.3$$

$$VY_4 + (1-V)(0.46) = 0.4$$

The equations are thus reduced to five equations with five unknowns. From the last two equations

$$Y_3 = (0.38V - 0.08)/V$$

$$Y_4 = (0.46V - 0.06)/V$$

The values of Y_i are substituted in the summation equation,

$$0.23 + 0.34 + (0.38V - 0.08)/V + (0.46V - 0.06)/V = 1$$

which is solved for V:

$$V = 0.3415$$

The remaining unknown mole fractions are now calculated directly from Equation set 4.1. The results are given in Table 4.2. The overhead and bottoms rates are calculated directly:

$$D = (100)(0.3415) = 34.15 \text{ mole/hr}$$

$$B = 100 - 34.15 = 65.85 \text{ mole/hr}$$

Table 4.2 Results for Example 4.1B

i	Z_i	Y_i	X_i
1	0.10	0.2300	0.0326
2	0.20	0.3400	0.1274
3	0.30	0.1457	0.3800
4	0.40	0.2843	0.4600

C. Since Y_2 and X_2 are both given for component 2, V may be calculated from Equation 4.1:

$$0.20 = 0.30V + (1 - V)0.15$$

$$V = 0.3333$$

The missing mole fractions are calculated from Equations 4.1 and 4.2 and listed in Table 4.3. The product rates are

$$D = (100)(0.3333) = 33.33 \text{ mole/hr}$$

$$B = 100 - 33.33 = 66.67 \text{ mole/hr}$$

Table 4.3 Results for Example 4.1C

i	Z_i	Y_i	X_i
1	0.10	0.20	0.05
2	0.20	0.30	0.15
3	0.30	0.06	0.42
4	0.40	0.44	0.38

4.2.2 Derived Variable Specifications

A number of variables of practical importance in separation terminology may be derived based on the primary variables. Derived variables include component rates and recoveries defined as follows:

Component distillate rates:

$$D_i = DY_i = FVY_i \tag{4.4}$$

Component bottoms rates:

$$B_i = BX_i = F(1 - V)X_i \tag{4.5}$$

Component recoveries in the distillate:

$$r_i^D = \frac{DY_i}{FZ_i} = \frac{VY_i}{Z_i} = \frac{D_i}{FZ_i} \tag{4.6}$$

Component recoveries in the bottoms:

$$r_i^B = \frac{BX_i}{FZ_i} = \frac{(1-V)X_i}{Z_i} = \frac{B_i}{FZ_i} \tag{4.7}$$

The separation specifications defined in the general form in Equation 4.3 may be written in terms of derived variables:

$$g_i = FVY_i - D_i = 0$$

$$g_i = F(1-V)X_i - B_i = 0$$

$$g_i = VY_i/Z_i - r_i^D = 0$$

$$g_i = (1-V)X_i/Z_i - r_i^B = 0$$

where D_i, B_i, r_i^D, and r_i^B are specified values.

The C specifications required to define the separation may be a combination of the above g functions or the above functions plus primary variable specifications.

It is important that the specifications be independent and feasible. For instance, one may specify the D_i's or the B_i's or the r_i^D's or the r_i^B's ($i = 1, \ldots, C$). If, however, D_i is specified, specifying B_i as well is redundant since D_i and B_i are not independent; they are related by material balance,

$$FZ_i = D_i + B_i$$

Possible independent derived and primary specifications include one of the variables D_i, B_i, r_i^D, r_i^B, Y_i, or X_i for each component. Specifying both D_i and Y_i, or a similar combination, is allowed for one component only. Since $D_i = FVY_i$, specifying both D_i and Y_i determines V. Once V is fixed, no other D_i, Y_i pair or the equivalent may be specified.

In general, specifications may be written in terms of primary variables in the form of constraining functions, Equation 4.3. To illustrate, consider a four-component mixture where the four specified variables are D_1, r_2^D, B_3, and r_4^B. The constraining equations would be

$$VY_1 = D_1/F$$

$$VY_2 = Z_2 r_2^D$$

$$(1-V)X_3 = B_3/F$$

$$(1-V)X_4 = Z_4 r_4^B$$

The terms on the right-hand side of the equations are known quantities. These four equations are independent and, together with four material balances, Equation 4.1, and one summation, Equation 4.2, they make up the nine equations $(2C + 1)$ required to solve for Y_1 through Y_4, X_1 through X_4, and V.

Consider a problem where the specified variables for the mixture separation are, for example, D_1, r_1^D, B_3, and r_4^B. The constraining equations would be

$$VY_1 = D_1/F$$

$$VY_1 = Z_1 r_1^D$$

$$(1 - V)X_3 = B_3/F$$

$$(1 - V)X_4 = Z_4 r_4^B$$

The variables on the right-hand side are known quantities. These four equations are not independent, however. In fact, the first two equations are identical if the specified values of D_1 and r_1^D are such that

$$D_1/F = Z_1 r_1^D$$

If this equation is not satisfied, conflicting specifications result. In either case, the system of equations cannot be solved.

Besides having to be independent, the specifications must be feasible. The obvious feasible limits for component product rate specifications are

$$D_i \leq FZ_i$$

and

$$B_i \leq FZ_i$$

For component recovery specifications, the feasible limits are

$$r_i^D \leq 1$$

$$r_i^B \leq 1$$

Example 4.2 Component Rates and Recovery Specifications

The mixture of Example 4.1 is to be separated into distillate and bottoms products to meet the following sets of specifications:

A. $D_1 = 8.5$ mole/hr
     $r_2^D = 0.72$
     $B_3 = 23.0$ mole/hr
     $r_4^B = 0.75$

B. $r_1^D = 0.82$
$r_2^D = 0.73$
$r_3^D = 0.21$
$r_4^D = 0.26$

Calculate product rates and compositions.

Solution. The compositions may be calculated by systematically solving Equations 4.1, 4.2, and the specifications. However, many steps may be bypassed because the problem lends itself to equation decoupling.

A. The feed component flow rates are first calculated:
$F_i = FZ_i$
$F_1 = 10.0$ mole/hr
$F_2 = 20.0$ mole/hr
$F_3 = 30.0$ mole/hr
$F_4 = 40.0$ mole/hr
The recovery specifications are converted to product rate specifications:

$$D_2 = F_2 r_2^D = (20.0)(0.72) = 14.4 \text{ mole / hr}$$

$$B_4 = F_4 r_4^B = (40.0)(0.75) = 30.0 \text{ mole / hr}$$

The remaining component flow rates are calculated by simple component material balance:

$$D_3 = F_3 - B_3 = 30.0 - 23.0 = 7.0 \text{ mole/hr}$$

$$D_4 = F_4 - B_4 = 40.0 - 30.0 = 10.0 \text{ mole/hr}$$

$$B_1 = F_1 - D_1 = 10.0 - 8.5 = 1.5 \text{ mole/hr}$$

$$B_2 = F_2 - D_2 = 20.0 - 14.4 = 5.6 \text{ mole/hr}$$

Product rates:

$$D = \sum_i D_i = 8.5 + 14.4 + 7.0 + 10.0 = 39.9 \text{ mole / hr}$$

$$B = \sum_i B_i = 1.5 + 5.6 + 23.0 + 30.0 = 60.1 \text{ mole / hr}$$

Distillate fraction:

$$V = 39.9/100.0 = 0.399$$

Product compositions:

$$Y_1 = 8.5/39.9 = 0.2130$$

$$Y_2 = 14.4/39.9 = 0.3609$$

$$Y_3 = 7.0/39.9 = 0.1754$$

$$Y_4 = 10.0/39.9 = 0.2506$$

$$X_1 = 1.5/60.1 = 0.0250$$

$$X_2 = 5.6/60.1 = 0.0932$$

$$X_3 = 23.0/60.1 = 0.3827$$

$$X_4 = 30.0/60.1 = 0.4992$$

B. The distillate fraction is first calculated directly from the distillate recoveries. Equation 4.6 is rewritten as

$$VY_i = Z_i r_i^D$$

Summation of this equation over all components results in a convenient expression for calculating V:

$$V = \sum_i Z_i r_i^D$$

$$V = (0.10)(0.82) + (0.20)(0.73) + (0.30)(0.21) + (0.40)(0.26) = 0.395$$

Product rates:

$$D = (100)(0.395) = 39.5 \text{ mole/hr}$$

$$B = 100 - 39.5 = 60.5 \text{ mole/hr}$$

The distillate mole fractions are calculated from Equation 4.6:

$$Y_i = Z_i r_i^D / V$$

$$Y_1 = (0.10)(0.82)/0.395 = 0.2076$$

$$Y_2 = (0.20)(0.73)/0.395 = 0.3696$$

$$Y_3 = (0.30)(0.21)/0.395 = 0.1595$$

$$Y_4 = (0.40)(0.26)/0.395 = 0.2633$$

The bottoms mole fractions are calculated from Equation 4.7 by substituting $1 - r_i^D$ for r_i^B:

$$X_i = Z_i \left(1 - r_i^D\right) \big/ (1 - V)$$

$$X_1 = 0.1(1 - 0.82)/(1 - 0.395) = 0.0298$$

$$X_2 = 0.2(1 - 0.73)/(1 - 0.395) = 0.0893$$

$$X_3 = 0.3(1 - 0.21)/(1 - 0.395) = 0.3917$$

$$X_4 = 0.4(1 - 0.26)/(1 - 0.395) = 0.4892$$

4.2.3 General Specifications

Although column separation constraints usually consist of primary and derived variable specifications as discussed in Sections 4.2.1 and 4.2.2, in general, specifications could be any function of these variables. One could define the g functions of Equation 4.3 as, for instance, sums, differences, or ratios of component rates, recoveries, etc. The only restrictions on the specified functions, at least from the mathematical standpoint, are that they be independent and feasible.

It is possible, for example, to specify the sum and ratio of the rates of two components in a given product. These would constitute two independent specifications for the two components. An additional specification on either of them would not be allowed since both have already been implicitly specified.

Depending on the functional form of the specifications, the system of equations comprising Equations 4.1, 4.2, and the specifications may or may not be linear. Nonlinear equations are solved by iterative methods such as the Newton–Raphson method. The solution method is illustrated in the following example.

Example 4.3 Mixed Specifications

It is required to separate a ternary mixture into two products that meet certain separation specifications. The feed consists of 25 mole/hr component 1, 33 mole/hr component 2, and 42 mole/hr component 3. The three specifications are as follows:

$$\text{Mole fraction 2 in the bottoms: } X_2 = 0.2833$$

$$\text{Mole fraction 3 in the bottoms: } X_3 = 0.5333$$

$$\text{Recovery of 1 and 2 in the overhead: } (D_1 + D_2)/(F_1 + F_2) = 0.5172$$

Calculate the product rates and compositions.

Solution. The feed compositions are

$$Z_1 = 0.25$$

$$Z_2 = 0.33$$

$$Z_3 = 0.42$$

Material balance equations:

$$VY_1 + (1 - V)X_1 = 0.25$$

$$VY_2 + (1 - V)X_2 = 0.33$$

$$VY_3 + (1 - V)X_3 = 0.42$$

Summation equation:

$$X_1 + X_2 + X_3 = 1$$

Specifications:

$$X_2 = 0.2833$$

$$X_3 = 0.5333$$

$$\frac{D_1 + D_2}{F_1 + F_2} = \frac{FVY_1 + FVY_2}{F_1 + F_2} = 0.5172$$

These are seven equations with seven unknowns: X_1, X_2, X_3, Y_1, Y_2, Y_3, V. The last specification was converted to a function of the primary variables by substituting Equation 4.4.

The first two specifications and the summation equation are solved directly:

$$X_1 = 0.1834$$

$$X_2 = 0.2833$$

$$X_3 = 0.5333$$

The values of X_1 and X_2 are plugged into the material balance equations for components 1 and 2 and into the third specification:

$$VY_1 + 0.1834(1 - V) = 0.25$$

$$VY_2 + 0.2833(1 - V) = 0.33$$

$$100(Y_1 + Y_2)(25 + 33) = 0.5172$$

The material balance equation for component 3 may be solved separately once V is determined. The above three simultaneous equations are first solved for Y_1, Y_2, and V. They are nonlinear, involving products VY_1 and VY_2. Although they may be solved by elimination, the method used here is Newton–Raphson's multivariable iterative technique. First, rewrite the equations in the form

$$g_1(Y_1, Y_2, V) = VY_1 - 0.1834V - 0.0666 = 0$$

$$g_2(Y_1, Y_2, V) = VY_2 - 0.2833V - 0.0467 = 0$$

$$g_3(Y_1, Y_2, V) = VY_1 + VY_2 - 0.3 = 0$$

The iteration requires starting values for the variables, Y_1^0, Y_2^0, and V^0. Superscript 0 refers to iteration 0, or initial estimates. In subsequent iterations, k, k + 1, ..., the values of the variables are improved by adding adjustments to current values:

$$Y_1^{k+1} = Y_1^k + \Delta Y_1^{k+1}$$

$$Y_2^{k+1} = Y_2^k + \Delta Y_2^{k+1}$$

$$V^{k+1} = V^k + \Delta V^{k+1}$$

Convergence is achieved when the adjustments ΔY_1^{k+1}, ΔY_2^{k+1}, and ΔV^{k+1} become smaller than a preset tolerance. The adjustments are calculated from the equations

$$\frac{\partial g_1}{\partial Y_1} \Delta Y_1^{k+1} + \frac{\partial g_1}{\partial Y_2} \Delta Y_2^{k+1} + \frac{\partial g_1}{\partial V} \Delta V^{k+1} + g_1 = 0$$

$$\frac{\partial g_2}{\partial Y_2} \Delta Y_1^{k+1} + \frac{\partial g_2}{\partial Y_2} \Delta Y_2^{k+1} + \frac{\partial g_2}{\partial V} \Delta V^{k+1} + g_2 = 0$$

$$\frac{\partial g_3}{\partial Y_1} \Delta Y_1^{k+1} + \frac{\partial g_3}{\partial Y_2} \Delta Y_2^{k+1} + \frac{\partial g_3}{\partial V} \Delta V^{k+1} + g_3 = 0$$

The functions g_1, g_2, g_3, and their partial derivatives are evaluated at the current values of the variables Y_1^k, Y_2^k, and V^k. The expressions for partial derivatives are

$$\frac{\partial g_1}{\partial Y_1} = V \qquad \frac{\partial g_1}{\partial Y_2} = 0 \qquad \frac{\partial g_1}{\partial V} = Y_1 - 0.1834$$

$$\frac{\partial g_2}{\partial Y_1} = 0 \qquad \frac{\partial g_2}{\partial Y_2} = V \qquad \frac{\partial g_2}{\partial V} = Y_2 - 0.2833$$

$$\frac{\partial g_3}{\partial Y_1} = V \qquad \frac{\partial g_3}{\partial Y_2} = V \qquad \frac{\partial g_3}{\partial V} = Y_1 + Y_2$$

The starting values of the variables are guessed or estimated. Good initial estimates help insure convergence, although equations that are not highly nonlinear are not too sensitive to the initial estimates. These values should at least be bracketed on the basis of physical realities. For instance, in the present example, Y_1, Y_2, and V should be between 0 and 1. The estimates may further be refined by inspecting the specified values of X_2 and X_3. These specifications indicate that $X_1 < Z_1$ and $X_2 < Z_2$. This implies that we should have $Y_1 > Z_1$ and $Y_2 > Z_2$. However, in order to check the ability of the iterative technique to recover from "bad" initial estimates, assume the following arbitrary starting values:

$$Y_1^0 = 0.25$$

$$Y_2^0 = 0.25$$

$$V^0 = 0.25$$

These values are used to calculate the functions g and their derivatives, yielding the following equations for calculating the adjustments:

$$0.25\Delta Y_1 \qquad + 0.0666\Delta V - 0.04995 = 0$$

$$0.25\Delta Y_2 - 0.0333\Delta V - 0.05502 = 0$$

$$0.25\Delta Y_1 + 0.25\Delta Y_2 + 0.5000\Delta V - 0.175 = 0$$

The solutions to these equations are

$$\Delta Y_1 = 0.1598$$

$$\Delta Y_2 = 0.2401$$

$$\Delta V = 0.1501$$

The updated values of the variables are

$$Y_1^1 = 0.4098$$

$$Y_2^1 = 0.4901$$

$$V^1 = 0.4001$$

Based on these values, new g functions and derivatives are computed and the following equations are obtained:

$$0.4001\Delta Y_1 \qquad + 0.2264\Delta V + 0.024 = 0$$

$$0.4001\Delta Y_2 + 0.2068\Delta V + 0.036 = 0$$

$$0.4001\Delta Y_1 + 0.4001\Delta Y_2 + 0.8999\Delta V + 0.060 = 0$$

The solutions are

$$\Delta Y_1 = -0.06$$

$$\Delta Y_2 = -0.09$$

$$\Delta V = 0$$

The updated variables are

$$Y_1^2 = 0.3498$$

$$Y_2^2 = 0.4001$$

$$V^2 = 0.4001$$

The procedure is repeated again, generating the equations

$$0.4001\Delta Y_1 \qquad\qquad\qquad + 0.1664\Delta V = 0$$

$$0.4001\Delta Y_2 + 0.1168\Delta V = 0$$

$$0.4001\Delta Y_1 + 0.4001\Delta Y_2 + 0.7499\Delta V = 0$$

for which the solutions are

$$\Delta Y_1 = 0$$

$$\Delta Y_2 = 0$$

$$\Delta V = 0$$

Convergence has been reached and the latest values of the variables are the solution. The other variables are easily calculated from the primary variables and are given in Table 4.4.

Table 4.4   Results for Example 4.3

i	Z_i	Y_i	X_i	F_i	D_i	B_i
1	0.25	0.3498	0.1834	25	14	11
2	0.33	0.4001	0.2833	33	16	17
3	0.42	0.2501	0.5333	42	10	32

NOMENCLATURE

B	Bottoms stream designation or molar flow rate	F_i	Component molar flow rate in the feed
B_i	Component molar flow rate in the bottoms	r	Component recovery
C	Number of components	V	Distillate mole fraction
D	Distillate or overhead designation or molar flow rate	X	Component mole fraction in the bottoms
D_i	Component molar flow rate in the distillate or overhead	Y	Component mole fraction in the distillate or overhead
F	Feed stream designation or molar flow rate	Z	Component mole fraction in the feed

SUBSCRIPTS

i, j   Component designation
s   Specified value

SUPERSCRIPTS

B   Bottoms
D   Distillate
k   Iteration number

PROBLEMS

4.1 A five-component stream splits into a distillate product with composition Y_i and a bottoms product with composition X_i. Which of the following sets of specifications uniquely define the separation, without taking into account vapor–liquid equilibrium or energy balance constraints?

a. X_1, X_2, X_3, X_4, X_5
b. X_1, X_2, X_3, Y_4, Y_5
c. X_1, X_2, X_3, Y_3, Y_4
d. X_1, X_2, X_3, Y_2, Y_3
e. X_1, X_2, X_3, X_4, V
f. X_1, X_2, X_3, Y_3, V

4.2 A stream F with composition Z_i, $i = 1, \ldots, C$, the number of components, is separated into a distillate product D with composition Y_i and a bottoms product B with compositions X_i. It is proposed to specify the separation with X_i specifications for $i = 1, \ldots, J$, and Y_i specifications for $i = J+1, \ldots C$. What are the feasibility conditions for these specifications? Are the following specifications feasible?

i	Z_i	Y_i	X_i
1	0.34		0.25
2	0.17		0.34
3	0.22	0.45	
4	0.27	0.19	

4.3 The feed stream of Problem 4.2 is to be separated in a distillation column according to the following specifications:

i	Feed, Z_i	Distillate, Y_i	Bottoms, X_i
1	0.34		0.30
2	0.17	0.22	
3	0.22	0.19	
4	0.27		0.37

Downstream process design requires distillate and bottoms data. Calculate the flow rates and compositions of these streams for a fixed feed rate of 1000 mole/hr.

4.4 Based on material balance considerations only, which of the following sets of specifications uniquely define the separation of a four-component stream into a distillate and a bottoms product? (See text for terminology).

a. D_1, D_2, B_3, B_4
b. D_1, D_2, D_3, r_1^B
c. B_1, Y_1, D_2, B_3
d. B_1, X_1, D_3, Y_3

4.5 Calculate the rates and compositions of the distillate and bottoms of a distillation column with the following feed and separation specifications:

i	Feed, Z_i	Distillate	Bottoms
1	0.27	$D_1 = 175.00$	$X_1 = 0.20$
2	0.35		$B_2 = 200.00$
3	0.21	$D_3 = 90.00$	
4	0.17		
Flow rate, lbmole/hr	1000.00		

4.6 It is proposed to design a distillation column for coarse separation of a four-component mixture between components 2 and 3. (The components are numbered by decreasing volatility). The combined mole fraction of components 1 and 2 in the distillate should be 0.65, and the ratio of the mole fraction of component 3 to 4 in

the bottoms should be 1.2. It is also required to recover 80% of component 1 in the distillate while maintaining its mole fraction in the bottoms at 0.07. Assuming the vapor–liquid equilibrium properties of this mixture are consistent with the separation specifications, calculate the distillate and bottoms compositions.
Feed composition:

Component, i	Z_i
1	0.12
2	0.37
3	0.29
4	0.22

CHAPTER 5

Binary Distillation: Principles

Binary distillation is a multistage process for separating a mixture of two components using the principles of vapor–liquid equilibria. Although binary distillation is not a very common occurrence in practice (impurities are normally present along with the two primary components to be separated), developing methods for solving it is beneficial in two ways. In the first place, binary distillation is helpful in representing the principles of distillation in general by using a simplified model. Second, many real-life distillation problems may be approximated by a binary system. The simplifying assumptions and the graphical techniques described in this chapter were necessary before the advent of computers and the fast and rigorous programs for solving columns. Why, then, bother with learning the simplified antiquated methods? The graphical methods have the advantage of providing a qualitative, as well as quantitative, albeit perhaps approximate, representation of the problem. They allow the engineer to visualize the effects of the different variables and their feasible ranges before proceeding to the rigorous numerical solution. The simplified methods should not be used as a substitute but in conjunction with rigorous methods. Computer programs can, in fact, enhance the use of graphical techniques for evaluating column performance trends because the very same data needed to construct the diagrams are usually readily available from the simulation output.

This chapter starts with a discussion of a *column section*, defined as a group of adjacent equilibrium stages with no side feeds or draws and with no heat gains or losses. The column section serves as a model for formulating the necessary assumptions and for developing analytical and graphical methods for representing binary multistage equilibrium separation.

The concepts developed for a column section are then applied to the solution of a total column, and a graphical technique is implemented on the Y–X diagram.

The simplifying assumptions that are necessary for the Y–X solution method are then relaxed to include enthalpy balances. A graphical solution of this more rigorous model, which requires enthalpy–composition data, is presented.

Solution steps are described for cases with different sets of column specifications. Detailed analysis of column performance viewed through example applications will be covered in the next chapter.

5.1 COLUMN SECTION

For the purpose of discussion in this chapter, a column section is defined as any part of a distillation or other multistage separation process that includes no side feeds or draws and no heaters or coolers. A column section does necessarily have a liquid feed at the top and a vapor feed at the bottom. It also has a vapor product at the top and a liquid product at the bottom. It operates adiabatically, i.e., no heat is transferred across the system boundary to or from the surroundings, except the enthalpy transferred with the streams themselves.

A simple absorber is itself a column section. However, in the case of binary systems, it is not practical to have a column section as a unit operation on its own; it must be part of a total column. The reason for this is that, in order to operate such a column section, one would need binary liquid and vapor feeds with disparate compositions. Generating such streams would have required some separation process in the first place.

5.1.1   Development of the Model

A column section model is shown in Figure 5.1. It consists of N equilibrium stages designated by subscript j. Liquid and vapor molar flow rates leaving stage j are designated as L_j and V_j, respectively. X_j and Y_j are mole fractions of the more volatile component in L_j and V_j, respectively.

The column section may be solved by simultaneous solution of the component mass balance and enthalpy balance equations and the vapor–liquid equilibrium relations. Additional equations include the temperature, pressure, and composition dependence of the equilibrium coefficients and enthalpies. The equations for stage j are as follows:

Component balance on the lighter component:

$$L_{j-1}X_{j-1} + V_{j+1}Y_{j+1} = L_jX_j + V_jY_j \qquad (5.1)$$

A similar equation may be written for the less volatile component. Instead, an overall material balance is written:

$$L_{j-1} + V_{j+1} = L_j + V_j \qquad (5.2)$$

Equilibrium relations:

$$Y_j/X_j = K_{j1} \qquad (5.3)$$

$$(1 - Y_j)/(1 - X_j) = K_{j2} \qquad (5.4)$$

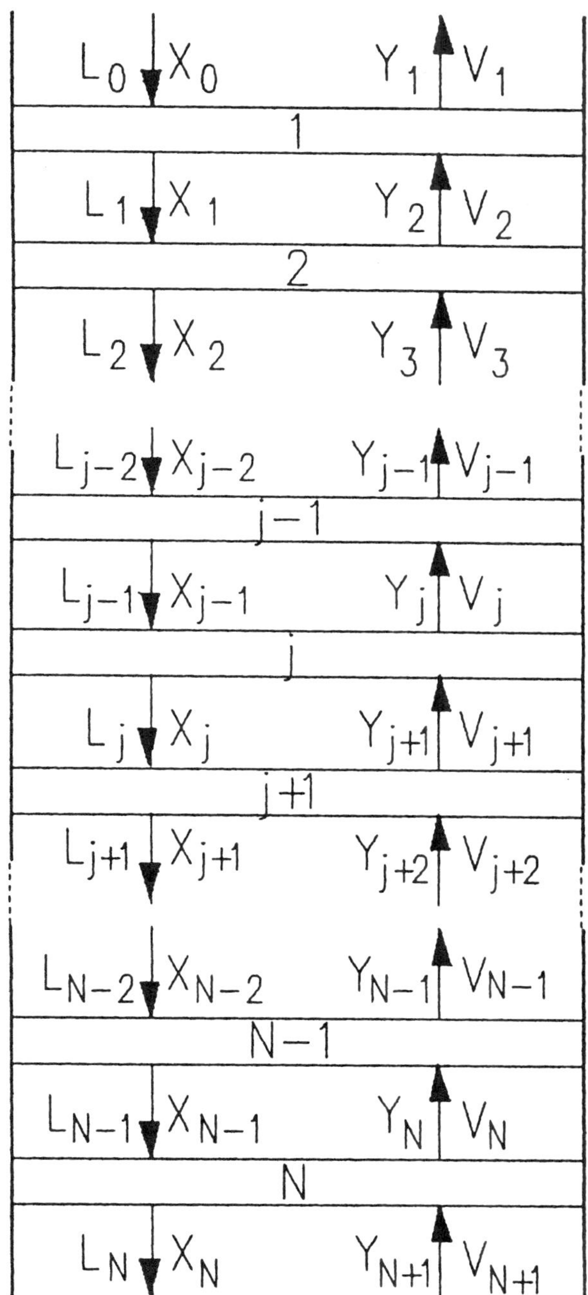

Figure 5.1 Schematic of a column section.

The second subscript of the equilibrium coefficients refers to components 1 and 2, the more volatile and less volatile components.

Enthalpy balance:

$$L_{j-1}h_{j-1} + V_{j+1}H_{j+1} = L_jh_j + V_jH_j \tag{5.5}$$

Data or correlations for predicting equilibrium coefficients and enthalpies may be represented as

$$K_{j1} = f(P_j, T_j, X_j, Y_j) \tag{5.6}$$

$$K_{j2} = f(P_j, T_j, X_j, Y_j) \tag{5.7}$$

$$H_j = f(P_j, T_j, Y_j) \tag{5.8}$$

$$h_j = f(P_j, T_j, X_j) \tag{5.9}$$

And, finally, the stage pressure equals the sum of the partial pressures:

$$P_j = P_{j1} + P_{j2} \tag{5.10}$$

For each stage j, the inlet streams are assumed to be of known rates, compositions, and thermal conditions, while outlet stream parameters are unknown. Thus, $L_{j-1}, V_{j+1}, X_{j-1}, Y_{j+1}, h_{j-1},$ and H_{j+1} are known quantities. The column or stage pressures are usually fixed and are therefore known, which renders Equation 5.10 redundant. This leaves nine equations for each stage with nine unknowns: $L_j, V_j, X_j, Y_j, h_j, H_j, T_j, K_{j1},$ and K_{j2}.

By extending this analysis to an N-stage column section, parameters $L_0, V_{N+1}, X_0, Y_{N+1}, h_0,$ and H_{N+1}, as well as the column section pressure, are assumed to be known quantities. That is, the feed rates, compositions, and thermal conditions and the column pressure are known. The column section is described by 9N equations with 9N unknowns. Since the number of equations equals the number of unknowns, the column section is completely specified if the feeds, number of stages, and stage pressures are fixed.

Assumptions and Simplifications

At this point we are not concerned with developing methods for rigorous solution of the above system of equations. Discussion of computational methods for the general case of multicomponent systems is covered in Chapter 13. This chapter considers a simplified model that lends itself to graphical solution and provides a tool for qualitative understanding of the operation of a distillation column. The model has a relatively low level of complexity because of its binary nature and also because of other simplifying assumptions.

The model shows that, if certain assumptions hold, then the vapor and liquid flows, V_j and L_j, are constant throughout the column section. If the sensible heat of the vapor and liquid is neglected when compared to the latent heat, i.e., if $h_{j-1} \approx h_j$ and $H_{j+1} \approx H_j$, Equation 5.5 becomes

$$L_{j-1}h_j + V_{j+1}H_j = L_j h_j + V_j H_j$$

Also, if the stream enthalpy equals the sum of component enthalpies, i.e., it exhibits ideal solution behavior, the above equation becomes

$$\sum_{i=1}^{2} \left[L_{j-1,i}h_{ji} + V_{j+1,i}H_{ji} \right] = \sum_{i=1}^{2} \left[L_{ji}h_{ji} + V_{ji}H_{ji} \right]$$

Parameters L_{ji} and V_{ji} refer to component molar flow rates in the liquid and vapor phases. The second subscript identifies the component. The component liquid and vapor enthalpies are related by the equation

$$H_{ji} = h_{ji} + \lambda_{ji}$$

where λ_{ji} is the heat of vaporization of component i on stage j. A key assumption at this point is that the two components have equal molar heats of vaporization,

$$\lambda_{ji} = \lambda_{j2} = \lambda_j$$

The enthalpy balance equation becomes

$$\sum_{i=1}^{2} \left[L_{j-1,i}h_{ji} + V_{j+1,i}h_{ji} + V_{j+1,i}\lambda_j \right] = \sum_{i=1}^{2} \left[L_{ji}h_{ji} + V_{ji}h_{ji} + V_{ji}\lambda_j \right]$$

The material balance equation (Equation 5.1) is now written for each component in terms of component flow rates, and these equations are multiplied by the component liquid enthalpy:

$$L_{j-1,1}h_{j1} + V_{j+1,1}h_{j1} = L_{j1}h_{j1} + V_{j1}h_{j1}$$

$$L_{j-1,2}h_{j2} + V_{j+1,2}h_{j2} = L_{j2}h_{j2} + V_{j2}h_{j2}$$

These two equations are subtracted from the enthalpy balance equation. The resulting equation, after cancelling λ_j, is

$$V_{j+1,1} + V_{j+1,2} = V_{j1} + V_{j2}$$

or

$$V_{j+1} = V_j$$

And, combining with Equation 5.2,

$$L_{j-1} = L_j$$

Thus, the assumptions of enthalpy additivity and equality of heats of vaporization of the two components in a binary lead to the approximation of constant molar overflow of liquid and vapor in a column section. The implication is that for each mole of component 1 vaporizing on a given stage, one mole of component 2 condenses.

Substituting

$$V = V_j = V_{j+1}$$

and

$$L = L_{j-1} = L_j$$

in Equation 5.1 yields the following:

$$Y_{j+1} = (L/V)X_j + Y_j - (L/V)X_{j-1} \tag{5.11}$$

This equation is the end result of combining Equations 5.1, 5.2, and 5.5, together with the simplifying enthalpy assumptions. Next, Equations 5.3 and 5.4 are combined into one equation, and the system of Equations 5.1 through 5.5 is reduced to two equations.

Each of Equations 5.3 or 5.4, aided by a vapor–liquid equilibrium coefficient prediction method such as Equations 5.6 or 5.7, could be used to establish the vapor–liquid equilibrium relationships. However, further simplification is possible if Equations 5.3 and 5.4 are used in combination with each other as follows:

$$K_{j1}/K_{j2} = (Y_j/K_j) (1 - X_j)/(1 - Y_j)$$

Rearranging the equation gives

$$Y_j = \frac{X_j \left(K_{j1}/K_{j2} \right)}{1 + X_j \left(K_{j1}/K_{j2} - 1 \right)} \tag{5.12}$$

If the relative volatility,

$$\alpha_{12} = K_1/K_2$$

is assumed constant, and if its value is known, Equation 5.12 may be used to generate the entire Y–X equilibrium curve.

Equations 5.11 and 5.12 may be used to obtain either an analytical or a graphical solution for the column section. Equation 5.12 relates the vapor and liquid compositions on the same tray. Equation 5.11 relates the composition of liquid descending from a given tray to the composition of vapor rising from the tray below.

5.1.2 Analytical Solution

As stated earlier in Section 5.1.1, if the external feeds to a column stage or section and its pressure are defined, the stage or section performance is defined. Equations 5.11 and 5.12 are now examined from this perspective. Consider first a single stage. If the external feeds are defined, then L, V, X_{j-1} and Y_{j+1} are known. This leaves two unknowns, X_j and Y_j, between Equations 5.11 and 5.12; hence, the problem is completely specified.

If the external feeds in an N-stage column section are defined, L, V, X_0, and Y_{N+1} are known. There are 2N unknowns (X_j and Y_j, $j = 1, ..., N$) and 2N equations (5.11 and 5.12, $j = 1, ..., N$) and the problem is completely defined. It is implied that N is also known, as well as the column pressure, which is assumed fixed. Thus the performance of a column section is defined if N, L, V, X_0, and Y_{N+1} and the pressure are known.

With a fixed number of stages N, parameters L, V, X_0, and Y_{N+1} may be substituted by other parameters as specified variables. For instance, one may specify, N, L, X_0, Y_{N+1}, and Y_1 instead of V, and the column section would still be completely specified. The vapor feed rate is relaxed to satisfy the vapor product composition specification, Y_1.

If the number of stages is relaxed, another parameter must be specified in order to define the column section. If, for example, L, V, X_0, and Y_{N+1} are specified, but N is not, the column section is not defined. If X_N or Y_1 is specified, the column section would be defined and N would have to be calculated to meet the specification. Table 5.1 lists the various possible ways of specifying a column section, together with the corresponding unknowns or dependent variables for each case.

Note that the system of simultaneous Equations 5.11 and 5.12 may easily be decoupled and solved independently and sequentially for $j = 1, ... N$. The computational steps for an analytical solution of a column section are as follows for the case where N, L, V, X_0, and Y_1 are specified:

1. Calculate X_1 from Equation 5.12 ($j = 1$).
2. Calculate Y_2 from Equation 5.11 ($j = 1$).

Repeat steps 1 and 2 for $j = 2, ..., N$. Cases where other sets of variables are specified may require a trial and error procedure.

5.1.3 Graphical Representation on the Y–X Diagram

Equations 5.11 and 5.12 are plotted on a Y–X diagram in Figure 5.2. Curve A is the equilibrium curve representing Equation 5.12 or any other set of vapor–liquid equilibrium data. Curve B is the operating line, or material balance relationship,

Table 5.1 Parameters in a Fixed-Pressure Binary Column Section

Specifications	Unknowns
Class 1	
N, L, V, X_0, X_N	$X_j(j = 1, ..., N - 1)$, $Y_j(j = 1, ..., N + 1)$
N, L, V, X_0, Y_{N+1}	$X_j(j = 1, ..., N)$, $Y_j(j = 1, ..., N)$
N, L, V, Y_1, Y_{N+1}	$X_j(j = 0, ..., N)$, $Y_j(j = 2, ..., N)$
N, L, V, Y_1, X_N	$X_j(j = 0, ..., N - 1)$, $Y_j(j = 2, ..., N + 1)$
Class 2	
N, L, V, X_0, Y_1	$X_j(j = 1, ..., N)$, $Y_j(j = 2, ..., N + 1)$
N, L, V, X_N, Y_{N+1}	$X_j(j = 0, ..., N - 1)$, $Y_j(j = 1, ..., N)$
Class 3	
N, L (or V), X_0, Y_1, X_N	V (or L), $X_j(j = 1, ..., N - 1)$, $Y_j(j = 2, ..., N + 1)$
N, L (or V), X_0, Y_1, Y_{N+1}	V (or L), $X_j(j = 1, ..., N)$, $Y_j(j = 2, ..., N)$
N, L (or V), X_N, Y_{N+1}, X_0	V (or L), $X_j(j = 1, ..., N - 1)$, $Y_j(j = 1, ..., N)$
N, L (or V), X_N, Y_{N+1}, Y_1	V (or L), $X_j(j = 0, ..., N - 1)$, $Y_j(j = 2, ..., N)$
Class 4	
L, V, X_0, Y_1, X_N	N, $X_j(j = 1, ..., N - 1)$, $Y_j(j = 2, ..., N + 1)$
L, V, X_0, Y_1, Y_{N+1}	N, $X_j(j = 1, ..., N)$, $Y_j(j = 2, ..., N)$
L, V, X_N, Y_{N+1}, X_0	N, $X_j(j = 1, ..., N - 1)$, $Y_j(j = 1, ..., N)$
L, V, X_N, Y_{N+1}, Y_1	N, $X_j(j = 0, ..., N - 1)$, $Y_j(j = 2, ..., N)$
Class 5	
L (or V), X_0, X_N, Y_1, Y_{N+1}	N, V (or L), $X_j(j = 1, ..., N - 1)$, $Y_j(j = 2, ..., N)$

expressed by Equation 5.11. The binary column section may be solved on this diagram by graphically duplicating the steps in the analytical procedure.

Any point on curve A represents compositions of vapor and liquid leaving an equilibrium stage. Points on curve B relate the compositions of passing vapor and liquid streams between stages. It can be shown that curve A must lie above curve B by comparing Y_A and Y_B, the values of Y on curves A and B respectively, corresponding to a given X. Y_A is the vapor composition on some stage, while Y_B is the vapor composition on the tray below. Since Y is the mole fraction of the more volatile component, Y is larger on trays that are higher up in the column. Therefore, Y_A is greater than Y_B.

As indicated by Equation 5.11, the operating line is a straight line if L/V is constant. Plotting it requires knowledge of the slope L/V and a point on the line, or two points on the line. Note that a point on the operating line has coordinates X_j and Y_{j+1} such as the inlet and outlet compositions at either end of the column section — X_0 and Y_1 or X_N and Y_{N+1}.

If only one point on the operating line or only its slope is known, the number of equilibrium stages must be known in order for the column section to be defined. The operating line would then be determined by trial and error to match the specified number of stages.

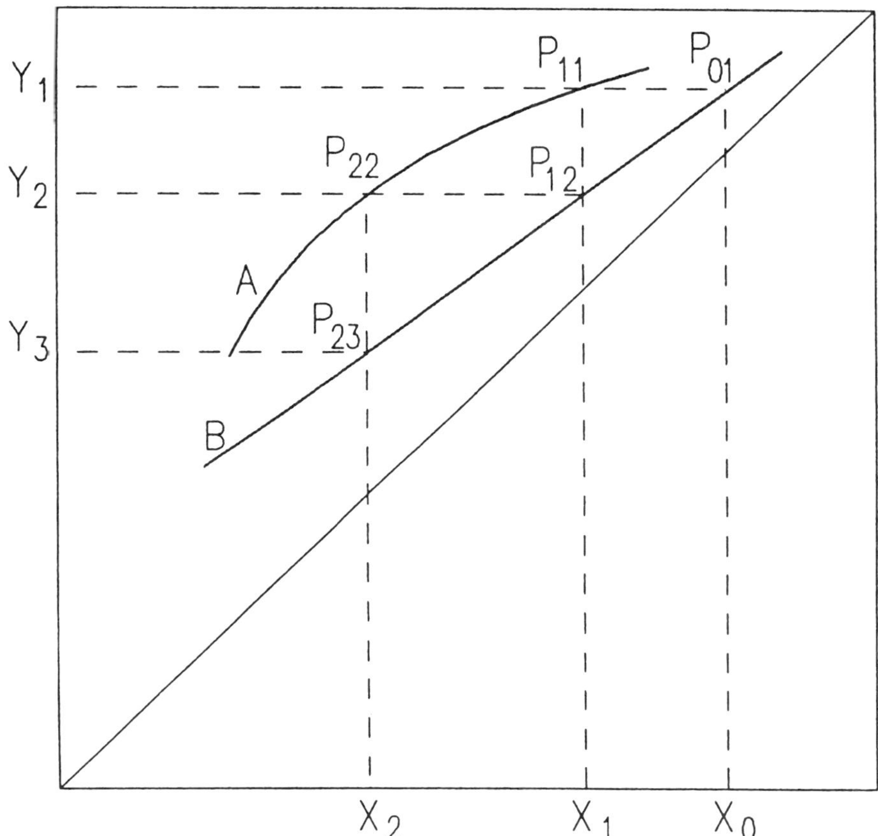

Figure 5.2   Equilibrium stages in a binary column section.

Constructing Equilibrium Stages

The equilibrium curve and the operating line drawn on a Y–X diagram (Figure 5.2) together describe the performance of a column section. Horizontal line $P_{01}P_{11}$ represents the transition from the region above stage 1 to stage 1 proper, and vertical line $P_{11}P_{12}$ represents the transition from stage 1 to the region between stages 1 and 2. Thus, point P_{01} corresponds to liquid composition entering stage 1 and vapor composition leaving stage 1, P_{11} corresponds to liquid and vapor compositions leaving stage 1, and P_{12} represents liquid composition from stage 1 to stage 2 and vapor composition from stage 2 to stage 1, etc. Points along connecting lines $P_{01}P_{11}$ and $P_{11}P_{12}$ do not represent actual compositions since the vapor and liquid compositions are assumed to undergo discrete changes from stage to stage rather than continuous variation. The graphical construction of stages is applied next to the different classes of column section specifications as listed in Table 5.1.

Class 1. The operating line slope L/V is known and the line is bounded by one coordinate on each end. Held at a fixed slope, the operating line is moved to a position where the number of steps between it and the equilibrium curve equals the given number of stages. In Figure 5.2, assume X_0 and X_2 are given as well as the slope and the number of stages, $N = 2$. Curve B is moved to a position that generates two steps between curves A and B, bounded by X_0 and X_2. Once curve B is determined, other compositions such as Y_1 and Y_3 can be read from the diagram.

Class 2. The operating line is defined by the specified values of the slope and one end of the line. The other end is determined by stepping off the specified number of stages. Referring to Figure 5.2, if the compositions of the liquid entering and vapor leaving the top of the column section are given, point $P_{01}(X_0, Y_1)$ is determined. Since the slope L/V is also given, the operating line, curve B, can be drawn. If the number of stages is given as 2, two steps are drawn between curves A and B down to point $P_{23}(X_2, Y_3)$, which determines the compositions of liquid leaving and vapor entering the bottom of the column section.

Class 3. Here, one end of the operating line is fixed and the other end is bounded by one coordinate. The operating line is free to rotate around the fixed point since the slope is not given. The correct position is determined by finding the slope that results in a number of steps equal to the given number of stages.

Class 4. This class of specifications fixes one end of the operating line and its slope. The other end is bounded by one coordinate. The required number of stages is found by constructing steps starting at the fixed point and ending at the intersection of the specified coordinate with the operating line. If the point of intersection does not coincide with the completion of a step, the last step should extend past the intersection.

Class 5. Finally, if two points on the operating line are specified, the line is fixed. The number of stages required for the specified separation is determined by stepping stages between the operating line and the equilibrium curve, starting at one of the specified points and ending at the other. Again, if the last step does not coincide with the given point, it is extended past it.

5.2 TOTAL COLUMN

A total column differs from a column section in that the former may have side feeds and products and coolers and heaters. Specifically, a conventional distillation column has one feed stream, an overhead (distillate) product and a bottoms product, and a condenser and a reboiler, as shown schematically in Figure 5.3.

Vapor rising from the top of the column is partially condensed, the vapor and liquid are separated, the vapor is taken out as a distillate or overhead product at its dew point, and the liquid is returned to the column as reflux. This *partial condenser* acts as an additional equilibrium stage since the vapor and liquid interact on it and are then separated. It is counted as stage 1, followed by the column top tray as stage 2, etc. The vapor from the column could be totally condensed, part of the condensate taken as liquid distillate and the remaining part refluxed to the column. With this mode of operation, the condenser is a *total condenser* and does not act as an equilibrium stage. No further change in the composition of the vapor leaving the

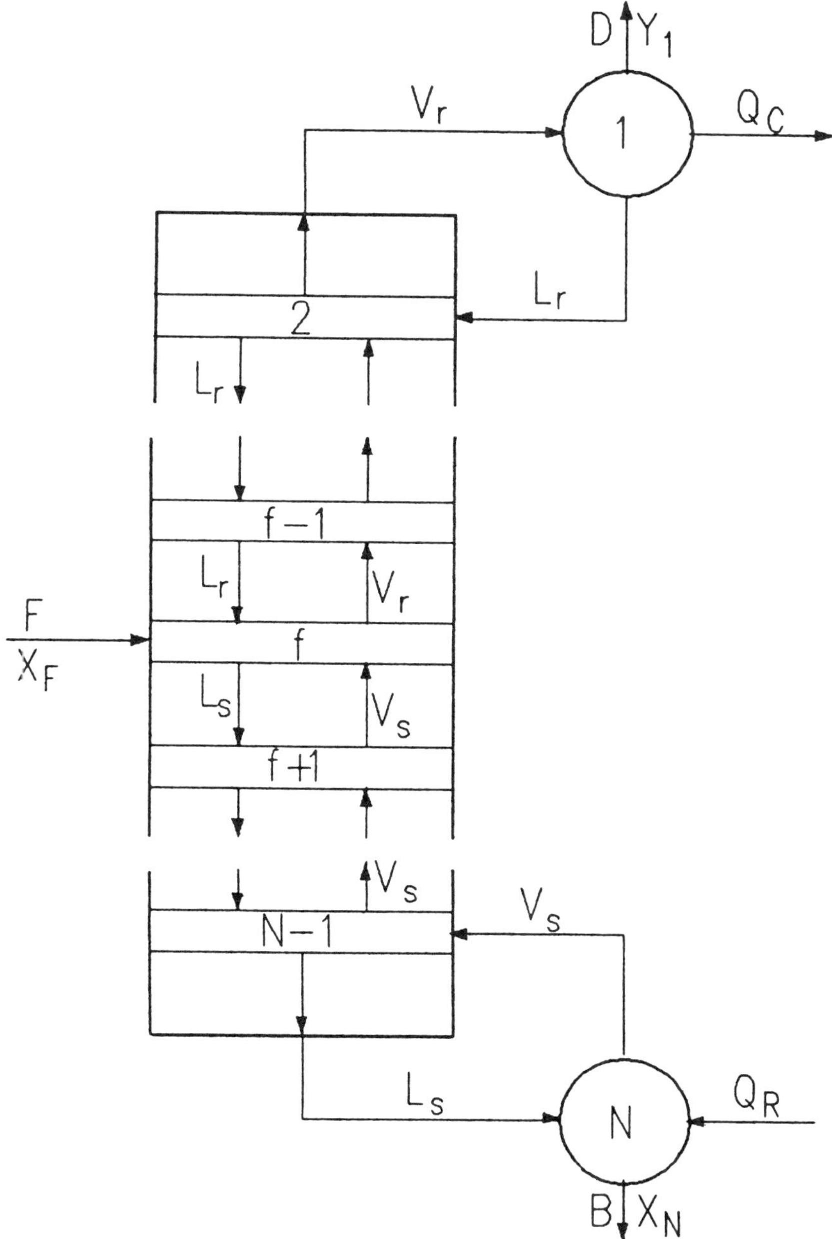

Figure 5.3 Schematic of a distillation column.

column top tray is achieved with a total condenser since no additional phase separation takes place. The column model discussed in this section is configured with a partial condenser.

Liquid from the column bottom tray is sent to the reboiler, stage N, where it is partially vaporized. The vapor is sent back to the column bottom tray, stage N-1, and the liquid is taken out as a bottoms product at its bubble point. In this configuration the reboiler is a *partial reboiler*, acting as an additional equilibrium stage.

The feed to the column could be at any thermal condition: superheated or saturated vapor, mixed phase, or saturated or subcooled liquid. The feed tray could be any tray between (and sometimes including) the condenser and reboiler. Regardless of its thermal condition, the feed is assumed to mix and equilibrate with the liquid and vapor on the feed tray.

The techniques developed earlier for solving column sections are now extended to include total columns. A mathematical model is set up which is used to derive graphical methods of solution on a Y–X diagram. Solution methods are described for different sets of specifications.

5.2.1 Mathematical Model

The assumptions and simplifications made for a column section apply to a total column as well. The implications of these assumptions for a column are that liquid and vapor molar flows from tray to tray are constant within each section of the column. As in the case of the column section, the component and enthalpy balances reduce to operating line equations, which must be solved in conjunction with the vapor–liquid equilibrium equation or data. Each section within a column is represented by a different operating line.

Top-Section Operating Line

The top section, or *rectifying section*, is that part of the column above the feed and including the condenser (stage 1). Referring to Figure 5.3, a component material balance on any part of such section that includes stages 1 through j (for j less than f, the feed tray) gives

$$V_r Y_{j+1} = L_r X_j + D Y_1$$

or

$$Y_{j+1} = (L_r/V_r)X_j + (D/V_r)Y_1 \tag{5.13}$$

for j = 1, ..., f − 1. D is the distillate molar rate, and Y_1 is its composition. The liquid and vapor molar flows, L_r and V_r are assumed constant throughout the rectifying section.

Bottom-Section Operating Line

The bottom section, or stripping section of the column, is the part that includes the reboiler (stage N) and all stages below the feed stage. A component material balance on any part of the stripping section that includes stages j through N (for j greater than f, the feed stage) gives

$$L_s X_{j-1} = V_s Y_j + B X_N$$

where B and X_N are the bottoms molar rate and composition, respectively. The equation is rearranged:

$$Y_j = (L_s/V_s)X_{j-1} - (B/V_s)X_N \tag{5.14}$$

where j = f + 1, ..., N. The liquid and vapor molar flows, L_s and V_s, are assumed constant throughout the stripping section.

Feed Stage

The feed stream, *F*, with composition X_F enters the column on stage f, mixes with the vapor and liquid entering that stage from adjacent stages, and reaches equilibrium. A component material balance on stage f gives

$$FX_F + L_r X_{f-1} - L_s X_f = V_r Y_f - V_s Y_{f+1} \tag{5.15}$$

A heat balance on the feed stage determines the relative flow rates of liquid and vapor above and below the feed stage, namely, $L_r, L_s, V_r,$ and V_s. As the feed stream enters the column, it changes from its initial thermal conditions (T_F, P_F) to the feed tray conditions (T_f, P_f). Let the amount of heat associated with the conversion of 1 mole of feed at T_F and P_F to saturated vapor at T_f and P_f be ΔH_F. Setting an arbitrary enthalpy reference point of 0 for saturated vapor at T_f and P_f and assuming as before a uniform molar heat of vaporization λ, a heat balance on the feed stage gives

$$(L_s - L_r)\lambda = F\Delta H_F$$

Defining the quantity q = $\Delta H_F/\lambda$, the following equation is obtained:

$$L_s = L_r + qF \tag{5.16}$$

Equation 5.16 is combined with an overall material balance on the feed stage to give

$$V_r = V_s + (1 - q)F \tag{5.17}$$

If the feed is a saturated vapor at T_f and P_f, then $\Delta H_F = 0$ and q = 0. The vapor rate leaving the feed tray is simply the sum of vapor from the bottom column section and the feed, $V_r = V_s + F$. Since no external liquid is involved, the liquid rate entering the tray from the upper section equals the liquid rate leaving it — $L_r = L_s$.

If the feed is a saturated liquid at T_f and P_f, then $\Delta H_F = \lambda$ and q = 1. The liquid rate leaving the feed stage is the sum of liquid from the upper column section and the feed, $L_s = L_r + F$. Since the external feed contains no vapor, the vapor rate leaving the tray equals the vapor rate entering it from the bottom column section — $V_r = V_s$.

For a mixed phase feed that is part vapor and part liquid at T_f and P_f, the liquid portion joins the liquid leaving the feed tray and the vapor portion joins the vapor leaving it. The amount of liquid is qF and that of the vapor is $(1 - q)F$. For a partially vaporized feed, q is equal to the liquid fraction $(0 < q < 1)$.

Note in the above cases that V_r is greater than or equal to V_s, and L_s is greater than or equal to L_r. That is, the feed has no tendency to "dry up" either the liquid or the vapor.

A feed that is superheated at T_f and P_f has a negative ΔH_F and, therefore, a negative q. A superheated feed will increase the vapor rate above the feed tray $(V_r > V_s)$ and decrease the liquid rate below the feed tray $(L_s < L_r)$. If the feed is sufficiently superheated, it may dry up the liquid in the lower column section, making L_s approach 0.

If the feed is a subcooled liquid at T_f and P_f, $\Delta H_F > 1$ and $q > 1$. In this case L_s is greater than L_r, and V_r is less than V_s. Thus, if the feed is sufficiently subcooled, it may dry up the vapor in the upper section (V_r approaches 0).

Analytical Solution

The equations describing total column operation include vapor–liquid equilibrium relations, Equation 5.12; component balances in the top and bottom sections, Equations 5.13 and 5.14; feed stage component balance, Equation 5.15; feed stage heat balance and overall material balance, expressed as Equations 5.16 and 5.17; and overall column component balance, Equation 5.18:

$$FX_F = DY_1 + BX_N \tag{5.18}$$

The problem definition is such that the feed is of fixed rate, composition, and thermal conditions. Also, the column pressure and the number of stages are known. The condenser and reboiler duties are not presently of concern; calculating them would require heat balance equations for the condenser and reboiler. Aside from these, the system of equations includes the following:

Equation 5.12	N	equations
Equation 5.13	f–1	equations
Equation 5.14	N–f	equations
Equation 5.15	1	equation
Equation 5.16	1	equation
Equation 5.17	1	equation
Equation 5.18	1	equation
Total 2N + 3		equations

The unknowns are $X_j (j = 1, \ldots, N)$, $Y_j (j = 1, \ldots, N)$, L_r, L_s, V_r, V_s, and B (or D), adding up to a total of $2N + 5$ unknowns. Thus, in order to define the column operation, two parameters must be specified. If the number of stages is not fixed, three parameters would have to be specified.

The method for solving the above set of equations depends on which parameters are specified. Although the simplified binary distillation model discussed in this chapter is mainly intended for developing a graphical solution, an analytical solution is first presented as an exercise for better understanding the interrelations among the different variables. An algorithm is provided only for one set of specifications — the distillate and bottoms compositions, Y_1 and X_N. Graphical solutions for these and other specifications are presented in the next section.

The computational steps are as follows:

1. Given Y_1, X_N, N, F, X_F, T_F, and P_F.
2. A column overall material balance, $F = D + B$, combined with Equation 5.18 gives the following:

$$FX_F = (F - B)Y_1 + BX_N$$

   All the parameters in this equation are known except B. Solve for B, then compute $D = F - B$.
3. Assume a reflux rate, L_r, and calculate $V_r = D + L_r$.
4. Calculate X_1 from Equation 5.12 with $j = 1$.
5. Calculate Y_2 from Equation 5.13 with $j = 1$.
6. Repeat steps 4 and 5 with $j = 2, \ldots, f - 1$. The last compositions calculated are X_f and Y_f.
7. From vapor–liquid equilibrium versus temperature data, calculate T_f that corresponds to $K_f = Y_f/X_f$.
8. Calculate ΔH_F, the difference between the molar feed enthalpy at initial conditions T_F and P_F and the enthalpy of a mole of saturated vapor at T_f and P_f.
9. Calculate q from its definition, $q = \Delta H_F/\lambda$.
10. Calculate L_s and V_s from Equations 5.16 and 5.17.
11. Calculate Y_{f+1} from Equation 5.15.
12. Calculate X_{f+1} from Equation 5.12 with $j = f + 1$.
13. Calculate Y_{f+2} from Equation 5.14 with $j = f + 2$.
14. Repeat steps 12 and 13 for $j = f + 2, \ldots, N$. The last compositions calculated are Y_N and X_N.
15. If X_N calculated matches X_N specified, the problem is solved. Otherwise, start over at step 3 with a new assumed reflux L_r.

The Description Rule

By taking an inventory of equations and variables for the above column, it was determined that two or three variables (depending on whether or not the number of stages is fixed) must be specified in order to define the column performance. An alternative to this mathematical analysis is to apply the *description rule* (Hanson et al., 1962), which can be a less tedious method for determining the number of independent variables based on a practical evaluation of the process.

The description rule states that the number of independent variables that are required to define a process uniquely is equal to the number of variables set by construction of the process equipment plus the number of variables that can be directly and independently controlled during operation of the process. The rule is simply a statement of the nature of the actual process which the mathematical model must represent.

Focusing again on the column described above, the engineer has the freedom during the design phase to decide on the number of trays as well as the feed location. These 2 degrees of freedom are used up at construction time. With an existing column, the operator starts up the process by pumping in the feed stream to the column. The process devices may include, in addition to the pump, a valve at the pump discharge and a feed preheater. If these pieces of equipment are considered part of the system, they provide the operator with the means to control the feed rate, temperature, and pressure by independently manipulating the pump speed, the valve setting, and the steam rate to the preheater. This particular subsystem therefore requires three independent variables to specify it. The setup does not include a device for controlling the feed composition and, hence, this parameter is not included in the list of independent variables.

If the box enclosing the system includes just the feed inlet and the existing column, the feed conditions are set and no devices are available to manipulate them. With a fixed feed and column configuration, the operator has the task of operating the column to achieve the desired separation. The variables available for controlling the separation are the operating devices that can be physically and independently manipulated. One variable that must be set is the column pressure, which is usually controlled by manipulating the vapor or vent flow out of the column. The column pressure is normally dictated by the temperature level of the condenser cooling medium and is not, in general, used to control the separation.

Two more operating variables exist — the reboiler heating medium rate (steam) and the condenser cooling medium rate (water). Therefore, two more variables must be set in order to specify the column operation. The steam and water rates themselves may be set or, alternatively, they may be controlled by two other variables such as the desired separation specifications.

Note that the column may have additional controllers such as condenser and reboiler level controls. The levels, however, are not independent variables since they must be maintained constant to ensure steady-state operation of the column. The liquid levels, which have no effect on the separation, may be controlled by manipulating the distillate and bottoms products rates.

In conclusion, for an existing column with a fixed feed, two variables are required to define the column performance with regard to separation — the condenser and reboiler duties.

5.2.2 Graphical Solution on the Y–X Diagram

The material balance equations, or operating lines (Equations 5.13 and 5.14) together with the feed stage heat balances (Equations 5.16 and 5.17), and the equilibrium relation, (Equation 5.12) form the basis for the McCabe–Thiele (1925) graphical solution of binary distillation on a Y–X diagram.

The phase equilibrium data, or Equation 5.12, are first plotted on a Y–X diagram (Figure 5.4). The operating lines, assumed straight lines, are then plotted on the basis of known points, slopes, and/or number of stages, depending on what variables are specified. Together, these curves completely define the column operation, and a graphical solution readily follows.

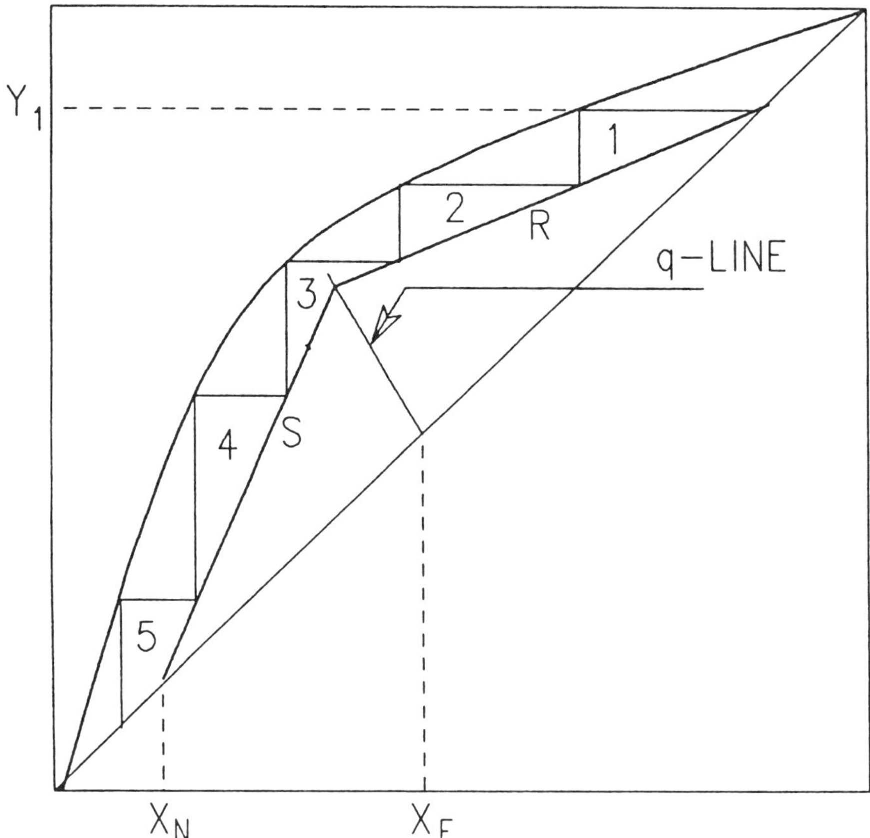

Figure 5.4   Performance of a binary distillation column.

Each operating line is the locus of discrete points, (X_1, Y_2), (X_2, Y_3), etc., relating the compositions of liquid and vapor counterflowing between stages. For the purpose of plotting them, the operating lines, Equations 5.13 and 5.14, are considered continuous in Y and X. Their slopes are L_r/V_r and L_s/V_s, respectively. Extrapolating them toward the diagonal determines one fixed point for each line. On the diagonal, since $Y = X$, Equation 5.13 becomes

$$V_r X = L_r X + DY_1$$

It is also known that $D = V_r - L_r$ which, when substituted in the above equation, gives $Y = X = Y_1$. That is, the rectifying section operating line intersects the diagonal at the distillate composition. Similarly, it may be shown that the stripping section operating line intersects the diagonal at $Y = X = X_N$, the bottoms composition.

Another point that determines the operating lines is their intersection point. Equations 5.13 and 5.14, again considered to represent Y as a continuous function of X, are both extrapolated toward the feed stage. Equation 5.13 is valid above that stage, and Equation 5.14 is valid below it. The operating line of the total column is made up of two straight lines, R and S (Figure 5.4). Since R corresponds to stages above the feed and S corresponds to stages below the feed, the transition from R to S, or their intersection, occurs at the feed stage. At the point of intersection, Y_{j+1} and X_j in Equation 5.13 are identical to Y_j and X_{j-1} in Equation 5.14. Subtracting these equations gives

$$Y(V_r - V_s) = X(L_r - L_s) + DY_1 + BX_N$$

which, when combined with Equations 5.16, 5.17, and 5.18, gives

$$Y = Xq/(q - 1) - X_F/(q - 1) \qquad (5.19)$$

Equation 5.19 is known as the q-line equation. For an optimum feed tray the q-line, which is determined by the feed composition and thermal conditions, must fall on the intersection of the operating lines. The q-line is a straight line with a slope of $q/(q - 1)$ and an intersection point with the diagonal at $Y = X = X_F$. This may be shown by setting $Y = X$ in Equation 5.19 and solving for X.

Feed Thermal Condition at T_f, P_f		q-Line Slope
Superheated Vapor	q < 0	0< slope < 1
Saturated Vapor	q = 0	0
Mixed Phase	0 < q < 1	< 0
Saturated Liquid	q = 1	Infinity
Subcooled Liquid	q > 1	> 1

The values of q for different feed thermal conditions at feed tray temperature and pressure, T_f and P_f, were discussed in Section 5.2.1. The corresponding values of the q-line slope are summarized on the previous page. Example q-lines are shown in Figure 5.5.

Representing a Total Column

Once the operating lines are plotted, the equilibrium stages are stepped off between the operating line and equilibrium curve as described in Section 5.1.3. If the number of stages is fixed, the operating lines are determined by trial and error to match the specified number of stages.

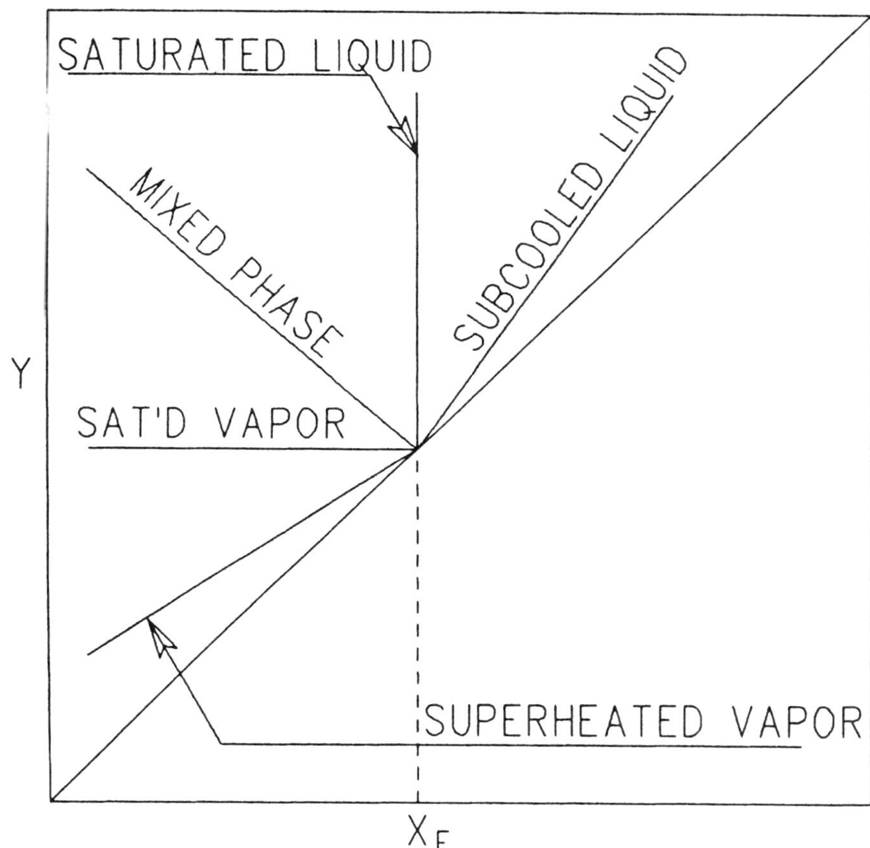

Figure 5.5   Plots of q-lines for different feed thermal conditions.

The mathematical model described in Section 5.2.1 indicates that two parameters must be specified in order to define the performance of a column with fixed pressure, number of stages, feed tray location, and feed rate, composition, and thermal conditions. Three specifications are required if the number of stages is unknown. Graphical methods for solving distillation columns depend on column specifications. A partial condenser with vapor distillate and a partial reboiler with liquid bottoms are assumed. As such, the condenser and reboiler are both counted as equilibrium stages. The column solution is described for the following sets of specifications.

Separation and Reflux Ratio Specified

The separation is specified in terms of the composition of one component in the distillate and bottoms, Y_1 and X_N. (The subscripts refer to stages. When component subscripts are dropped, implied reference is to the more volatile component.) The separation may alternatively be specified in terms of parameters that are functions of Y_1 and X_N. The reflux ratio expresses relative flow rates of liquid and vapor within the column. It may be specified as L_r, V_r, L_s, or V_s, or a function thereof such as L_r/V_r.

The number of stages is unknown; therefore, three independent specifications are required to define the column operation. Consider the case where the specifications are Y_1, X_N, and L_r/V_r. The first step in a graphical solution is to plot the equilibrium curve corresponding to the given column pressure on a Y–X diagram based on vapor–liquid equilibrium data (Figure 5.4). Next, the q-line is plotted on the basis of the feed composition and its thermal conditions at feed tray temperature and pressure. The feed tray pressure is known, but the feed tray temperature must be assumed for the purpose of calculating the q-line slope. The rectifying section operating line, R, has one point fixed at the intersection of a horizontal line through Y_1 and the diagonal. The slope of R is L_r/V_r as indicated by Equation 5.13. With one fixed point and the slope defined, the operating line, R, is next drawn on the diagram. The stripping section operating line, S, is drawn by joining X_N on the diagonal to the intersection of R with the q-line.

This completes the operating line for the entire column. The equilibrium stages are next stepped off between the operating line and the equilibrium curve beginning at Y_1, as shown in Figure 5.4. (Construction of the stages could just as well be started at X_N.) The diagram shows that four full equilibrium stages plus a fraction of a stage are required to achieve the specified separation at the given reflux ratio. A fraction of a theoretical stage could translate into one or more actual trays, depending on their efficiency. Fractional stages are generally rounded up, bringing the total in this case to five. The resulting separation would be slightly sharper than specified, i.e., the bottoms composition would be somewhat less than X_N. The diagram also shows that the feed should be introduced at the third stage from the top. The top stage is the condenser and the bottom stage is the reboiler.

The assumed feed tray temperature may now be checked and, if necessary, the q-line redrawn using the updated slope. The above procedure is then repeated for the new q-line slope. To check the feed tray temperature, it is necessary to make use of the vapor–liquid equilibrium data from which the equilibrium curve was drawn.

To complete the column solution, the product rates are computed from a total material balance and one component balance around the column:

$$F = D + B$$

$$FX_F = DY_1 + BX_N$$

The values of L_r and V_r are determined from the given L_r/V_r ratio and a material balance on the condenser, $D = V_r - L_r$. V_s and L_s may be calculated from Equations 5.16 and 5.17.

Distillate Composition, Reflux Ratio, and Number of Stages Specified

The equilibrium curve and the q-line are plotted as above. Since Y_1 and L_r/V_r are given, the rectifying section operating line may be drawn. The point of intersection of this line with the q-line fixes one point on the stripping section operating line. The slope of this line must be determined by trial and error to meet the specified number of stages. The stripping section operating line is rotated around its pivotal point on the intersection of the q-line and the operating line and, for each position, equilibrium stages are stepped off between the operating lines and the equilibrium curve. The correct position is found when the stepped number of stages matches the specified number. The bottoms composition, X_N, is then determined by dropping a vertical line through the intersection of the lower operating line with the diagonal. The solution also determines the feed location.

A similar technique may be used if the bottoms composition and L_s/V_s are specified instead of Y_1 and L_r/V_r.

Separation and Number of Stages Specified

The separation is specified in terms of Y_1 and X_N or quantities that are functions thereof, such as recoveries. The intersections of a horizontal line through Y_1 and the diagonal and a vertical line through X_N and the diagonal determine one point on each operating line. The q-line is drawn on the basis of feed data as before. This line is the locus of intersection points of the two operating lines. The correct intersection point is determined by trial and error until the stepped number of stages equals the specified number. The closer the intersection point to the equilibrium curve, the larger the number of stages. The resulting operating lines may be used to determine the feed location, L_r/V_r, L_s/V_s (or reflux ratio), tray compositions, etc.

Reflux Ratio, Product Rates, and Number of Stages Specified

If the reflux ratio, or L_r/D, and the product rate, D, are specified, the rectifying operating line slope, L_r/V_r may be calculated. The feed composition and thermal conditions are also known, which allows the q-line to be plotted and L_s and V_s to be calculated from Equations 5.16 and 5.17. Thus, the slopes of the two operating lines are known as well as the locus of their intersection points.

The column is solved by trial and error by moving the operating lines with their intersection point along the q-line, keeping their slopes fixed (Figure 5.6). The number of stages for each position is determined by stepping off stages as usual. The position that results in the specified number of stages is the required solution. Information obtained from the solution includes the separation achievable with the given reflux ratio and number of stages and the appropriate feed location.

Columns with Multiple Feeds, Side Draws, and Side Heaters/Coolers

The methods developed for graphical construction of equilibrium stages on Y–X diagram for single-feed, two-product distillation columns can be generalized to include columns with multiple feeds, side draws, and side heaters, or coolers. When these discontinuities exist, each column section operating line has, in general, a different L/V ratio or slope. A column section operating line is defined if its L/V ratio and one point on the line are known or if two points on it are known.

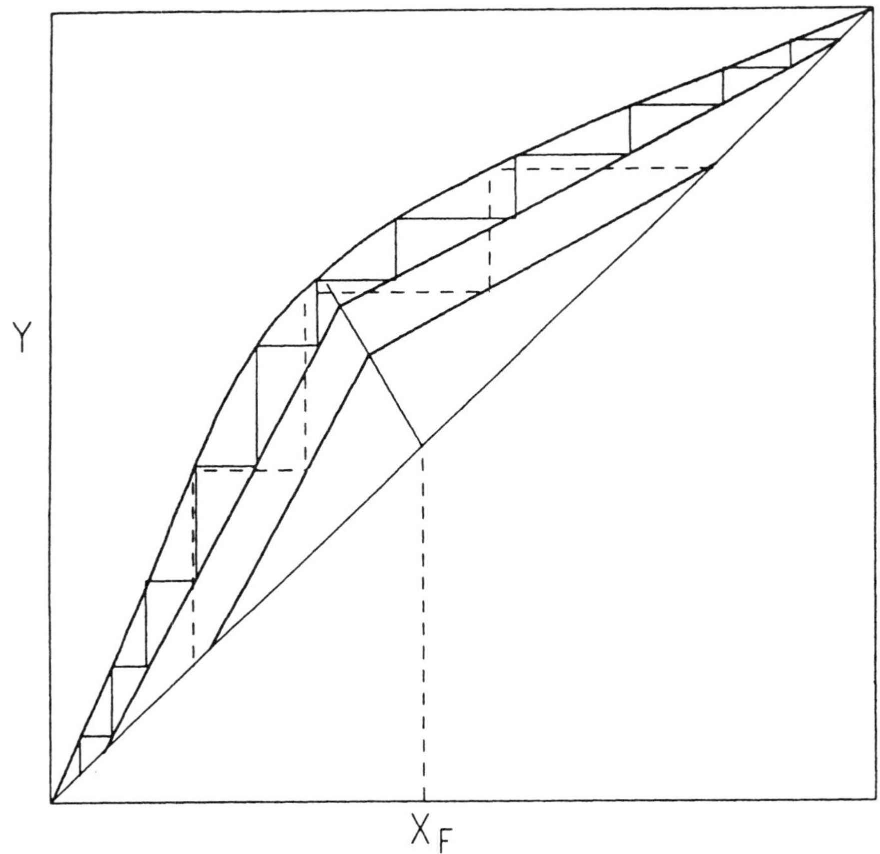

Figure 5.6   Binary distillation: given number of stages, reflux ratio, and product rate.

The way the operating lines are actually constructed depends on how the particular column is specified. Figure 5.7 shows a hypothetical column with two feeds, a liquid draw, a vapor draw, a partial condenser and a reboiler, a side cooler, and a side reboiler. Also shown is the Y–X diagram and operating lines for this column. If we assume the distillate rate and composition and the reflux ratio are known, the calculations can be started at the top. A vertical line drawn through the distillate composition intersects the diagonal at a point that defines the upper end of the top operating line. The L/V ratio, calculated from the distillate rate and reflux ratio, determines the slope of the line. Equilibrium stages are then stepped off between this operating line and the equilibrium curve. This is continued until the side cooler stage is reached.

The effect of a side cooler or heater on the L/V ratio is determined from an enthalpy balance on stage j, the stage where heat duty q_j is added. For a cooler, q_j is negative. The enthalpy balance is written as

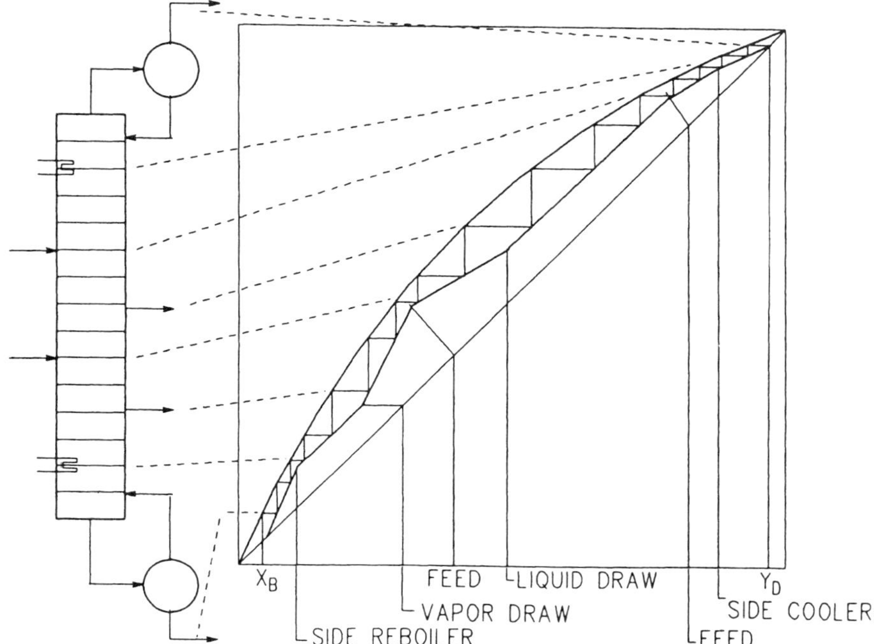

Figure 5.7 Column with multiple feeds, side draws, and side heaters/coolers.

$$q_j + L_{j-1}h_{j-1} + V_{j+1}H_{j+1} = L_jh_j + V_jH_j$$

When the McCabe–Thiele assumptions are applied, the sensible heat is neglected in comparison with the latent heat, which is assumed identical for both components in the binary. With these assumptions, the enthalpy balance is simplified, resulting in the equation

$$-q_j = H(V_{j+1} - V_j) + h(L_{j-1} - L_j)$$

where H and h are average vapor and liquid molar enthalpies, respectively. The material balance equation on tray j,

$$L_{j-1} - L_j = V_j - V_{j+1}$$

is combined with the above equation to give

$$-q_j = (H - h)(V_{j+1} - V_j) = (H - h)(L_j - L_{j-1})$$
$$= \lambda(V_{j+1} - V_j) = \lambda(L_j - L_{j-1})$$

The operating line slope below the side cooler, L_j/V_{j+1} is related to the liquid and vapor flows above the side cooler stage by the following equation, based on a material balance around stage j:

$$\frac{L_j}{V_{j+1}} = \frac{L_{j-1} + \left(L_j - L_{j-1}\right)}{V_j + \left(L_j - L_{j-1}\right)}$$

For a side cooler, the enthalpy balance on stage j indicates that

$$L_j - L_{j-1} > 0$$

Therefore, if $L_{j-1}/V_j > 1$, then $L_j/V_{j+1} < L_{j-1}/Vj$, the slope of the operating line above the side cooler. Also, if $L_{j-1}/V_j < 1$, then $L_j/V_{j+1} > L_{j-1}/V_j$.

If the slope of the operating line below the side cooler stage is known or can be calculated, the line is constructed directly. If the number of stages between the side cooler and the upper feed is known, the slope of the operating line must be varied to match the number of stages.

The upper feed is represented graphically by drawing a vertical line through its composition to intersect the diagonal where the q-line is drawn with the appropriate slope corresponding to the feed thermal conditions. The intersection of the q-line with the previous operating line determines the upper end of the next operating line. With a known slope, or L/V ratio, this line is now drawn, and the stages between the upper feed and the next discontinuity, the liquid draw, are stepped off.

The operating line above the liquid draw ends at its intersection with the q-line of the liquid draw. Since this stream is always a saturated liquid, its q-line must be vertical. Hence, a vertical line constructed at the draw tray represents this side stream and determines its composition.

The liquid flow below the liquid draw tray is reduced by an amount equal to the side draw rate, S. Thus, if the side draw is taken at tray j,

$$L_j = L_{j-1} - S$$

Under the present simplifying assumptions, the vapor flow does not change at the liquid draw tray. Therefore, the L/V ratio below the liquid draw is less than the ratio above it, and the slope of the operating line below the draw is reduced. The next discontinuity occurs at the lower feed, whose q-line is constructed in the usual manner. The slope of the operating line between the liquid draw and the lower feed is adjusted to match the number of stages in that section.

The slope of the operating line between the lower feed and the vapor draw is again determined so that the stepped stages equal the known number of stages in that section. The q-line of the vapor draw is horizontal since it is a saturated vapor. Its composition is determined by dropping a vertical line from the intersection of the q-line with the diagonal.

The next section operating line has a smaller slope since the vapor flow is reduced above the draw tray j by the vapor draw rate, S:

$$V_j = V_{j+1} - S$$

And, since the tray liquid flow is assumed constant,

$$L_j/V_{j+1} < L_{j-1}/V_j$$

The operating line slope is determined so that it matches the number of stages between the vapor draw and the side reboiler.

The side reboiler has the opposite effect of the side cooler on the operating line slope. Thus, if $L_{j-1}/V_j > 1$, then

$$L_j/V_{j+1} > L_{j-1}/V_j$$

and, if $L_{j-1}/V_j < 1$, then

$$L_j/V_{j+1} < L_{j-1}/V_j$$

If the number of stages between the side reboiler and bottom reboiler is known, the slope of the last section operating line is adjusted to match the number of stages. If the bottoms composition is known, the operating line is determined and the equilibrium stages are stepped off.

5.2.3 Tray Efficiency

Most column tray-by-tray calculations, including the graphical methods described in this chapter, are based on the concept of the equilibrium stage. Since actual column trays can generally only approach equilibrium stage performance to varying degrees, it is common practice to use tray efficiencies to relate actual performance to equilibrium conditions. There are different types of tray efficiencies, a detailed discussion of which is deferred to Chapter 16. The Murphree tray efficiency (Murphree, 1925) is introduced at this point because it can be readily incorporated in the Y–X graphical construction of stages.

Figure 5.8 shows a segment of an operating line and the equilibrium curve. Tray j, with a liquid composition X_j, receives vapor with composition Y_{j+1} from the tray below. If tray j were an equilibrium stage, the vapor leaving it would have a composition Y_j^*, determined by the equilibrium curve. The change in the vapor composition as it passes through tray j would be $Y_j^* - Y_{j+1}$. In an actual tray, the vapor leaving tray j has a composition Y_j, resulting in a change in vapor composition of $Y_j - Y_{j+1}$. The value of Y_j is a function of the tray efficiency. The Murphree tray vapor efficiency is defined as

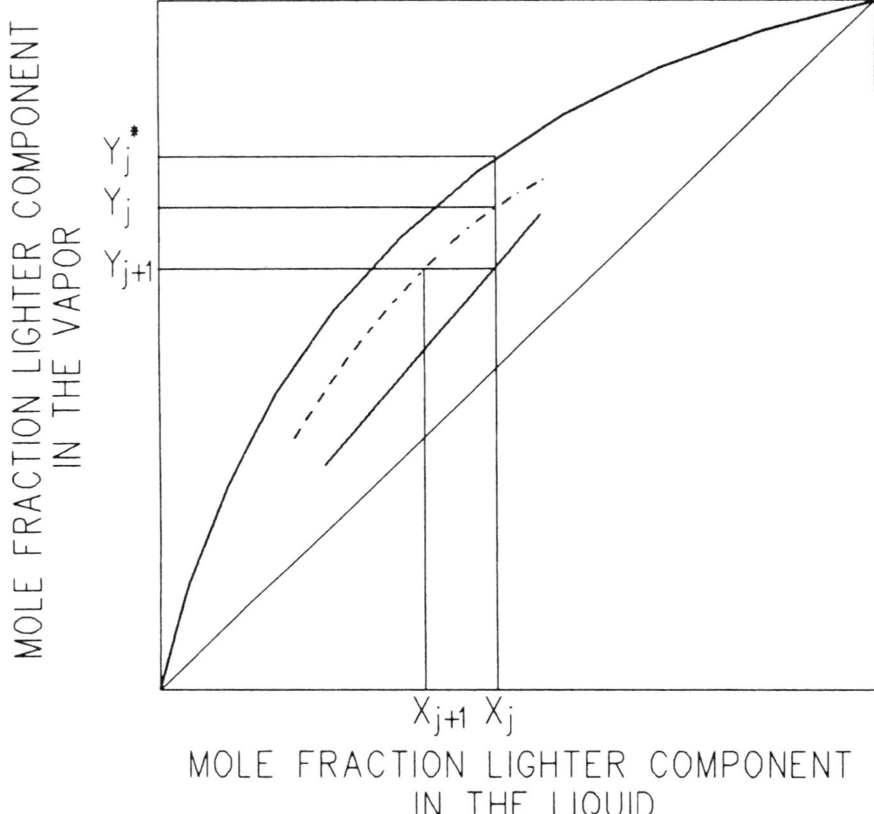

Figure 5.8   Implementing Murphree tray efficiency on the Y–X diagram.

$$E_v = \frac{Y_j - Y_{j+1}}{Y_j^* - Y_{j+1}}$$

The tray efficiency may generally be estimated on the basis of past experience with similar columns or from operating data of an existing column. Given the tray efficiency, a pseudo-equilibrium curve can be drawn, such as the dotted line in Figure 5.8. This curve is constructed based on the definition of the tray vapor efficiency by marking off vertical segments between the operating line and the equilibrium curve so that

$$Y_j - Y_{j+1} = E_v \left(Y_j^* - Y_{j+1} \right)$$

Once this curve is in place, the graphical construction of stages proceeds as before.

5.3 COLUMN SOLUTION WITH MASS AND ENTHALPY BALANCES

The graphical binary distillation solution technique described on the Y–X diagram in the previous section is based on a number of simplifying assumptions. Simplifications in the enthalpy balances were made, which lead to the approximation that the vapor and liquid flows are constant in each column section and that the operating lines can be represented by straight lines. A more rigorous solution must include accurate enthalpy balances. In order to accomplish this, enthalpy–composition data for the binary are needed, generally presented in the form of an enthalpy–composition diagram.

The method described here was originally presented by Ponchon (1921) and Savarit (1922). Besides the enthalpy–composition diagram, the method makes use of the lever rule for relating the rates of vapor and liquid products of a binary equilibrium stage to their compositions and enthalpies, as described below.

5.3.1 Single-Stage Mass and Enthalpy Balances

If a feed stream with mole rate F separates in an equilibrium stage into a vapor product with mole rate V and a liquid product with mole rate L, then by overall mass balance,

$$F = V + L \tag{5.20}$$

A component mass balance on either component is written as

$$FZ = VY + LX \tag{5.21}$$

where Z, Y, and X are the mole fractions of one of the components in the feed, vapor product, and liquid product, respectively. The mole fractions are not subscripted since the mass balance applies to either component, and no distinction between them is necessary at this point.

A total enthalpy balance on the equilibrium stage is written as

$$FH_F = VH + Lh \tag{5.22}$$

where H_F, H, and h are the enthalpy per mole of the feed, vapor product, and liquid product, respectively.

By combining Equations 5.20 and 5.21, and 5.20 and 5.22, the following equations are obtained:

$$V/F = (Z - X)/(Y - X) \tag{5.23}$$

$$L/F = (Y - Z)/(Y - X) \tag{5.24}$$

$$V/F = (H_F - h)/(H - h) \tag{5.25}$$

$$L/F = (H - H_F)/(H - h) \qquad (5.26)$$

Equation pairs 5.23 and 5.25 as well as 5.24 and 5.26 are now combined to give

$$\frac{H_F - h}{Z - X} = \frac{H - h}{Y - X} \qquad (5.27)$$

$$\frac{H - H_F}{Y - Z} = \frac{H - h}{Y - X} \qquad (5.28)$$

The enthalpies and compositions of the feed, vapor product, and liquid product are now plotted on an enthalpy–composition diagram, Figure 5.9. Point F represents the feed with coordinates (Z, H_F); point V represents the vapor product with coordinates (Y, H); and point L represents the liquid product with coordinates (X, h). It can be seen from Equations 5.27 and 5.28 that line segments LF and FV are actually segments on the same straight line since they have identical slopes and a common point, F. The fact that points F, V, and L lie on the straight line is a result of mass and enthalpy balances and can be used to determine unknown variables graphically. If, for instance, points L and F and composition Y are known, H can be determined independently of enthalpy–composition data and based solely on mass and enthalpy balances. Graphically, H is determined by joining L and F with a straight line, then extending it until it intersects the vertical line through Y. The intersection point corresponds to point V with coordinates (Y, H).

Looking back at Equations 5.23, 5.24, 5.25, and 5.26, it can be seen that the rates of F, L, and V obey the lever rule with respect to compositions and enthalpies. If, for instance, F is considered a *pivot point*, the product of vapor rate V times compositional line segment ZY equals the product of liquid rate L times compositional line segment XZ. Similar relations apply to enthalpy line segments H_FH and hH_F and also to line segments FV and LF. Also, points V and L could be selected as the pivot points instead of point F.

The above concepts amount to a graphical representation of the overall mass, component mass, and enthalpy balances around an equilibrium stage. A column solution may be developed by applying these graphical rules on an enthalpy–composition diagram.

5.3.2 Binary H–X Diagrams

As explained in Section 1.4, the enthalpy of a mixture is a function of temperature, pressure, and composition. These parameters determine the phase, so that, in vapor–liquid equilibrium calculations, the enthalpy is also implicitly a function of the phase. For a binary mixture at constant pressure, the equilibrium vapor and liquid temperatures vary with composition as represented by the dew point and bubble point curves (Figure 2.2). The enthalpy at each point may be plotted as a function of the composition, resulting in a saturated vapor enthalpy curve and a saturated

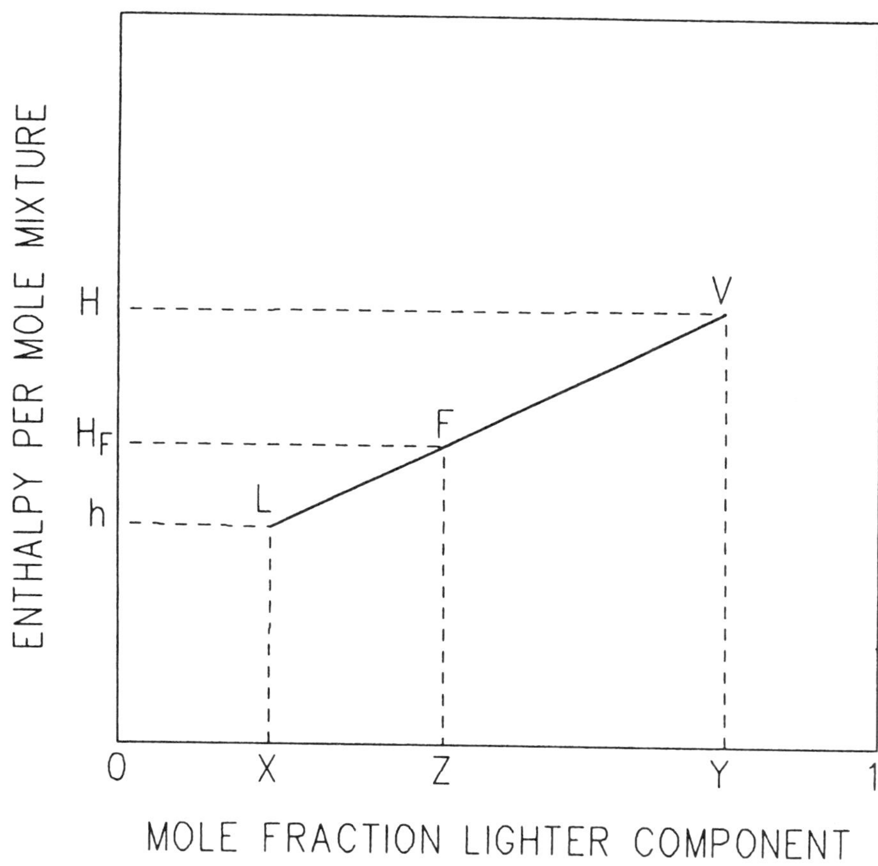

Figure 5.9 Mass and enthalpy balances on the H–X diagram.

liquid enthalpy curve, as shown in Figure 5.10. The composition is plotted as the mole fraction of the lighter component.

At temperatures above the dew point curve and below the bubble point curve, superheated vapor and subcooled liquid enthalpies are represented by isotherms as indicated.

Each point on the saturated liquid curve is associated with a point on the saturated vapor curve at equilibrium with it. The equilibrium vapor and liquid compositions may be obtained from Y–X or T–X diagrams. Saturated vapor and liquid points on the H–X diagram at equilibrium with each other are joined by straight lines called *tie lines*. The single-stage graphical representation described in Section 5.3.1 is an illustration of a tie line. If, for instance, X is known, L can be determined as a point on the saturated liquid curve with composition coordinate X. Point V must lie on the other end of the tie line on the saturated vapor curve. F can then be determined either from information on the relative rates of feed, liquid, and vapor or from its composition or enthalpy.

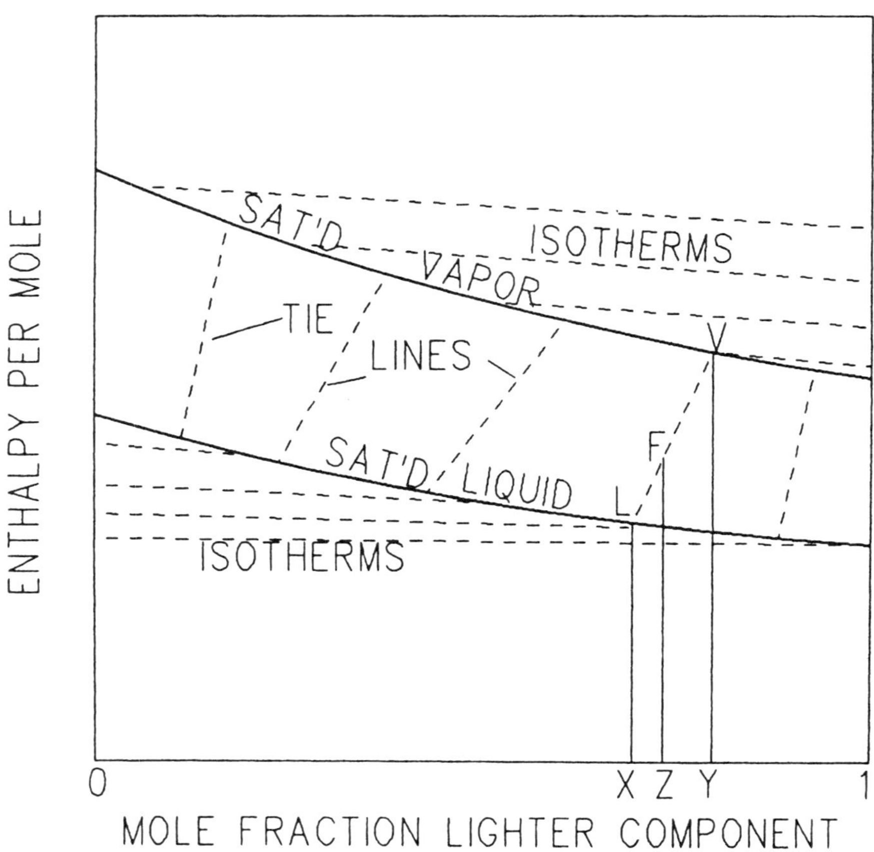

Figure 5.10 Binary enthalpy–composition diagram.

The H–X diagram may be constructed on the basis of either experimental data or computations using any of the prediction methods such as equations of states (Chapter 1).

5.3.3 Solving Distillation Columns on the H–X Diagram

Binary distillation columns are solved on the H–X diagram by graphically implementing the appropriate mass and enthalpy balances on an H–X diagram prepared for the given binary. The detailed solution steps somewhat depend on the way the column is specified.

A column having a completely defined single feed, an overhead product, a bottoms product, a condenser, a reboiler, and a fixed column pressure has 3 degrees of freedom if the number of trays is variable (Section 5.2). The column model depicted in Figure 5.3 meets these qualifications and is used here to describe the method. Note that the condenser in this model is a partial condenser, so that the distillate is a saturated vapor. With 3 degrees of freedom, the column requires three specifications to define its operation. Let the condenser duty, q_C, the distillate composition, Y_D, and the bottoms composition, X_B, be specified.

Starting the calculations at the top of the column, an enthalpy balance around the condenser is written as

$$V_2H_2 = L_1h_1 + DH_1 + q_C$$

$$= L_1h_1 + DH_1 + DQ_C = L_1h_1 + D(H_1 + Q_C)$$

$$= L_1h_1 + D'$$

where D' is the *difference point* for streams V_2 and L_1, q_C is the condenser duty per unit time, and Q_C is the condenser duty per mole distillate:

$$Q_C = q_C/D$$

A difference point on the enthalpy–composition diagram represents the enthalpy difference between two streams. It lies on the straight line joining the points representing these streams and extended to one side of them. Point D' represents the distillate with its enthalpy incremented by Q_C. Its coordinates are Y_D and $H_1 + Q_C$. The distillate itself is represented by point D, with coordinates Y_D and H_1. Since the distillate composition Y_D (or Y_1) and the condenser duty Q_C are known, D' can be located on the H–X diagram (Figure 5.11). The intersection of the vertical line through Y_D with the saturated vapor curve provides a handy way for determining H_1 at point D. A tie line drawn through this point crosses the saturated liquid curve at X_1 and h_1, the coordinates of the reflux L_1. A straight line joining D' and L_1 is the operating line corresponding to the mass and enthalpy balances on the condenser (stage 1). It represents the addition of streams D' and L_1 to generate stream V_2, the vapor leaving stage 2. By mass and enthalpy balances alone, V_2 could be represented by any point on the operating line. However, since V_2 must also lie on the saturated vapor curve, its coordinates are set by the intersection point at Y_2 and H_2.

Next, an enthalpy balance is written around the top two stages:

$$V_3H_3 = L_2h_2 + D(H_1 + Q_C)$$

Stream L_2 coordinates (X_2, h_2) are determined by the intersection of the tie line through (Y_2, H_2) and the saturated liquid curve. V_3 coordinates (Y_3, H_3) lie at the intersection of the operating line through D' and L_2 and the saturated vapor curve.

This procedure is repeated for all trays above the feed. For operating lines relating passing liquid and vapor streams, L_j and V_{j+1}, below the feed tray, the enthalpy balance is performed around the top j trays that include the feed:

$$Fh_F + V_{j+1}H_{j+1} = L_jh_j + DH_D + DQ_C$$

$$= L_jH_j + D'$$

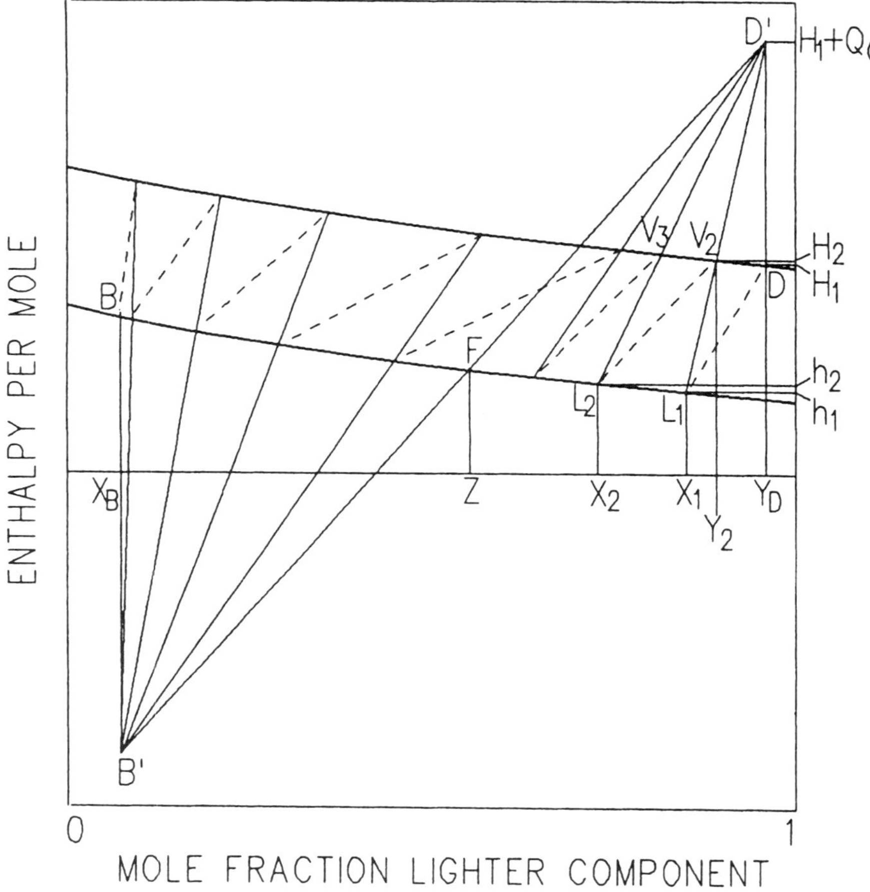

Figure 5.11 Solving a distillation column on the H–X diagram.

An enthalpy balance around the entire column relates the condenser and reboiler duties:

$$Fh_F = B(h_B - Q_B) + D(H_D + Q_C)$$

$$= B' + D'$$

These two equations are combined to derive an operating line equation in terms of B′:

$$V_{j+1}H_{j+1} = L_jh_j + B'$$

As this equation indicates, the operating lines for trays below the feed are constructed using B′ as a pivot (Figure 5.11). This point has coordinates X_B and $h_B - Q_B$, and may be located graphically from the overall enthalpy balance, represented by a straight line through D' and F. It is the intersection of this line with the vertical line through X_B, one of the specified variables. The feed composition Z and enthalpy H_F determine the coordinates of F. For a feed at its bubble point, F falls on the saturated liquid curve.

Continuing with the procedure of defining equilibrium stages down the column, the pivot point is switched from D' to B' when the feed tray is crossed. Alternatively, the procedure could be started at the reboiler, stepping off stages up the column and switching from B' to D' as the feed tray is crossed.

For an existing column, or if the feed tray is given, the graphical construction of equilibrium stages on the H–X diagram proceeds as described above, using D' as a pivot point above the feed and B' below it. In a design situation or in a case where the feed tray is to be determined, the optimum feed tray would be the target. For a given number of trays, the optimum feed tray is that which minimizes the reflux required to achieve a specified separation or maximizes the separation obtainable with a fixed reflux. When designing a new column, the optimum feed tray is that which minimizes the number of trays required for a specified separation, assuming other conditions such as reflux ratio and column pressure are fixed.

The optimum feed tray location is determined on the H–X diagram as the stage or tie line that crosses the overall column enthalpy balance line joining D' and B'. This fact, which is one of the interesting features of the H–X diagram method, may be verified graphically by constructing a few stages around the presumed optimum feed tray, as shown in Figure 5.12. The dotted lines represent tie lines, and the stage just above the feed is designated as tray m. The solid lines indicate the operating lines originating at D', the pivot point used above the feed. The overall column enthalpy balance line $D'B'$ crosses tie line n. If this is determined to be the feed stage, the next stage, k, should be constructed with B' as the pivot, using the indicated dashed line. This results in a liquid on stage k with composition X_k. If the stage below n is used as the feed tray, then this stage, k', is constructed with D' as the pivot. The resulting liquid is of composition X_k'. Thus, with the same number of stages, moving the feed from tray n to k' raises the liquid composition from X_k to X_k'. Since these are mole fractions of the lighter component, it can be seen that a poorer separation results from lowering the feed. A similar conclusion is reached by raising the feed; hence, the presumed optimum feed tray is indeed optimal.

Finally, the distillate and bottoms rates are calculated from the lengths of line segments $D'B'$, $D'F$, and $B'F$, using the lever rule. Other tray liquid or vapor rates may be calculated from segment lengths on the appropriate operating line. For instance, the reflux rate may be calculated from segments $D'V_2$ and V_2L_1. The reboiler duty is calculated as the difference between the enthalpy coordinates of B' and B.

5.3.4 Other Column Features Represented on the H–X Diagram

The H–X diagram offers a convenient and versatile, yet rigorous, means for preliminary evaluation of the performance of multistage separation prior to proceeding to numerical solutions. This graphical representation of columns is not hampered by the assumptions made in the Y–X method for the purpose of generating straight

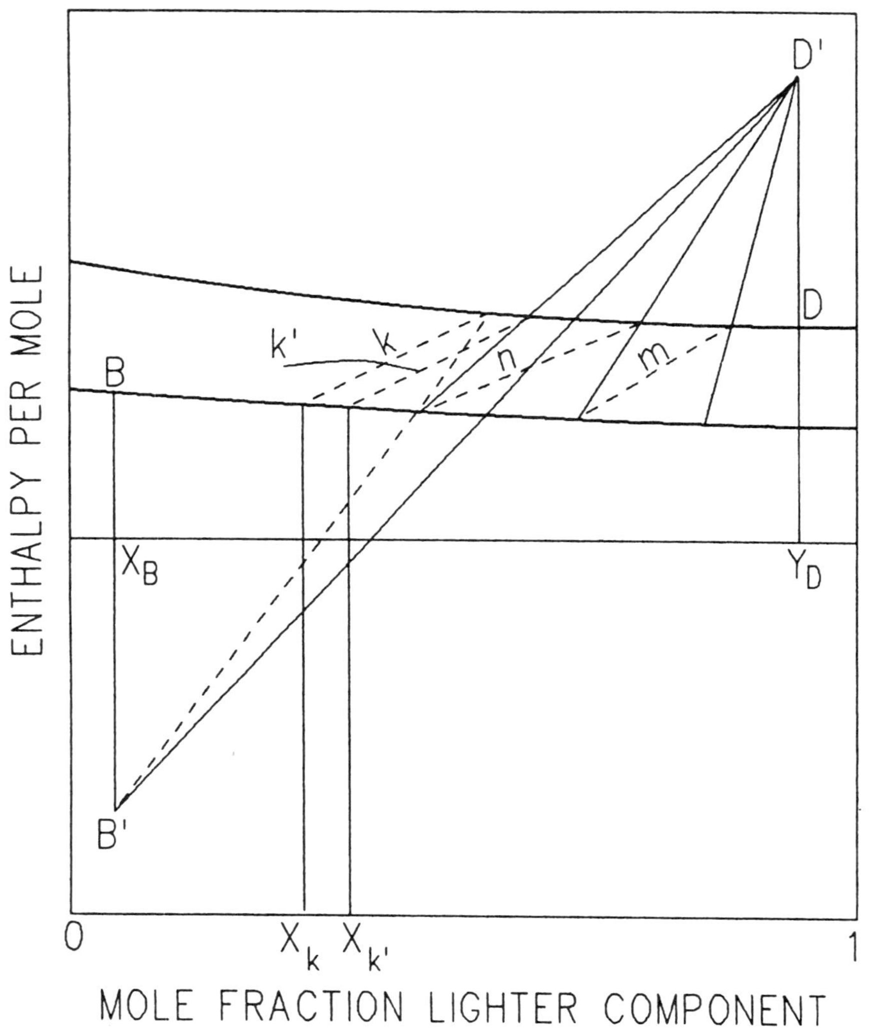

Figure 5.12 Effect of feed tray location on the separation.

operating lines. In view of this, the method is applicable to a wider range of columns, including nonideal distillation. It is still, however, essentially for binary systems and is contingent on the availability of H–X data. It lends itself to modeling a number of column features, some of which are described next.

Condenser Types

The column discussed in Section 5.3.3 is equipped with a partial condenser where part of the overhead vapor from the top tray (stage 2) is condensed and returned to the top tray as reflux. The vapor leaving the condenser is the distillate.

Part of the liquid in a partial condenser may be drawn out as an additional product — a liquid distillate — alongside the vapor distillate. These two products, coming from the same stage, are at equilibrium with each other. The enthalpy balance around a partial condenser with vapor and liquid distillates is stated as

$$V_2 H_2 = L_1 h_1 + D_v H_1 + D_L h_1 + q_C$$

where V_2 is the vapor rate from the top of the column, L_1 is the reflux rate, D_v is the vapor distillate, and D_L is the liquid distillate. Since D_v and D_L are at equilibrium, the points representing them on the H–X diagram fall on a tie line (Figure 5.13A). This line also represents a mass balance comprising the two distillate products. The total distillate D must, therefore, lie on this line. Its position on the line is determined by the relative rates of D_v and D_L. In terms of the total distillate, the enthalpy balance around the condenser is written as

$$V_2 H_2 = L_1 h_1 + D H_D + q_C$$
$$= L_1 h_1 + D\left(H_D + Q_C\right)$$
$$= L_1 h_1 + D'$$

Point D' has coordinates X_D and $H_D + Q_C$. The reflux, L_1, has the same composition and enthalpy per mole as D_L and is, therefore, represented by the same point. The operating line representing the condenser is obtained by joining D' and L_1.

In a total condenser, the entire overhead vapor is condensed; part of it is refluxed, while the remainder is taken as liquid distillate. The distillate point, D, could either be on the saturated liquid curve or below it, depending on whether the overhead vapor is condensed to the bubble point or is subcooled (Figure 5.13B). From an enthalpy balance around the condenser,

$$V_2 H_2 = L_1 h_1 + L_D h_D + q_C$$
$$= L_1 h_1 + L_D\left(h_D + Q_C\right)$$
$$= L_1 h_1 + D'$$

Note that L_1 and D' have the same composition, X_D, and, therefore, the operating line is vertical.

Multiple Feeds, Side Draws

The versatility of the H–X method is also demonstrated in the analysis of the performance of multiple-feed and multiple-product columns. The basic approach is unchanged: relate the passing vapor and liquid streams at the point of interest in the

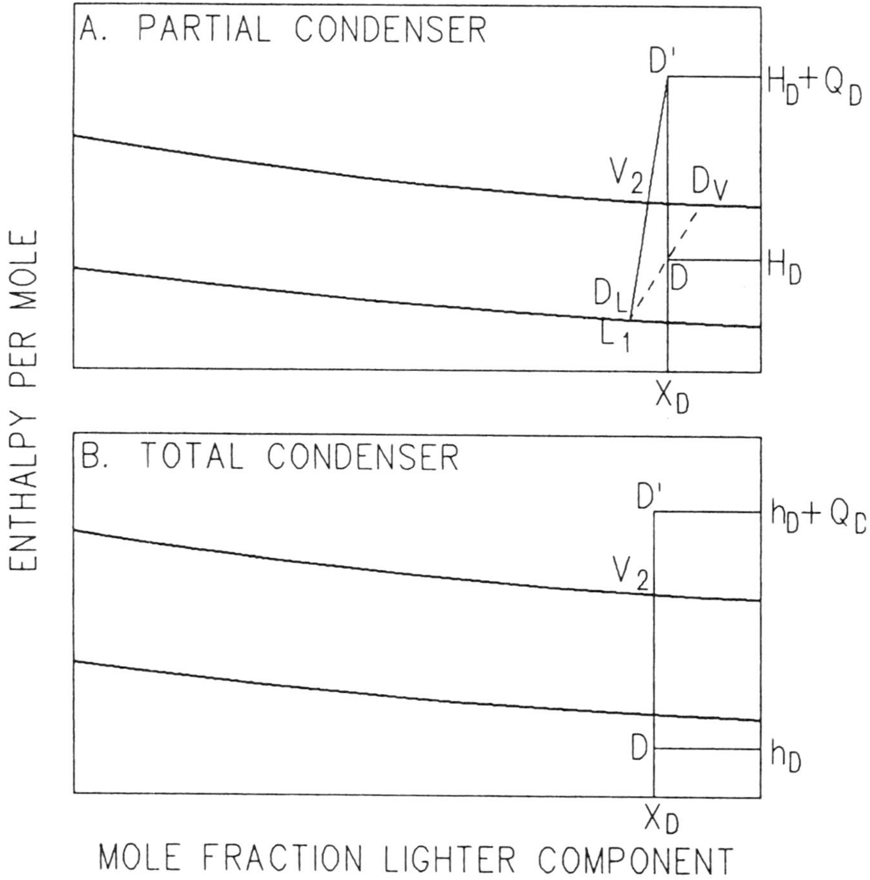

Figure 5.13 Modeling condensers on the H–X diagram.

column by performing an enthalpy balance, then implement the relationship graphically on the H–X diagram.

The situation may arise, for instance, where it is necessary to study the effect of a liquid side draw rate on the performance of the column below the draw tray. Consider any tray j above which the only heat or material transfer across the column boundary occurs at the condenser, distillate, and side draw. An enthalpy balance around the top part of the column down to and including tray j is written as

$$V_{j+1}H_{j+1} = L_j h_j + DH_D + q_C + Sh_s$$

$$= L_j h_j + D(H_D + Q_C) + Sh_s$$

Graphical construction of the trays above the draw proceeds as usual, using point D′ with coordinates X_D and $H_D + Q_C$ as the pivot (Figure 5.14). Starting at the tray

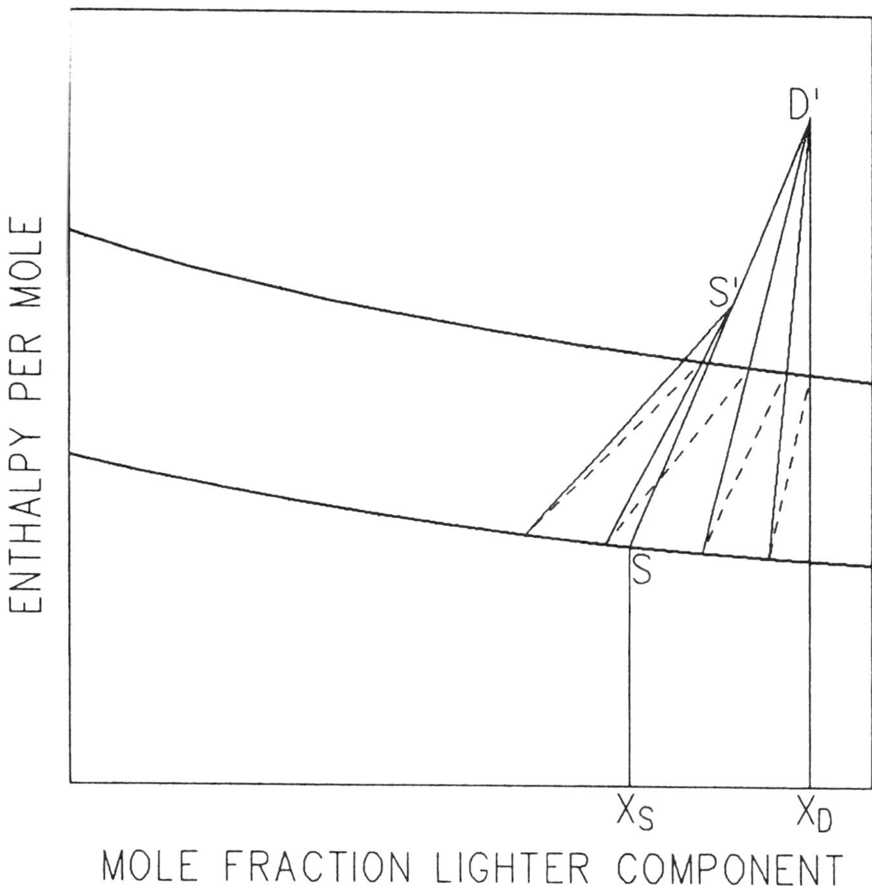

Figure 5.14 Modeling side draws on the H–X diagram.

just below the draw, the pivot point is switched to a point S' representing the combined effect of D' and S. The latter corresponds to the draw composition, X_s, and its enthalpy, h_s. Since the draw is a saturated liquid, S lies on the saturated liquid curve. Point S' lies on a straight line joining D' and S. Its position is determined by the relative rates of D' and S.

With a small draw rate, S' is not far from D' and its effect on fractionation below the draw tray is small. As the draw rate is increased, S' moves farther away from D' toward S. Comparing operating lines using S' as a pivot vs. those using D' indicates that fractionation below the draw tray is compromised as the draw rate is increased. If S is large enough, S' can reach a point where a tie line coincides with the operating line for the passing vapor and liquid streams below the draw tray. This implies that no additional fractionation takes place by adding more trays below that tray.

Analyzing the effect of a feed above a given tray is analogous to the case with a side draw. The enthalpy balance that includes the upper part of the column through any tray j below the feed takes the form

$$V_{j+1}H_{j+1} = L_j h_j + D(H_D + Q_C) - Fh_F$$

$$= L_j h_j + D' - Fh_F$$

Point D' is the usual pivot that corresponds to the distillate and the condenser. Point F has the coordinates of the feed composition and enthalpy, (X_F, h_F). The combined effect is represented by a new pivot point F', that corresponds to the difference D' $-$ Fh_F. To locate F', a straight line is drawn from F to D' and extended past D' to F' so that, by the lever rule,

$$F(FF') = D(D'F')$$

where (FF') and (D'F') are the line segment lengths.

Side Coolers, Heaters

Heat sources or sinks other than the condenser and reboiler may exist at different points in the column such as in the case of pumparounds. As an example, a column with a partial condenser may have a side cooler a few trays down the column. For trays above the side cooler, the enthalpy balance around passing vapor and liquid streams and the overhead is unchanged. Therefore, the graphical construction of trays starting at the top is not affected by the side cooler. When the side cooler is crossed, the enthalpy balance becomes

$$V_{j+1}H_{j+1} = L_j h_j + D(H_D + Q_C + Q_s)$$

$$= L_j h_j + D''$$

where Q_S is defined as q_S/D and q_S is the side cooler duty. When the side cooler tray is crossed, the pivot point is switched to D'' with coordinates X_D and $H_D + Q_C + Q_S$.

Tray Efficiency

Modifying the graphical construction of stages to account for tray efficiency can easily be incorporated in the H–X diagram method. Using the column model described in Section 5.3.3 with a partial condenser as an example, the graphical construction is started at the condenser by marking the distillate composition at Y_1 (Figure 5.15). Assuming the condenser performs as an equilibrium stage, the liquid reflux composition is X_1, as determined by the tie line. Next, Y_2, the vapor composition on stage 2, is set by the operating line. If stage 2 were an equilibrium stage,

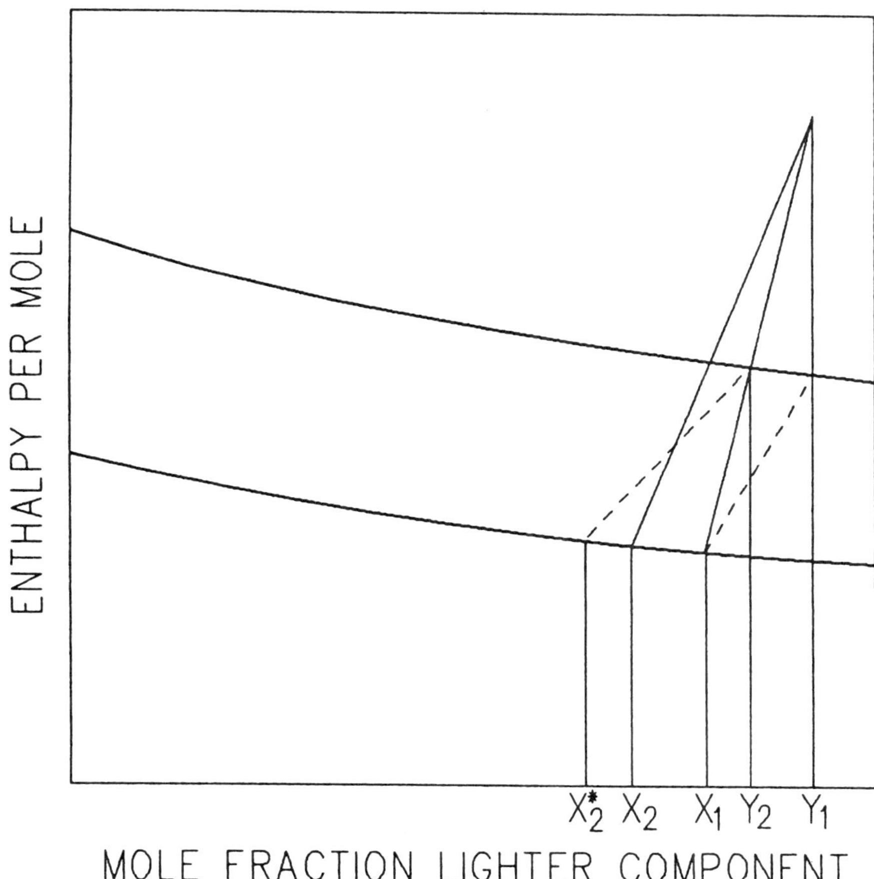

Figure 5.15 Implementing Murphree tray efficiency on the H–X diagram.

its liquid composition, X_2^*, would be determined by the tie line passing through Y_2. In an actual tray, the liquid composition, X_2, falls short of the equilibrium composition and the difference between X_1 and X_2 is some fraction of the difference between X_1 and X_2^*. The Murphree tray liquid efficiency is defined simply as the ratio of these two differences:

$$E_L = \frac{X_1 - X_2}{X_1 - X_2^*}$$

or, in general,

$$E_L = \frac{X_j - X_{j+1}}{X_j - X_{j+1}^*}$$

This definition of the Murphree tray efficiency should be compared with the Murphree tray vapor efficiency, defined in Section 5.2.3. The tray efficiency may generally be estimated on the basis of past experience with similar columns or from operating data of an existing column. The tray efficiency may then be used in the graphical construction of actual trays on the H–X diagram. The procedure follows the usual steps, except that the operating lines are drawn from the pivot point to the actual tray compositions, X_{j+1}, rather than the equilibrium compositions, X_{j+1}^*. The length of each incremental segment, $X_j - X_{j+1}$, is calculated from the tray efficiency:

$$X_j - X_{j+1} = E_L\left(X_j - X_{j+1}^*\right)$$

An analogous procedure is followed if the Murphree tray vapor efficiency is known. In this case it is more convenient to start the graphical construction at the reboiler.

NOMENCLATURE

B	Bottoms stream designation or molar flow rate	P	Pressure
D	Distillate or overhead designation or molar flow rate	Q	Heat duty per mole
		q	Heat duty per unit time
F	Feed stream designation or molar flow rate	S	Side draw designation or molar flow rate
H	Vapor molar enthalpy	T	Temperature
h	Liquid molar enthalpy	V_j	Vapor molar flow rate leaving tray j
K	Vapor–liquid distribution coefficient	V_{ji}	Component i molar flow rate in vapor leaving tray j
L_j	Liquid molar flow rate leaving tray j	X	Mole fraction in the liquid
		Y	Mole fraction in the vapor
L_{ji}	Component i molar flow rate in liquid leaving tray j	α	Relative volatility
		λ	Molar heat of vaporization
N	Number of equilibrium stages in a column		

SUBSCRIPTS

B	Bottoms or reboiler designation	i	Component designation
C	Condenser designation	j	Stage designation
F	Feed conditions designation	r	Rectifying section designation
f	Feed tray designation	s	Stripping section designation

REFERENCES

Hanson, D. N., J. H. Duffin, and G. F. Somerville, *Computation of Multistage Separation Processes*, New York, Reinhold Publishing Corporation, 1962.

McCabe, W. L. and E. W. Thiele, *Ind. Eng. Chem.*, 17, 605, 1925.

Murphree, E. V., *Ind. Eng. Chem.*, 17, 747, 1925.

Ponchon, M., *Tech. Moderne*, 13, pp. 20, 55, 1921.

Savarit, R., *Arts et Metiers*, 1922, 65.

CHAPTER **6**

Binary Distillation: Applications

The principles of binary distillation presented in Chapter 5 are applied in this chapter to predict column performance under different operating conditions. The object is to study performance trends as well as to solve numerical examples.

As a learning tool, the binary model is useful for qualitatively studying the characteristics of multistage separation. The model is used in this chapter to answer such questions as what effect the reflux ratio or product rate or number of trays has on separation or what is the minimum number of trays or minimum reflux ratio required to achieve a given separation, or over what ranges column performance specifications are feasible.

Although rigorous computer programs will provide design data for a particular separation process, understanding the process characteristics is necessary in order to evaluate its performance under varying operating conditions. Reams of multiple cases of computer output may cover all the expected operating ranges, but a visual representation of the process can be invaluable. Thus, the graphical methods can be applied to get the most out of computer simulation results.

Moreover, the availability of computer programs provides an added impetus to the use of graphical methods because these programs either plot the distillation diagrams directly or provide the data required to plot them.

Multicomponent separations can also be approximated with graphical methods by selecting appropriate key components to define the separation.

Using the graphical techniques developed in Chapter 5, Section 6.1 examines the effect on column performance of each parameter separately, and Section 6.2 considers the interactions of different parameters. Section 6.3 presents example applications where graphical methods are used in conjunction with simulation results to check the design data.

6.1 PARAMETERS AFFECTING COLUMN PERFORMANCE

The binary model, conveniently represented on the Y–X or H–X diagrams, demonstrates the effect on column performance of the reflux ratio, product rate, number of trays, feed location, and thermal conditions. In the cases that follow in this section, the feed is assumed of fixed flow rate and composition. The effect of column pressure, which influences column performance through its effect on the equilibrium curve, is not considered here and is held constant.

6.1.1   Effect of Reflux Ratio and Product Rates

The reflux ratio is a measure of the relative flows of liquid and vapor counter-flowing in the column. It is commonly defined as $R = L_1/D$, where L_1 is the reflux rate or rate of liquid flowing from the condenser back to the column and D is the distillate rate. The operating line of the column rectifying section given by Equation 5.13 has a slope of L_r/V_r, where L_r and V_r are the liquid and vapor flow rates in the rectifying section. Since the molar flow is assumed constant in each column section, $L_1 = L_r$ and $R = L_r/D$. Also, writing a material balance around the condenser, $V_r = D + L_r$, the operating line slope may be expressed in terms of the reflux ratio:

$$\frac{L_r}{V_r} = \frac{L}{D + L_r} = \frac{R}{1 + R} \tag{6.1}$$

As R increases, so does L_r/V_r, and as R goes from 0 to infinity, L_r/V_r goes from 0 to 1.

The effect of reflux ratio on the separation may be examined by looking at the McCabe–Thiele Y–X diagram. If the total number of stages and feed location are fixed, the number of stages in the rectifying section is also fixed. Line r_1 in Figure 6.1 represents the rectifying section operating line at some reflux ratio R_1. The corresponding number of stages in the rectifying section is four. If the reflux ratio is increased, the slope of r_1 increases. In order to maintain the same number of stages in the rectifying section, the new operating line, r_2, must intersect the diagonal at a higher concentration of the light component. The overhead product is therefore richer in the lighter component, and better separation is achieved at a higher reflux ratio when all other parameters are held constant.

The number of stages in the stripping section is also constant. Lines S_1 and S_2 in Figure 6.1 show that, as the reflux ratio increases, the slope of the stripping section operating line must decrease in order to maintain a fixed number of stages. With a lower L_s/V_s slope, the stripping section operating line intersects the diagonal at a lower concentration of the lighter component. Consequently, the bottoms is richer in the heavier component as the reflux ratio is increased, resulting in better separation.

As the reflux ratio approaches 0, L_r/V_r also approaches 0 and the rectifying section operating line becomes a horizontal line intersecting the equilibrium curve at its intersection with the q-line (Figure 6.1). The stripping section operating line becomes vertical since any other slope would imply an infinite number of stripping stages. At 0 reflux, the number of stages is indeterminate and has no effect on the

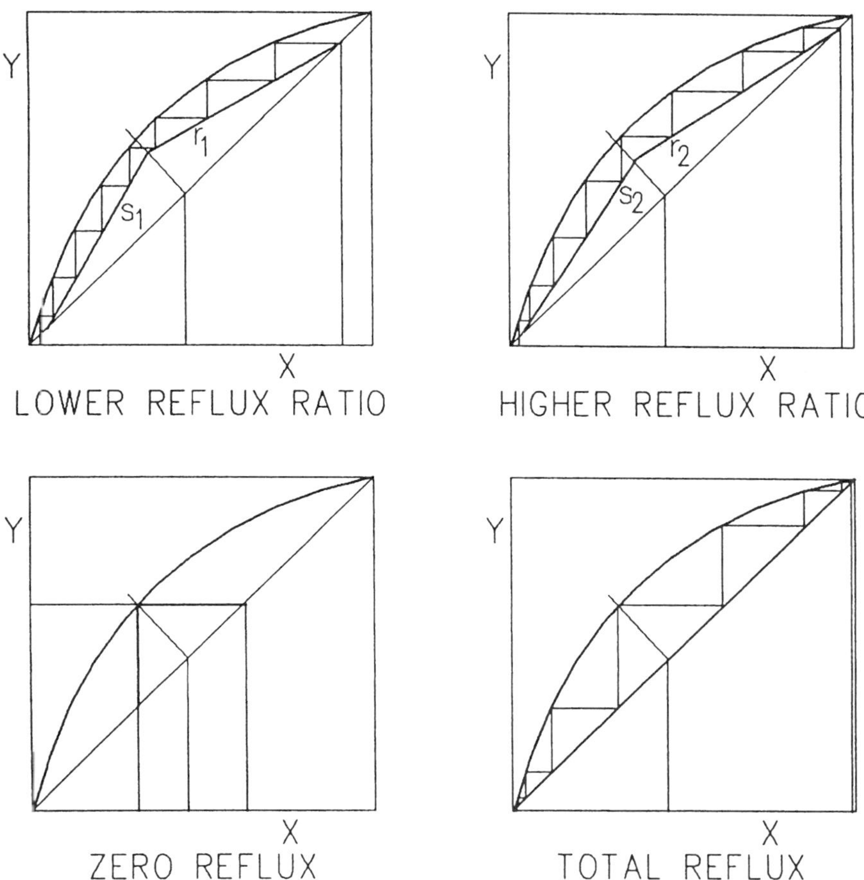

Figure 6.1   Effect of reflux ratio on column performance — Y–X diagram.

separation. A single stage would result in the same separation as any number of stages. This conclusion was arrived at in Section 3.2.3 on the basis of a two-stage model. The overhead and bottoms compositions at 0 reflux correspond to the Y and X coordinates of the intersection of the q-line with the equilibrium curve.

At total reflux, R goes to infinity and $L_r/V_r = 1$ so that both rectifying and stripping section operating lines coincide with the diagonal. The distance between the equilibrium curve and the operating lines is largest at total reflux, resulting in the largest size steps between equilibrium stages. Hence, maximum separation is obtained at total reflux, which is another conclusion arrived at earlier in Section 3.2.3 using the simplified two-stage model. Graphically, the separation obtainable at total reflux with a given number of stages may be determined on the Y–X diagram by stepping off the number of stages in each section, starting at the intersection of the q-line with the equilibrium curve (Figure 6.1).

The effect of reflux ratio on column performance may also be inspected using the H–X diagram. Given a fixed distillate rate, a higher reflux ratio implies larger vapor and liquid traffic in the rectifying section, requiring a higher vapor condensation rate in the condenser. A higher condenser duty, q_C (or $Q_C = q_C/D$), results in a higher pivot point D′, on the H–X diagram and greater slopes for the operating lines. This causes the stepping off of stages to take larger strides, thereby enhancing the separation (Figure 6.1A).

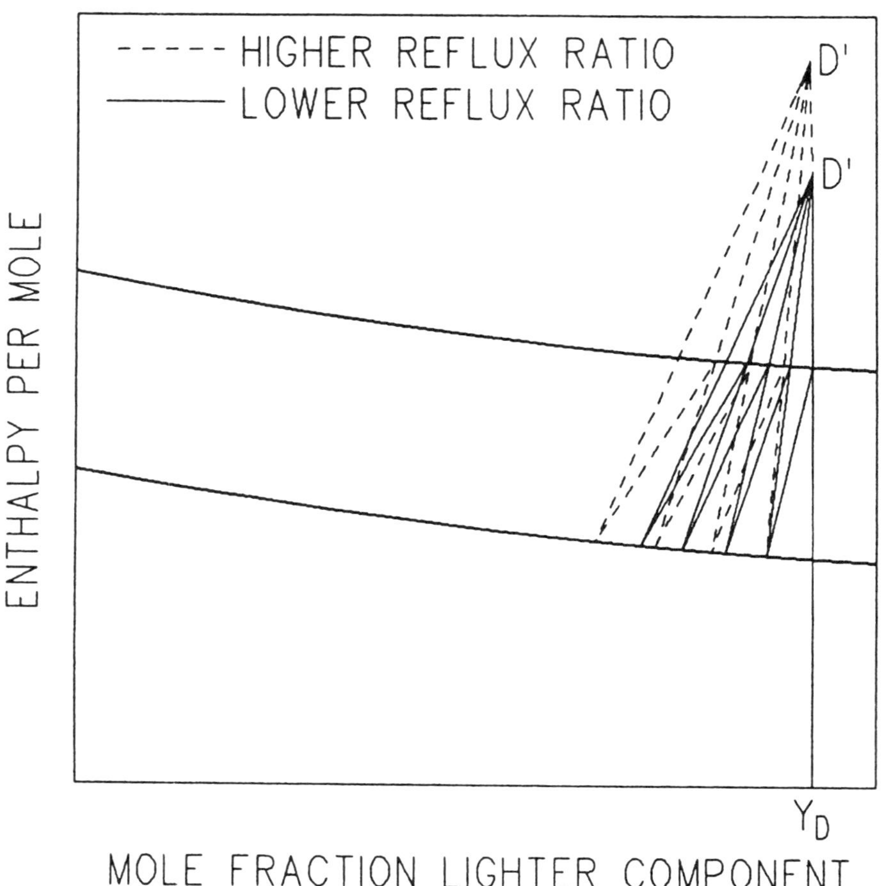

Figure 6.1A Effect of reflux ratio on column performance — H–X diagram.

At total reflux L/D and Q_C go to infinity, causing the operating lines to become vertical. These conditions determine the minimum number of stages required to achieve a given separation.

Product Rates

The complete definition of column performance when the number of stages is fixed requires two specifications (Section 5.2.2). In studying the effect of reflux ratio on separation, other parameters, including product rates, are assumed constant. In Y–X graphical representation, a fixed product rate is implied by fixing the q-line slope. This slope depends on the feed thermal conditions at feed tray temperature and pressure (Section 5.2.2). The column pressure is fixed, but its temperature depends on the product rates. Therefore, in order to maintain a fixed product rate, the q-line slope must be held constant. The actual product rates corresponding to a given q-line slope are calculated by material balance once the product compositions have been determined.

The qualitative effect of product rates on product purities may be examined graphically by comparing Y–X diagrams with a vertical q-line and a horizontal q-line (Figure 6.2). For a feed with given thermal conditions, the q-line is vertical if the feed tray temperature is such that the feed is saturated liquid at this temperature. To convert the same feed to saturated vapor with a horizontal q-line requires a higher temperature feed tray. A vertical q-line column thus corresponds to a generally colder column with a lower overhead rate and a higher bottoms rate. The converse is true for a horizontal q-line column. In the former case, the rectifying and stripping steps are shifted to the right of the Y–X diagram, resulting in higher purity overhead and lower purity bottoms. The shift is to the left for a hotter column with high overhead and low bottoms rates, resulting in lower purity overhead and higher purity bottoms.

LOWER OVERHEAD RATE   HIGHER OVERHEAD RATE

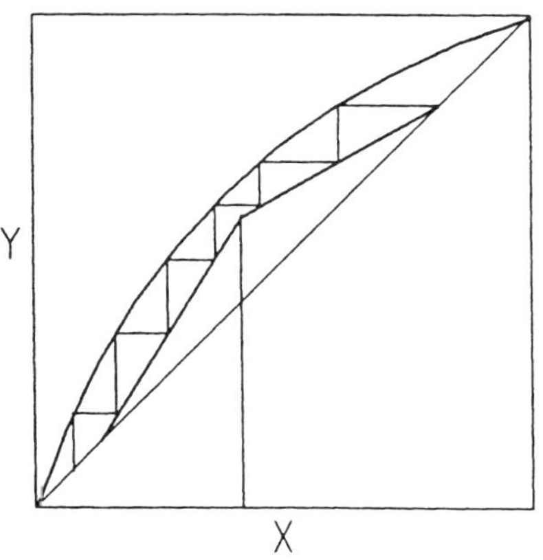

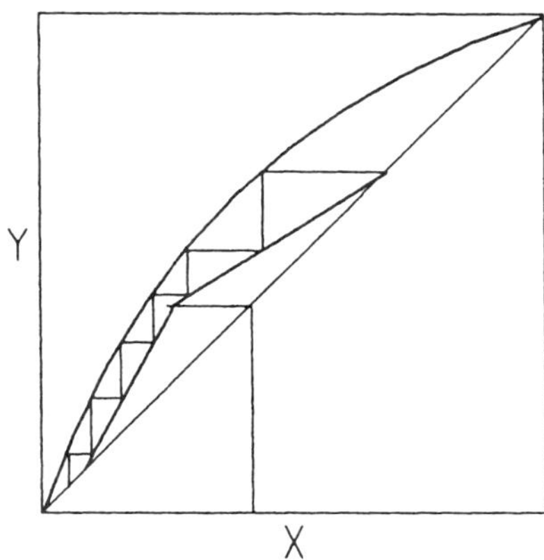

Figure 6.2 Effect of product rates on column performance.

Example 6.1 Benzene–Toluene Column: Reflux Ratio Specified

A feed stream made up of 40% mole benzene and 60% mole toluene is to be separated into benzene-rich and toluene-rich products using a distillation column. The column has ten equilibrium stages including a partial condenser and a partial reboiler and is operated at 25 psia. The feed stream, with a flow rate of 100 moles/hr, is at its bubble point at 25 psia and is placed on the fourth stage from the top. The problem is to determine the compositions of the two products at different reflux ratios.

Vapor–liquid equilibrium data for the benzene–toluene system are provided in Table 6.1 at 25 psia.

Table 6.1   Vapor–Liquid Equilibrium Data for the Benzene–Toluene Binary at 25 psia

X (Benzene)	Y (Benzene)	T, F
0.00	0.00	267
0.10	0.19	259
0.20	0.35	252
0.30	0.49	245
0.40	0.60	239
0.50	0.70	233
0.60	0.77	227
0.70	0.84	222
0.80	0.90	218
0.90	0.95	213
1.00	1.00	209

Solution. The McCabe–Thiele graphical method is applied using the Y–X diagram. The equilibrium curve is first plotted from the given data (Figure 6.3). A temperature coordinate is also included in the diagram (nonlinear) to determine the condenser, reboiler, and tray temperatures.

The next step is to plot the q-line, which intersects the diagonal at the feed composition. The slope of the q-line is computed on the basis of the feed thermal conditions at the feed tray temperature, which is dependent on the product rates. It is assumed that the product rate (distillate or bottoms) is such that the feed is almost a saturated liquid at feed tray conditions so that the q-line is close to vertical.

The operating lines are drawn next. Assuming a reflux ratio R = 1, the slope of the rectifying section operating line is $L_r/V_r = 0.5$ (Equation 6.1). The rectifying section operating line is determined by trial and error by drawing lines with a slope of 0.5 between the q-line and the diagonal. The correct line is that which results in four rectifying stages, which is the specified number of theoretical stages above the feed tray (Figure 6.3).

The stripping section operating line must pass through the intersection of the q-line with the rectifying section operating line. The slope of the former is varied until the number of stages in the stripping section is six.

The compositions of the products are read from the intersections of the operating lines with the diagonal. The results are tabulated in Table 6.2 for different reflux ratios. Also shown are the condenser and reboiler temperatures as read from Figure 6.3, and the product flow rates, calculated from a component material balance:

$$FX_F = DY_D + (F - D)X_B$$

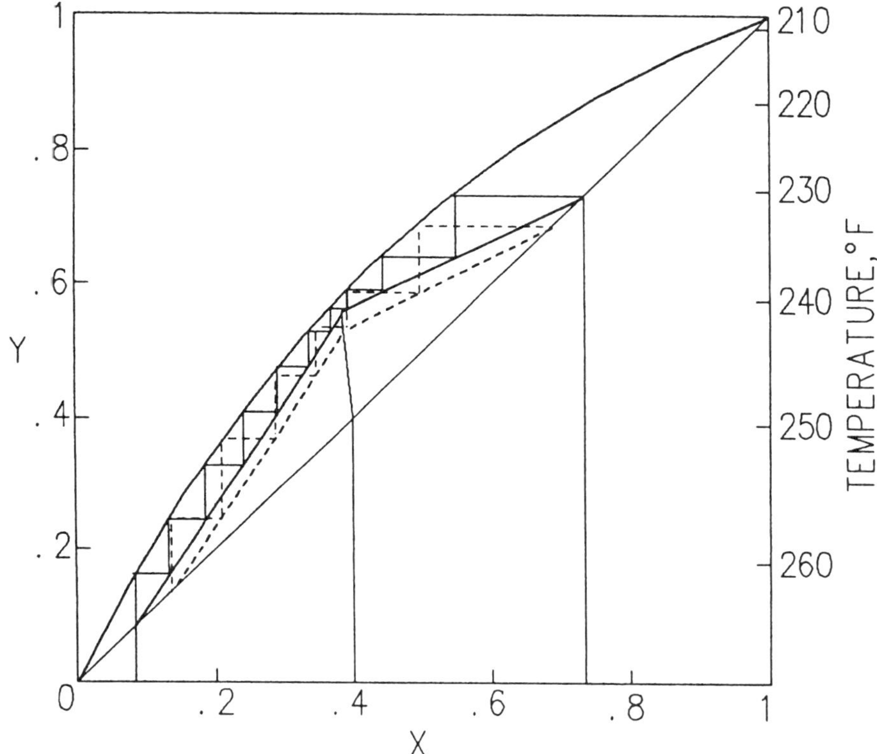

Figure 6.3   Y–X diagram for benzene–toluene.

Table 6.2   Results of Example 6.1

Reflux Ratio	0	1	10	Infinity
Distillate Composition Y (Benzene)	0.540	0.735	0.796	0.950
Bottoms Composition X (Benzene)	0.340	0.070	0.004	0.002
Distillate Rate, Mole/Hr	30	50	50	40
Condenser Temperature, °F	242	230	226	213
Reboiler Temperature, °F	242	261	266	267

The product rates at high and low reflux ratios are sensitive to the q-line slope, which varies slightly depending on feed tray conditions. As a result, certain variations in the product rates are observed.

6.1.2   Effect of Number of Stages and Feed Location

The effect of the number of stages on separation may be readily inspected using the binary Y–X diagram. Other parameters, namely the reflux ratio and the product rates, are assumed constant. The condition of constant product rate is implied by fixing the q-line slope.

The equilibrium curve and q-line are first drawn as usual on a Y–X diagram. The slope of the rectifying section operating line is known from the specified reflux ratio. For a given number of stages in the rectifying section, the correct operating line is found by trial and error aimed at matching the given number of stages while maintaining a constant slope. As the operating line gets closer to the equilibrium line, it intersects the diagonal at a higher concentration of the light component, and the number of rectifying stages increases.

The stripping section operating line is drawn by joining the q-line intersection with the rectifying operating line to the diagonal. Again, the correct stripping operating line is one that matches the number of stripping stages. Stripping section operating lines closer to the equilibrium curve imply more stripping stages and lower concentrations of the light component in the bottoms.

In conclusion, with other parameters held constant, more stages in a given section result in higher purity of the corresponding product, and vice versa. Consequently, if the total number of stages is held constant, but the feed location is raised, higher purity bottoms and lower purity distillate products are expected. The reverse is true if the feed location is lowered.

Using the H–X diagram, if the feed tray is not fixed, its optimum location can be determined as discussed in Section 5.3.3.

Example 6.2 Benzene–Toluene Column: Number of Stages and Feed Location Specified

The benzene–toluene mixture of Example 6.1 is to be separated into approximately 50 moles/hr distillate and 50 moles/hr bottoms products. The column is to be operated at a reflux ratio of 1. Determine the effect of the number of stages and feed location on the purity of the products.

Solution. Based on the results of Example 6.1, a distillate rate of about 50 moles/hr corresponds to a near vertical q-line, i.e., a bubble point feed. The equilibrium curve is drawn from data given in Table 6.1.

The operating lines are drawn as described above. Figure 6.3 illustrates the effect of varying the number of stages in each section, and Table 6.3 summarizes the results. The first two points in the table represent columns with the same number of stripping stages (four) but with different numbers of rectifying stages (three and four). The results show that the bottoms concentrations are about the same for the two points, but the distillate has a higher purity when more rectifying stages are

Table 6.3   Results of Example 6.2; Effect of Number of Stages and Feed Location on Separation

Number of Stages	7	8	10	12
Feed Location (From Top)	3	4	4	4
Distillate Composition Y (Benzene)	0.690	0.759	0.761	0.761
Bottoms Composition X (Benzene)	0.0681	0.0704	0.0394	0.0185

available. The columns corresponding to the last three points have the same number of rectifying stages (four) but increasing numbers of stripping stages (four, six, and eight). With more stripping stages, the distillate compositions do not change much, but the bottoms purity goes up.

6.1.3   Number of Stages vs. Reflux Ratio

If the separation is specified, one could calculate the number of stages required with a given reflux ratio or the reflux ratio required with a given number of stages. Higher reflux ratios are required with fewer trays, and lower reflux ratios are required with columns having more trays. Using multiple calculations, a curve could be generated of the required reflux ratio vs. the number of trays available for a specified separation. Each point on this curve would correspond to a combination of the number of trays and reflux ratio that satisfies the specified separation. In a design situation, the selection of a particular combination is based on, among other factors, economic considerations of capital expenditure vs. operating cost.

The separation is specified by fixing the distillate and bottoms compositions, Y_D and X_B. The product compositions determine the product rates, which are calculated by material balance. The product compositions are used to locate one end point of each operating line on the Y–X diagram. The feed conditions determine the q-line (Figure 6.4).

If the reflux ratio is specified, the slope of the rectifying section operating line is known. Its intersection with the q-line defines another point on the stripping section operating line. The equilibrium stages are then stepped off as usual between the operating lines and the equilibrium curve. The optimum feed location is the stage that lies on both operating lines, i.e., the stage through which the q-line passes.

If the reflux ratio is increased, the slope of the rectifying section operating line increases, the operating lines move farther away from the equilibrium curve, and the required number of stages decreases. At total reflux $L_r/V_r = 1$ and the operating lines coincide with the diagonal. Total reflux corresponds to the minimum number of stages required to achieve a specified separation. With fewer stages this separation is impossible regardless of the reflux ratio.

As the reflux ratio is lowered, the L_r/V_r slope decreases and the operating lines move closer to the equilibrium curve, thereby requiring more stages for the same separation. If the reflux ratio is lowered to a point where one of the operating lines

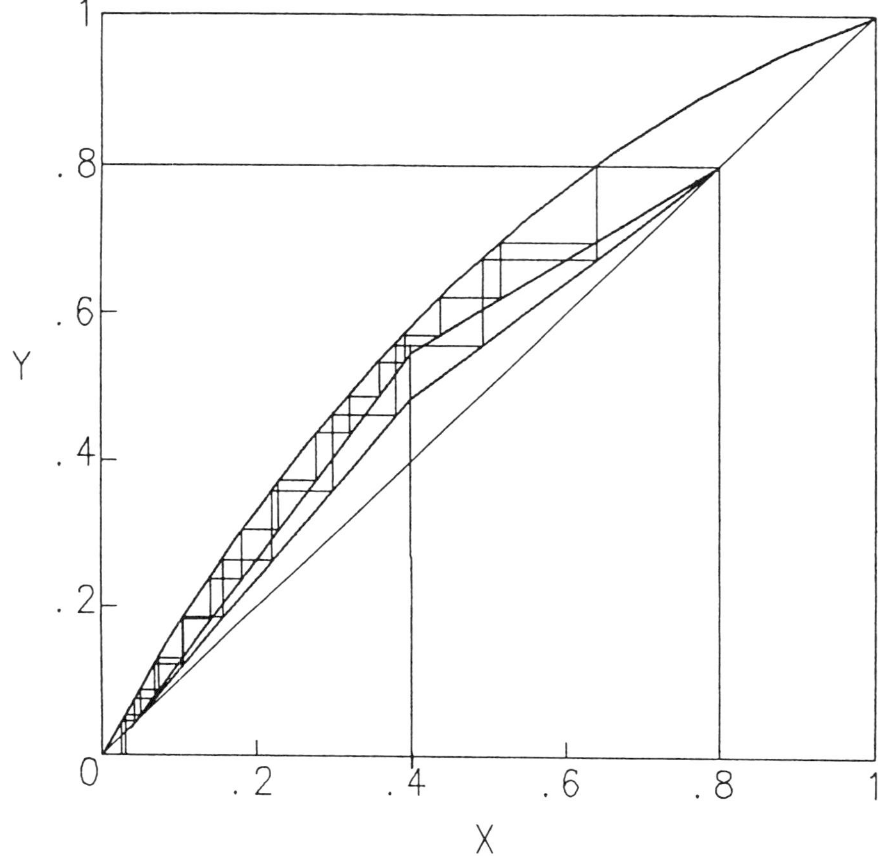

Figure 6.4   Product purities specified.

touches the equilibrium curve, the number of stages becomes infinite. This corresponds to the minimum reflux ratio, below which the specified separation cannot be met with any number of stages. The point where the operating line touches the equilibrium curve, or the *pinch point*, can occur in the vicinity of the feed stage at the intersection of the operating lines, or anywhere else, depending on the shape of the equilibrium curve.

Looking at the H–X diagram (Figure 6.1A), as the reflux ratio is reduced, the pivot point D′ moves down toward the saturated vapor enthalpy curve. A point is reached at which the operating line originating at D′ coincides with one of the H–X diagram tie lines. When this happens, it becomes impossible to change the composition from stage to stage, theoretically requiring an infinite number of stages to achieve the specified separation. This position of D′ defines the minimum reflux ratio.

If the number of stages is fixed, the problem is to calculate the reflux ratio required to meet the specified separation. In this situation two cases may arise — the feed location is fixed or is to be determined.

If the feed location is fixed, the number of rectifying stages is known and the reflux ratio is calculated by adjusting the L_r/V_r slope until the number of stages is matched. The intersection of the rectifying section operating line with the q-line is then joined to the bottoms composition projection on the diagonal. The resulting number of stripping stages may or may not match the existing number, meaning that the feed location may or may not be optimum. If the required number of stripping stages exceeds the actual, the bottoms specified composition cannot be met. If the required number is less than the actual, then too many unnecessary stages exist below the feed.

In the case where the feed location is free to move up or down the column, the problem becomes to determine the optimum feed location as well as the reflux ratio. The calculation proceeds as above, starting with an assumed number of rectifying stages. If the resulting number of stripping stages matches the existing number, a solution has been reached. Otherwise, a new trial is started with an altered number of rectifying stages. The procedure is repeated until a match is obtained for the number of stages.

The H–X diagram technique for determining the number of stages and optimum feed location for a specified separation was described in Section 5.3.3. It was implied that the reflux ratio was also specified since the condenser and reboiler duties were given. If the number of stages is specified, but the condenser and reboiler duties are allowed to vary, it becomes necessary to repeat the above procedure with different pivot points (corresponding to different condenser and reboiler duties) until the number of stages is matched.

Example 6.3 Benzene–Toluene Column: Separation Specified

It is required to separate the benzene–toluene mixture of Example 6.1 into a benzene-rich distillate with 0.80 mole fraction benzene and a toluene-rich bottoms with 0.05 mole fraction benzene. The separation is to be made using a distillation column with 15 theoretical stages that include a partial condenser and a partial reboiler. Calculate the reflux ratio required to achieve the specified separation and determine the optimum feed location. What effect would lowering the number of stages to ten have on the reflux ratio and the optimum feed location?

Solution. The product rates consistent with the specified separation are calculated by a material balanced on benzene:

$$FX_F = DY_D + BX_B$$

$$(100)(0.40) = D(0.80) + (100 - D)(0.50)$$

$$D = 46.7 \text{ mole/hr}$$

Using the Y–X diagram, the q-line is assumed vertical, corresponding to a saturated liquid feed. The procedure outlined in the above section amounts to sliding the intersection of the operating lines up and down the q-line until the number of stepped stages equals 15. The reflux ratio is then calculated from the slope of the rectifying section operating line. From Figure 6.4, $L_r/V_r = 0.55$ and from Equation 6.1,

$$R = \frac{L_r/V_r}{1 - L_r/V_r} = 1.22$$

The feed location, according to the diagram, is between the fourth and fifth stages from the top. This feed location is optimum because it requires the least amount of reflux for the given total number of stages. If the feed were to be introduced at a higher stage, a smaller number of rectifying stages would be available, thus requiring a higher reflux ratio to achieve the same distillate purity. The stripping section would have more stages than necessary. If the feed is sent to a lower stage, fewer stripping stages would be available and a higher boilup rate would be required to achieve the same bottoms purity. The higher boilup rate results in a higher reflux ratio and more fractionation than necessary in the rectifying section.

Figure 6.4 shows that with ten stages the optimum feed location is the third stage and the L/V ratio is 0.66, corresponding to a reflux ratio of 1.94.

6.2 PARAMETER INTERACTIONS IN FIXED CONFIGURATION COLUMNS

A column with a fixed number of stages and feed location may be operated over certain ranges of conditions. It may be operated to satisfy given separation requirements or product rates or component recoveries. The column tray size and type, and condenser and reboiler capacities, set practical limits on throughput, reflux ratio, and condenser and reboiler duties and temperatures. Thermodynamics and mass and energy balances also place constraints on the operable ranges. From a rating standpoint for studying the performance of existing columns or from the standpoint of evaluating column design options to meet preset process specifications, the engineer must consider the ranges of different column parameters and their interactions. Each parameter has certain ranges of allowable values and, when one parameter assumes a certain value, the others become restricted to varying degrees in their allowable ranges.

6.2.1 Column Operable Ranges

A column operator can control the column performance by manipulating the reboiler and condenser duties. Consider starting up a column with a mixed-phase feed introduced at some intermediate tray between the condenser and reboiler. With no condenser or reboiler duties, the liquid flows down the column and out as bottoms, and the vapor flows up the column and out as overhead. The column thus acts as a flash drum.

The reboiler is now started, the bottoms is partially reboiled, and the vapor is sent back to the bottom of the column. The ascending vapor interacts with the descending liquid and, if sufficient vapor is generated at the reboiler, a net amount of vapor reaches the feed stage and joins the vapor portion of the feed. The total vapor flows out the top of the column as overhead product. If the condenser is not put in service, the column operates as a stripper, i.e., fractionation takes place on the trays between the feed and the bottoms but not above the feed. By increasing the reboiler duty, one can strip the bottoms of more of the lighter components, thus generating a bottoms product with a higher concentration of the heavy components. The overhead purity cannot be improved as long as the condenser is not in operation.

If the reboiler is turned off and the condenser is activated, the column operates as a rectifier. The condenser liquid is returned to the top of the column as reflux and fractionation takes place on the trays between the overhead and the feed, with no fractionation occurring below the feed. In this mode of operation, higher reflux results in higher concentration of the lighter components in the distillate, but the bottoms cannot be enriched in the heavier components as long as the reboiler is inactive.

The stripping and rectifying modes of operating the column represent limiting conditions of the operable range. Between these limits, the column has both a stripping section and a rectifying section. At reflux ratios approaching 0 with finite boilup ratios, the column acts primarily as a stripper and, at boilup ratios approaching 0 with finite reflux ratios, it performs mostly as a rectifier.

The column operable range is determined in part by the requirement that no tray be allowed to "dry up," i.e., liquid and vapor must exist on each tray to maintain phase equilibrium. This range may be defined by the limits over which the condenser and reboiler duties may vary. As such, the condenser and reboiler duties are considered the two independent variables required to define the column performance (Sections 3.2.3 and 5.2.1). Alternatively, other pairs of variables may be chosen as the independent variables defining the column performance and each set can vary within certain feasible limits. The following sections look at different possible pairs of independent column variables.

6.2.2 Feasible Ranges of Product Rates and Reflux Ratios

Consider the bottoms product rate and the reflux ratio as the independent variables. The feed to the column is assumed of mixed-phase vapor and liquid. The bottoms product rate could vary from 0 to the feed flow rate. At a bottoms rate approaching 0, the reboiler duty must be sufficient to vaporize all the liquid flowing down the column, leaving an infinitesimal amount of bottoms. The vaporized liquid joins the feed vapor, and the combined vapor flows up the column to the condenser. A minimum amount of reflux is required to maintain liquid and vapor phases on all the trays. The conditions of 0 bottoms rate and minimum reflux ratio are represented for the benzene–toluene system of Example 6.1 by point A in Figure 6.5. The reboiler and condenser duties may be increased simultaneously, raising the reflux ratio while keeping a fixed bottoms rate. Thus, the column is operable at any point above A. The overhead composition does not change with reflux ratio since practically all the feed leaves as overhead product. The bottoms product, although approaching 0 flow rate, becomes more concentrated in the heavy components as the reflux ratio is increased.

With higher bottoms flow rates, lower reboiler duties are required at minimum reflux since less liquid must be vaporized. For any given bottoms rate, a minimum

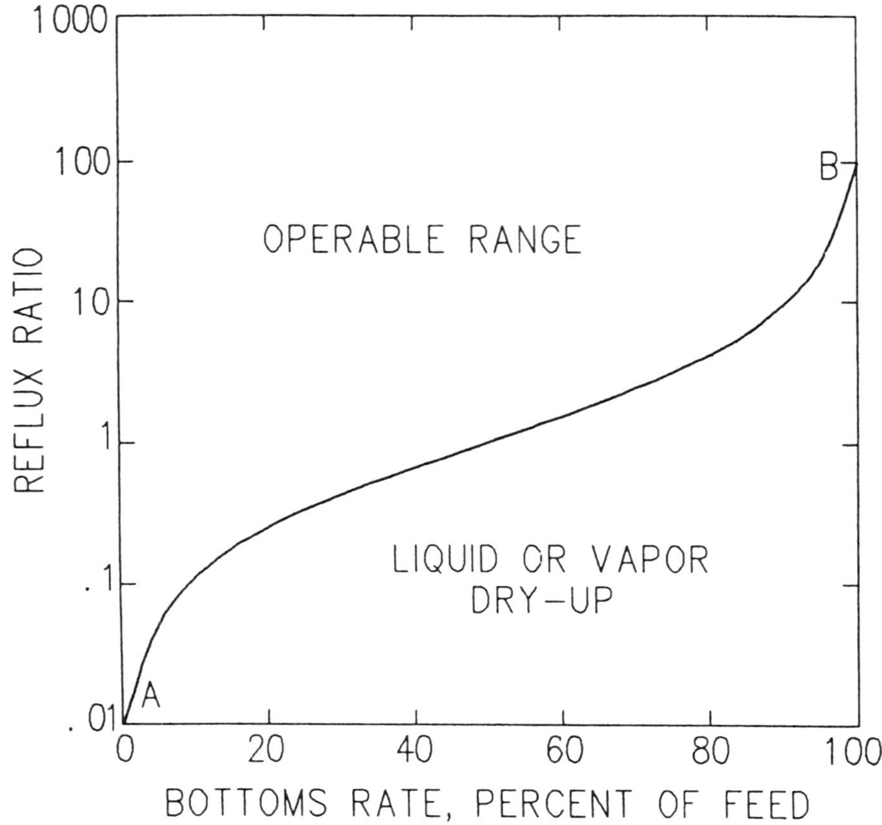

Figure 6.5 Column operable range.

amount of reflux is required to maintain both liquid and vapor phases on all trays. Curve AB in Figure 6.5 is a plot of the minimum reflux ratio. This minimum is not to be confused with the minimum reflux ratio required to bring about a specified separation with a given number of stages. The curve characteristics depend mainly on the feed thermal conditions. Below this curve either the liquid or the vapor dries up on some of the trays. The upper limit of the reflux ratio is determined by practical considerations such as duty limitations and column diameter.

Figure 6.5 shows that a column can operate over wide ranges of product rate and reflux ratio specifications. Expressed mathematically, this fact results from a high degree of independence between the two parameters. This independence may also be verified by inspecting the Y–X diagram. The effect of changing a product rate is equivalent to varying the q-line slope (Section 6.1.1). Changing the reflux ratio is expressed graphically by varying the operating lines' slopes. The q-line and the operating lines may be moved quite independently of each other, subject only to maintaining a fixed number of stages.

6.2.3 Feasible Ranges of Distillate and Bottoms Compositions

If the overhead and bottoms compositions (Y_D and X_B) are specified, the product rates are determined by material balance. The specified separation also determines the reflux ratio. (The number of stages and feed location are assumed fixed.)

The product compositions can vary only over portions of the entire composition range. For instance, the following inequalities must be observed: $Y_D > X_F$ and $X_B < X_F$, where X_F is the feed composition. Also, the upper limit of Y_D and the lower limit of X_B are bounded by the available number of stages operating at or close to total reflux. Moreover, the reflux ratio must be high enough to prevent column dry-up. Figure 6.6 charts the feasible ranges of Y_D and X_B combinations for the benzene–toluene binary.

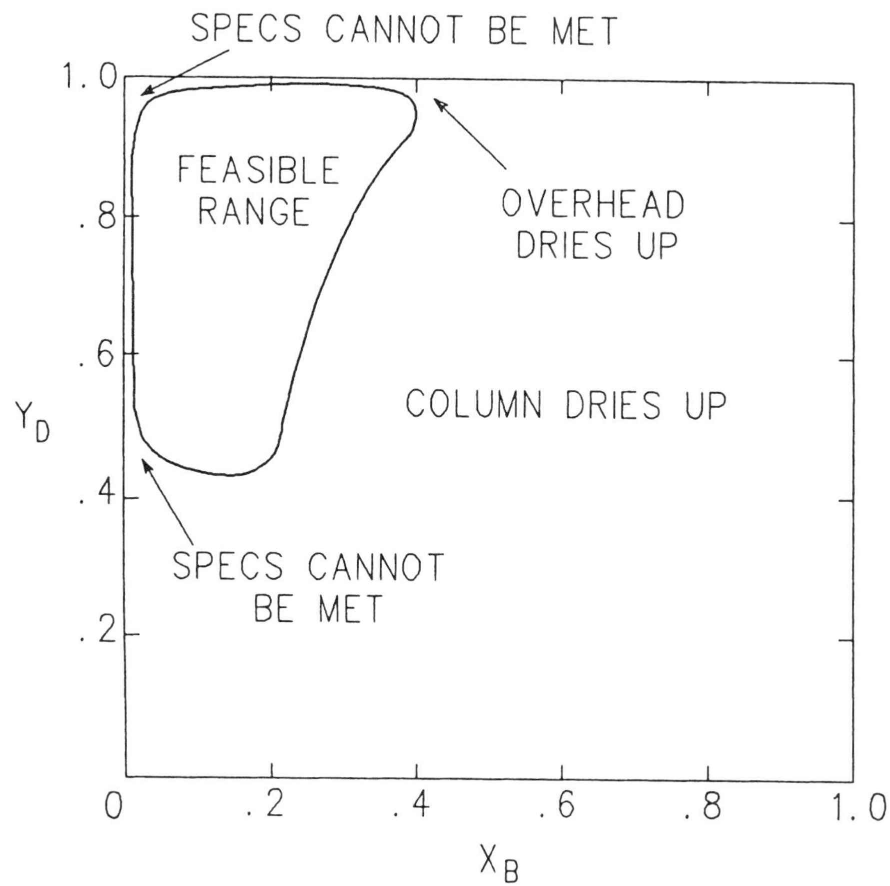

Figure 6.6 Ranges of feasible compositions.

The relatively small operable range of composition specifications results from the lesser degree of independence between Y_D and X_B. The higher the dependence between the specified variables, the smaller the operable range. Seen on a Y–X equilibrium diagram, the dependence between Y_D and X_B is underscored by the limited flexibility in finding a solution if Y_D and X_B, as well as the number of stages and feed location, are fixed. In this situation the q-line slope offers the only flexibility in solving the problem.

6.2.4 Feasible Ranges of Distillate Composition and Reflux Ratio

With a fixed number of stages and feed location, if the reflux ratio is specified, a distillate composition specification may be satisfied within a limited range by varying the product rates. On a Y–X equilibrium diagram, this is equivalent to fixing the rectifying section operating line slope and intersection with the diagonal and varying the q-line slope. The specified variables (Y_D and reflux ratio) are, therefore, not highly independent; they define a relatively small feasible region. This is illustrated for the benzene–toluene system in Figure 6.7.

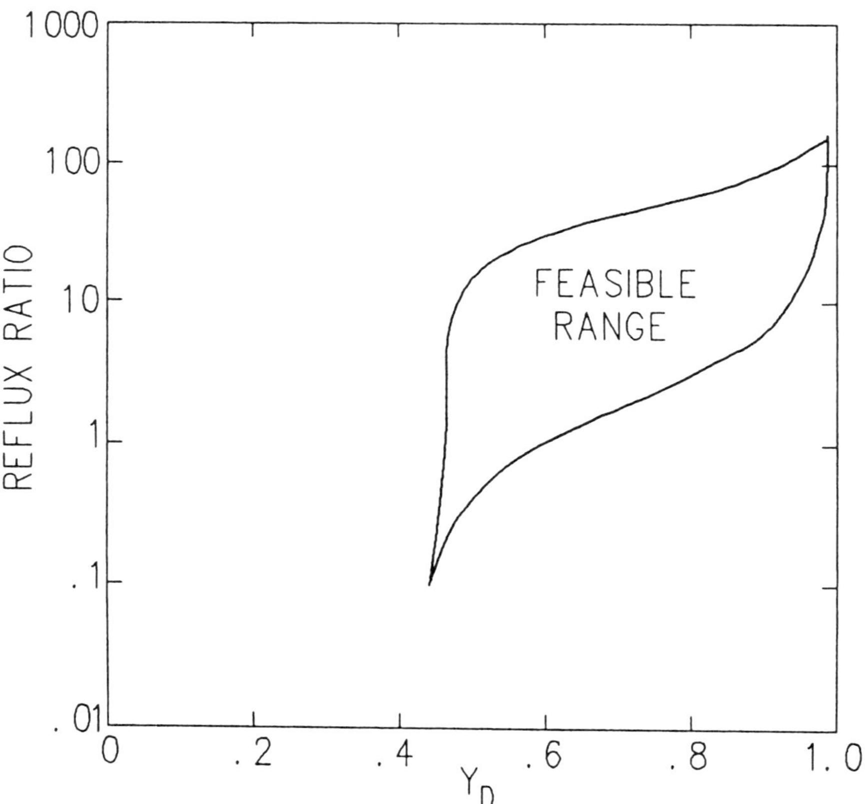

Figure 6.7   Feasible compositions and reflux ratios.

6.2.5   Feasible Ranges of Distillate Composition and Bottoms Rate

If the distillation bottoms product rate is fixed (thereby also fixing the distillate rate), a distillate composition specification may be satisfied within a certain range by varying the reflux ratio. Higher distillate purities may be obtained by increasing the reflux ratio, with maximum Y_D obtainable at total reflux. At the lowest possible reflux required to keep all column trays from drying up, a low Y_D is expected, which at the limit would approach the vapor composition obtainable from a single-stage flash.

Operating the column at a specified product rate and composition is equivalent to fixing the q-line slope and the Y_D-diagonal intersection of the rectifying section operating line on a Y–X equilibrium diagram. The required reflux ratio is determined from the slope of the rectifying stages. There is a limited degree of independence between the product rate and composition and, thus, a limited feasible range can be defined in terms of these two variables. This is shown in Figure 6.8 for the benzene–toluene binary.

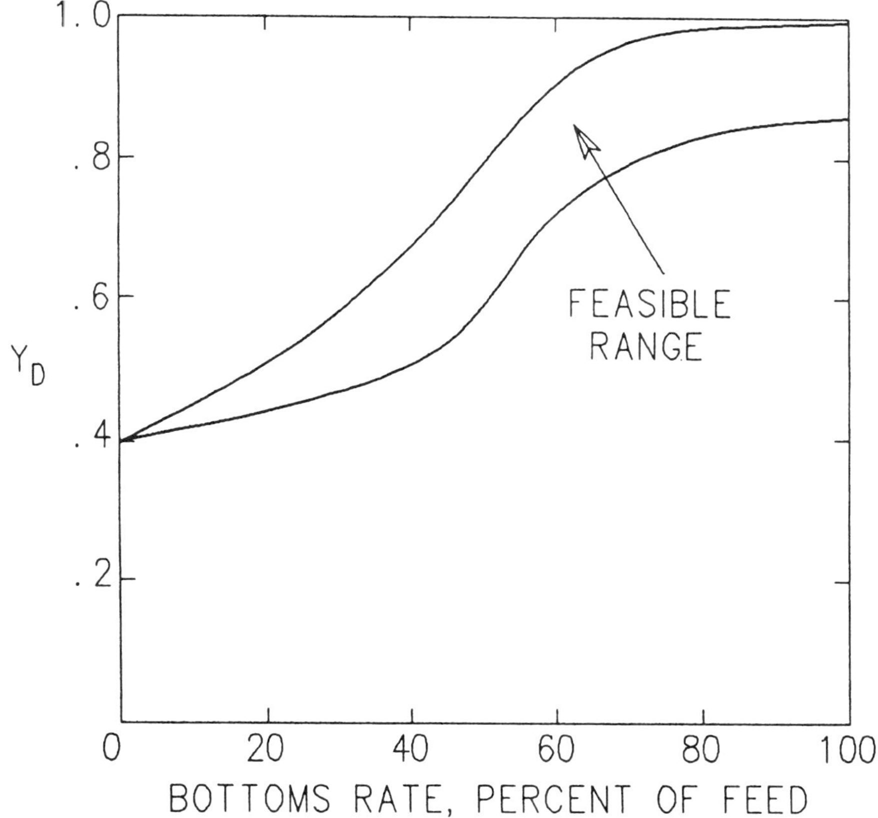

Figure 6.8   Feasible compositions and product rates.

6.3 DESIGN STRATEGIES GUIDED BY GRAPHICAL REPRESENTATION

The design of a separation process must take into account anticipated variations in operating conditions. The process is usually not expected to operate only at design conditions and, therefore, flexibility must be built into the design to ensure stable performance over the expected parameter ranges. The basic design typically relies on a rigorous computer simulation of the process. In conjunction with this, the graphical methods can aid understanding the functioning of the process and provide a visual backup to check design reasonableness and flexibility.

The following applications illustrate some of the factors considered in the validation of a column design using graphical representation.

Example 6.4 Methanol Dehydration

A distillation column is designed for separating methanol from a methanol–water mixture containing 60% mole methanol. The overhead methanol product must have a purity of 98% mole, and the methanol in the bottoms should not exceed a concentration of 4% mole. The column operating pressure is 25 psia with the feed entering the column at its bubble point. It is proposed to design the column at a reflux ratio of 1, or an L/V ratio of 0.5.

For this application a preliminary design is done graphically on a Y–X diagram. The equilibrium properties are provided by the curve in Figure 6.9. The graphical representation of the column is done in the usual manner for known values of X_D, X_B, and feed definition. The q-line is vertical since the feed is a saturated liquid. The rectifying section operating line is drawn with a slope of L/V = 0.5. The graphical design indicates that 13 theoretical stages are required to meet the desired specifications at a reflux ratio of 1 and that the optimum feed location is the tenth stage from the top.

The next step in the column design is to decide on a column diameter that will handle the expected reflux. With an assumed overall tray efficiency of 65%, the actual number of trays would be 13/0.65 = 20, including condenser and reboiler. A column could then be designed based on this reflux ratio and number of trays although economic considerations might suggest alternative trays vs. reflux combinations. Aside from this, an evaluation of the process design is in order before the actual construction is begun. An examination of the Y–X diagram (Figure 6.9) can reveal potential operating problems that might require reevaluating the design.

Figure 6.9 shows that the rectifying section operating line is close to the equilibrium curve, while the stripping section operating line is considerably farther away. It is harder to produce a given purity methanol in the overhead than a comparable purity water in the bottoms. This means that more rectifying trays are needed than stripping trays, i.e., a relatively low feed location is required.

The proximity of the rectifying section operating line to the equilibrium curve gives the column the tendency to "pinch out" in that region. That is, any slight change in some of the operating conditions could cause the operating line to move even closer to the equilibrium curve, requiring many more trays in that region. Since the number of trays is fixed, the changes in operating conditions could prevent the

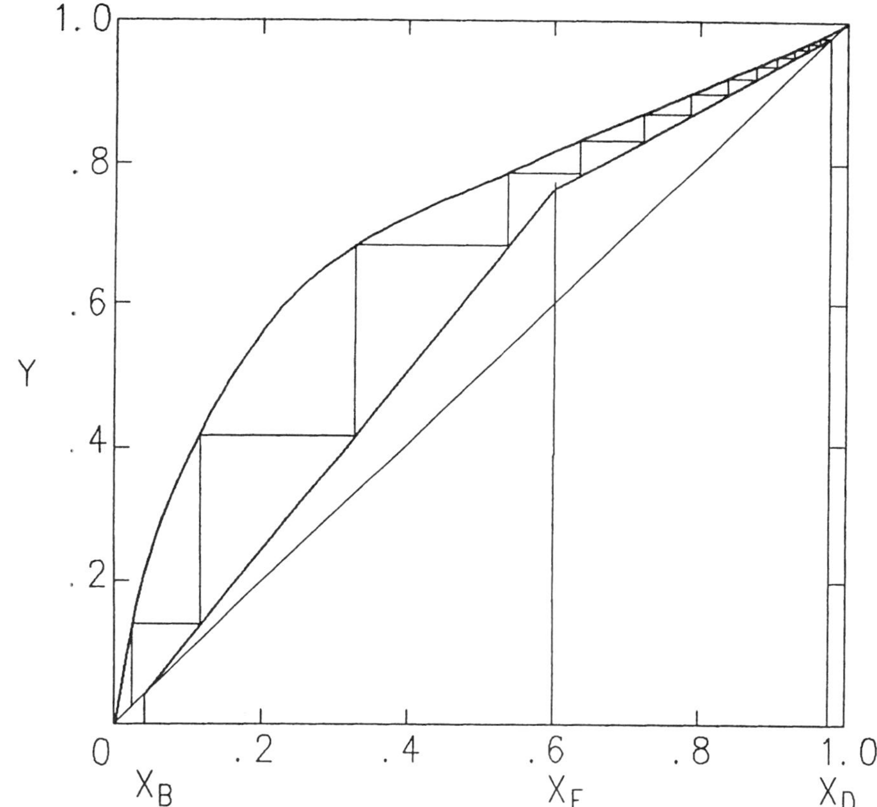

Figure 6.9   Methanol dehydration.

column from making the required separation at the design reflux. Increasing the reflux ratio may not be possible with the existing column diameter. A design that is pinched or that is close to the minimum reflux ratio is sensitive to small changes in the reflux. The column is not stable under these conditions because a slight drop in the reflux can cause a large reduction in the separation. Because of this, it is good practice to avoid designing columns close to the minimum reflux ratio. The methanol column should, therefore, be redesigned with a higher reflux ratio, such as 1.5.

The changes in operating conditions that can cause problems may be identified by referring again to Figure 6.9. If, for instance, the feed is partially vaporized, the q-line tilts to the left, requiring either more rectifying trays or a higher reflux ratio. A similar situation would occur if the methanol concentration in the feed drops slightly or the methanol product purity specification is made a bit more stringent.

By contrast, the stripping section of the column is quite stable since the operating line in that section is a good distance from the equilibrium curve. Lowering or raising reflux ratio by reasonably small amounts causes minor changes in the bottoms composition.

Example 6.5 Demethanizer Column

In this application the column is designed with a computer simulation program; then the computer output is used for plotting the distillation diagram to check the design. This example, which is based on two articles by Johnson and Morgan (1985, 1986), also shows how the principles of binary distillation can be applied to multicomponent mixtures.

In a typical computer-aided column design, the number of trays and feed location are assumed based on past experience, shortcut distillation, or a graphical method. The column is then solved rigorously on a simulator with the assumed configuration. Following this, the design is checked and perhaps optimized using additional computer runs. A graphical check of the initial simulation run prior to the optimization runs can help define the direction of these simulations and thereby avoid unnecessary runs.

The column in this application is a demethanizer in an expander gas plant that recovers ethane and heavier components from natural gas. The feed to the column contains 58% mole methane and lighter components. The product specifications are 98% mole methane and lighter components in the overhead and 3% mole methane and lighter components in the bottoms.

Using the Y–X Diagram. In the initial simulation, ten theoretical stages are assumed, with the feed going on the fourth stage from the top. The results from this run are used to construct a McCabe–Thiele diagram, shown schematically in Figure 6.10. The multicomponent mixture is represented on a binary Y–X diagram by lumping the light key component, methane, and lighter components into one pseudocomponent. The equilibrium curve is then plotted as Y vs. X for the pseudocomponent. These data are usually readily available from the simulation run. The stage compositions are then plotted from the simulation results, and the equilibrium stages are stepped off. Note that the graphical construction of equilibrium stages from simulation results does not require an operating line since the tray vapor and liquid compositions are known. In fact, the operating lines are generated after the stages have been stepped off by drawing the best straight line through the corners of the steps in each column section. The points obtained from simulation may not, in general, lie exactly on one straight line since the simplifying assumptions that are the basis for the straight line derivation are not used in the rigorous methods.

To complete the construction of the Y–X diagram from simulation results, the feed line must be drawn. The intersection with the diagonal of a straight line drawn through the feed composition determines one point on the q-line. One other point is determined by the feed equilibrium vapor and liquid compositions. If the feed is a saturated liquid, the equilibrium liquid composition is the same as the feed composition, and the equilibrium vapor composition is the bubble point composition on the equilibrium curve. In this case the q-line is vertical. For a saturated vapor feed, the equilibrium vapor composition is the same as the feed composition, the equilibrium liquid composition is the dew point composition, and the q-line is horizontal. For a mixed-phase feed, the q-line slope is determined by the feed thermal condition (Section 5.2.2). Note that, for a multicomponent mixture, the feed equilibrium vapor and liquid compositions from the simulation output may not lie exactly on the

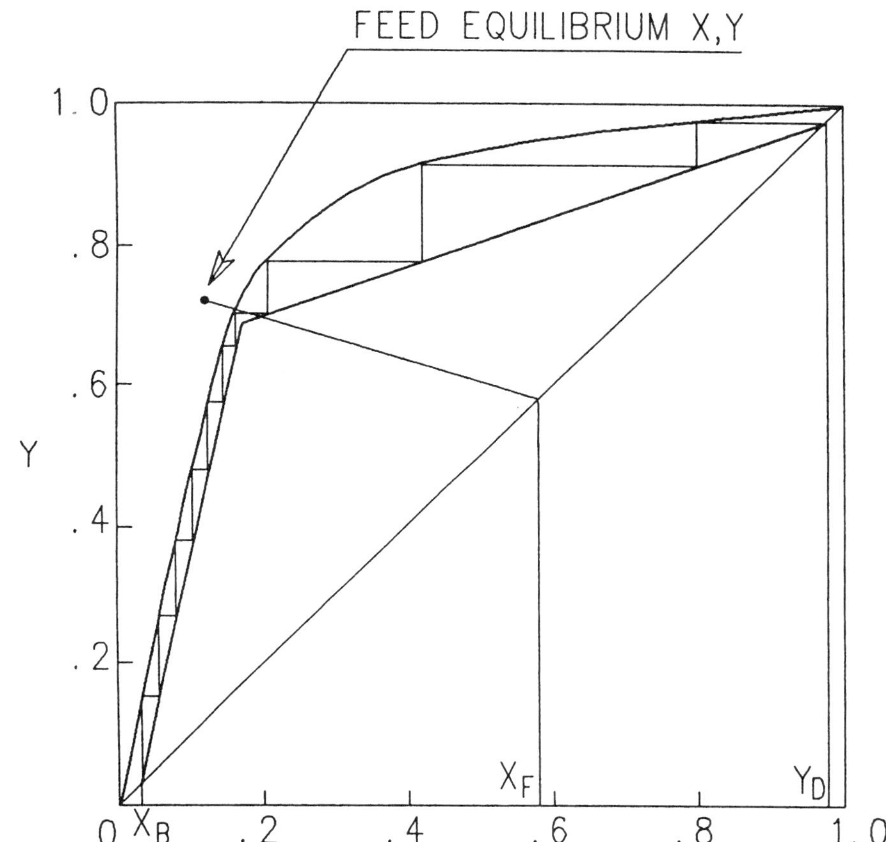

Figure 6.10 Demethanizer column — ten-tray design.

equilibrium curve because of discrepancies resulting from lumping the light components in one pseudocomponent.

A Y–X distillation diagram has thus been completed based on the same data used in the rigorous solution of the column. The Y–X diagram can now be used for a graphical check of the design. The following observations may be made after examining Figure 6.10.

The q-line passes through the same stage that straddles both operating lines. This verifies that the initial selection of the feed stage was correct. Also, the operating lines are adequately distanced from the equilibrium curve, indicating that the column is stable and should have no tendency to pinch out as a result of small changes in operating conditions. The separation progresses steadily from stage to stage, showing no overtraying anywhere in the column. The reflux ratio required to achieve the specified separation, although readily available from the computer output, can be calculated from the slope of the rectifying section operating line. All in all, the graphical representation indicates that the initial design in this case is basically sound, requiring no significant modifications.

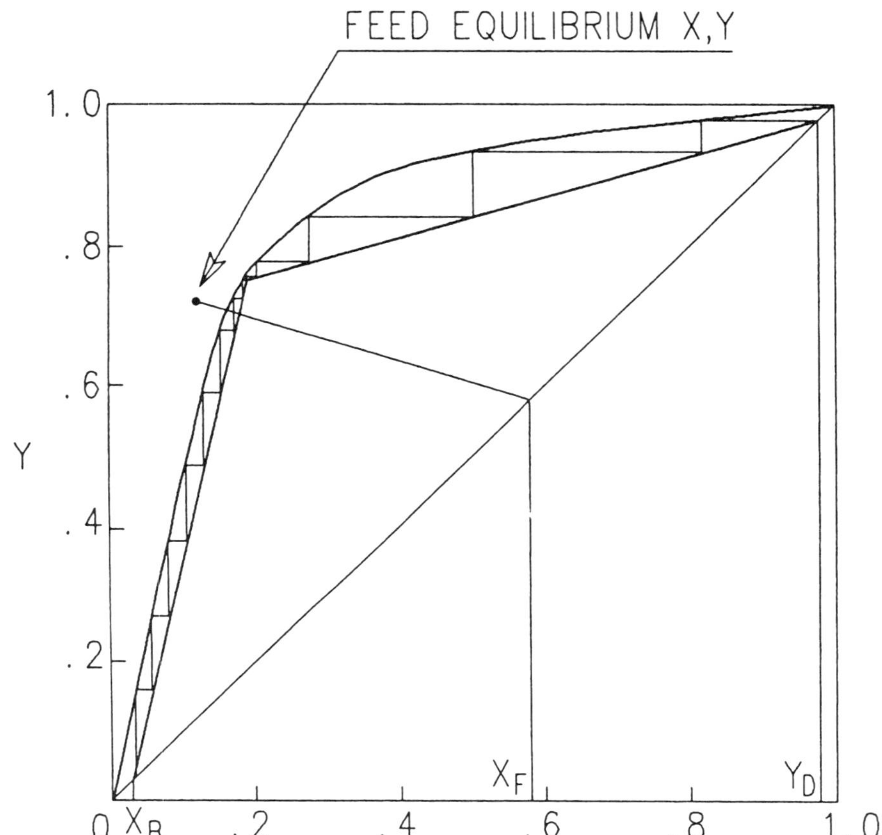

Figure 6.11 Demethanizer column — 13-tray design.

Figure 6.11 represents schematically another case where different design characteristics may be observed. The Y–X diagram is drawn based on the results of a simulation run using 13 stages, with the feed going on the fourth stage from the top. The stripping section operating line is distinctly closer to the equilibrium curve and, consequently, about two or three trays below the feed are achieving little separation. It is also noted that the intersection of the rectifying section operating line and the stripping section operating line is significantly above the q-line, meaning that the feed location is a few trays above optimum.

The effect of redistributing the reboiler duty between the bottom reboiler and a side reboiler a few trays up the column can also be studied on the Y–X diagram. As discussed in Section 5.2.2, a side reboiler in this case has the effect of increasing the slope of the operating line below the side reboiler. Given the other problem constraints, the side reboiler tends to move the operating line closer to the equilibrium curve. One result of this is the redistribution of the degree of separation above and below the side reboiler. Moreover, the proximity of the operating line to the equilibrium curve in the vicinity of the side reboiler may have the potential to cause a pinch point in that area.

Using the H–X Diagram. The construction of the Ponchon–Savarit H–X diagram for binary distillation calculations was described in Section 5.3. A multicomponent mixture can be represented on the binary diagram as in the McCabe–Thiele case, by lumping the light key and lighter components into one light pseudocomponent. Generating the H–X diagram when the tray compositions are not known proceeds by stepping off stages starting at the distillate or bottoms (Section 5.3). To use the H–X diagram for evaluating simulation results, the procedure is modified to take advantage of enthalpy vs. composition data, tray compositions, and other data from the simulation (Johnson and Morgan, 1985, 1986).

The tray vapor and liquid compositions and enthalpies, generally available from the computer output, are first used to construct the saturated vapor and liquid enthalpy–composition curves on a light pseudocomponent basis. The H–X method is especially suited for nonideal systems where the McCabe–Thiele assumptions may be in gross error. The H–X method takes into account the variation of enthalpy with composition (Section 5.3), which is reflected in the shape of the enthalpy curves. Ideal mixtures are characterized by straight, horizontal lines while nonideal mixture lines exhibit varying slopes and curvature.

The tie lines are drawn next by joining corresponding dew and bubble points on the saturated vapor and liquid curves. These points also correspond to individual tray vapor and liquid compositions. The pivot points are plotted next, based on the distillate enthalpy and condenser duty and the bottoms enthalpy and reboiler duty. The rectifying section operating lines are drawn by joining the distillate pivot point to the compositions on the saturated liquid curve corresponding to the trays above the feed; the stripping section operating lines are drawn by joining the bottoms pivot point to the compositions on the saturated vapor curve corresponding to the trays below the feed. Since a consistent set of data is used both for constructing the enthalpy–composition curves and for determining the tray compositions, each rectifying section operating line should intersect the saturated vapor curve at the vapor composition of the tray below. Likewise, each stripping section operating line intersects the saturated liquid curve at the liquid composition of the tray above.

The completed H–X diagram can now be used for a visual evaluation of the proposed column design. The feed location is optimum if its tie line crosses the line joining the distillate and bottoms pivot points (Section 5.3). If this condition is not satisfied, consideration should be given to moving the feed tray up or down as indicated. The spacing or density of the operating lines gives a visual measure of the degree of separation achieved per tray at various points in the column. The number of trays in the different column sections may then be considered for modification. Finally, the effect of varying the condenser and reboiler duties can be visualized on the H–X diagram by moving the pivot points up or down and observing the positions of the operating lines, etc.

NOMENCLATURE

B   Bottoms stream designation or molar flow rate

D   Distillate or overhead stream designation or molar flow rate

F   Feed stream designation or molar flow rate

L   Liquid molar flow rate in the column

R   Reflux ratio

V   Vapor molar flow rate in the column

X   Component mole fraction in the liquid

Y   Component mole fraction in the vapor

SUBSCRIPTS

B   Bottoms designation

D   Distillate designation

F   Feed designation

r   Rectifying section designation

s   Stripping section designation

REFERENCES

Johnson, J. E. and D. J. Morgan, *Chemical Engineering*, July 8, 1985.

Johnson, J. E. and D. J. Morgan, *Encyclopedia of Chemical Process Design*, New York, Marcel Dekker, 1986.

PROBLEMS

6.1 The K-values and enthalpies of components 1 and 2 in a binary mixture at 1 atm are given by the following equations:

$$\ln K_1 = 11.9 - 7200/T(^\circ R)$$

$$\ln K_2 = 12.1 - 7800/T(^\circ R)$$

Liquid enthalpies:

$$h_1 \ (Btu/lbmole) = 3000 + 45.0T(^\circ F)$$

$$h_2 \ (Btu/lbmole) = 3300 + 52.0T(^\circ F)$$

Vapor enthalpies:

$$H_1 \ (Btu/lbmole) = 18000 + 32.0(^\circ F)$$

$$H_2 \ (Btu/lbmole) = 21000 + 40.0(^\circ F)$$

Construct the Y–X and H–X diagrams for this binary. What assumptions are necessary for generating the enthalpy diagram from the given data alone?

6.2 A binary stream at the rate of 1000 lbmole/hr containing 35% mole of component 1 (the lighter component) is to be separated in a distillation column to produce 95% component 1 in the distillate and 90% component 2 in the bottoms. The column will have a partial reboiler and a partial condenser, and will operate at 1 atm. It is proposed to utilize an existing hot process stream in the plant as a heat source for the reboiler, which limits the reboiler duty to 25×10^6 Btu/hr. Using either the Y–X diagram or the H–X diagram, determine the number of theoretical trays required, the optimum feed location, the product rates, and the condenser duty. Assume feed thermal conditions that result in a saturated liquid at the feed tray pressure. Use thermodynamic data from Problem 6.1.

6.3 The distillation column described in Problem 6.2 is to be designed with the provision that the hot process stream will supply only 20×10^6 Btu/hr to the reboiler. It is proposed to supplement the column duty requirements by utilizing another process stream that can supply 10×10^6 Btu/hr. However this stream is at 200°F, and a temperature approach of 25°F is required between the process stream and the column temperature where the heat would be supplied. Determine the optimum locations of the heater and the feed and the total number of equilibrium stages. Calculate the condenser duty.

6.4 Two streams, each a binary mixture of components 1 and 2, are to be separated in a distillation column to produce a 95% mole component 1 distillate and a 5% mole component 1 bottoms. The flow rates and compositions of the feed streams are as follows:

	Stream 1	Stream 2
Component	Mole Fraction	
1	0.40	0.70
2	0.60	0.30
Flow rate, lbmole/hr	600.0	400.0

The column will operate at 1 atm with a total condenser at the bubble point and a reflux ratio of 3.0. The thermodynamic data in Problem 6.1 may be used for this mixture. The feeds are assumed to be saturated liquids at feed tray conditions.

Two alternative schemes are proposed: 1. Mix the feed streams and send the mixture to a single point in the column. 2. Feed each stream separately to different trays in the column. For each option calculate the required number of trays, the optimum feed locations, and the condenser and reboiler duties. Use either the Y–X or H–X diagram methods.

6.5 A distillation column was designed to separate a 1000 lbmole/hr binary mixture of 50% mole component 1 to produce a distillate of 95% mole component

1 and to recover in this product 90% of component 1 in the feed. The column was constructed to handle a liquid traffic of 2500 lbmole/hr. Due to upstream process changes the column feed composition dropped to 40% mole component 1 at the same total rate of 1000 lbmole/hr. While maintaining the required 95% distillate composition and operating at optimum performance, what recovery of component 1 is achievable, and can the column handle the required liquid traffic? The column operates with a partial condenser at 1 atm. The feed is a saturated liquid at feed tray conditions. Thermodynamic data given in Problem 6.1 may be used in this problem.

6.6 A column with eight equilibrium trays and a partial condenser and reboiler is used to separate a 60–40 mole percent mixture to produce a distillate containing 95% mole component 1. The feed may be introduced on either the 4th or 7th stage from the top. Assuming the feed location is optimum in both cases, calculate component 1 recovery in the distillate in each case. The feed enters the column as saturated liquid. Use thermodynamic data from Problem 6.1.

6.7 A distillation column is designed to separate a 70–30 mole percent mixture to produce a distillate product with 97.5% mole component 1 and a bottoms product with 20% mole component 1. The column has a partial condenser and reboiler and operates at 1 atm pressure. Determine the required number of theoretical trays and the optimum feed location at a reflux ratio of 1.5. Calculate the recovery of component 1 in the distillate. Calculate the recovery at the same distillate purity and reflux ratio if the feed is introduced 3 trays below the optimum feed tray. Use thermodynamic data from Problem 6.1.

CHAPTER **7**

Multicomponent Separation: Conventional Distillation

The principles of multistage separation were applied in the past chapters mainly to binary mixtures. The objective was to formulate the theories and methods of solution and prediction for an idealized system. This model was used to study the effect of different parameters on the outcome of a separation process and also to solve quantitatively certain binary problems.

Most practical separation problems, however, involve multicomponent mixtures, simply because mixtures occurring in nature are rarely binary. Certain industrial streams, the products of various processing steps, could include primarily two components to be further separated in downstream unit operations. Even such streams are in fact multicomponent mixtures on account of impurities carried along in the stream.

Calculating or predicting the performance of multicomponent, multistage columns is carried out rigorously by setting up the equations that relate the various interacting parameters, then solving the equations. Discussion of these methods in detail is deferred to Chapter 13. Although the mathematical models are accurate and rigorous, staring at an elaborate equation matrix hardly helps the engineer determine what parameters could affect what performance specifications and to what extent.

This chapter, as well as a few more that follow, is concerned primarily with a qualitative understanding of the factors that influence the separation process. The principles and ideas introduced in earlier chapters and applied to idealized and binary systems are generalized here to multicomponent separation.

This chapter centers around multicomponent distillation with conventional columns, i.e., columns with a single feed and overhead and bottoms products. In general, such columns have a condenser and a reboiler, although the special cases of rectifiers and reboiled strippers are also considered.

211

7.1 CHARACTERISTICS OF MULTICOMPONENT SEPARATION

A multicomponent mixture is separated in a fractionation column into an overhead product that is enriched in the lighter components and a bottoms product that is enriched in the heavier components. The product compositions depend on the extent of fractionation (or separation) taking place inside the column and on the product rates.

Assume the feed mixture is made up of components 1 through C with mole fractions Z_1 through Z_C. Assume also that the components in the feed are listed in order of decreasing volatility, such that component 1 has a lower boiling point than component 2, which has a lower boiling point than component 3, etc. If the column has a sufficient number of stages and enough reflux to result in perfect fractionation, the product compositions are determined entirely by the product rates. If, for instance, the overhead rate is less than or equal to FZ_1 (where F is the molar feed rate), the overhead composition would be pure component 1. As the overhead rate is increased, a certain amount of component 2 starts flowing with the overhead. Since perfect fractionation is assumed, the overhead composition is determined by material balance alone. If the overhead rate is D, the following relationships hold for values of D between FZ_1 and $F(Z_1 + Z_2)$:

$$FZ_1 = DY_1$$

$$Y_2 = 1 - Y_1$$

where Y_1 and Y_2 are mole fractions of components 1 and 2 in the overhead. Similar relationships may be derived for the overhead composition as its rate is increased beyond $F(Z_1 + Z_2)$. Thus, in this model, the point in the component list where separation takes place shifts as the overhead rate changes.

With a finite number of stages, where fractionation is not perfect, the separation point also shifts with the product rate. The above relationships, however, are no longer accurate. Nevertheless, for a given overhead or bottoms flow rate, there generally exists a pair of components i and i + 1 with different volatilities, between which the separation takes place. Since component i is more volatile than i + 1, most of i in the feed goes overhead, while most of i + 1 goes with the bottoms. It is also clear that the larger part of each of the components that are lighter than i flows with the overhead, while those heavier than i + 1 end up mostly in the bottoms. Component i is known as the *light key component*. It is the heaviest component that goes mostly in the overhead. Component i + 1 is the *heavy key* and is the lightest component that goes mostly in the bottoms.

It may be desired to separate components i and j that are not necessarily adjacent on the volatility list. Between them there may be irrelevant components with intermediate volatilities. In this case, if i is lighter than j, then i is the light key and j is the heavy key. The components in between are *distributed components*; they may flow either mostly with the overhead or with the bottoms.

7.2 FACTORS AFFECTING SEPARATION

The separation into lighter and heavier products in a multistage column is accomplished as a result of a composition gradient that develops along the column. The upper stages have higher concentrations of lighter components and the lower stages have higher concentrations of heavier components. The composition gradient is maintained by the counter flow of liquid and vapor streams with different compositions passing each other and interacting on each stage. In conventional distillation columns, the liquid and vapor streams within the column are generated internally by condensing vapor in the condenser and reboiling liquid in the reboiler. In absorbers and strippers, the column liquid and vapor traffic is maintained by feeding external liquid and vapor streams to the top and bottom of the column, respectively.

As the composition changes from stage to stage in the column, so does the temperature. If the feed to a distillation column is a narrow-boiling mixture, the compositional gradient along the column is linked mostly to the temperature gradient. The predominant factor influencing the temperature and composition profile in this type of column is the temperature dependence of the vapor–liquid equilibrium coefficients (K values). The contribution of the heat of vaporization/condensation to the temperature profile is secondary. In fact, in the idealized treatment of binary systems in previous chapters, the latent heat was cancelled out (Section 5.1). A simple, two-stage, binary model was also described which illustrates the phenomenon of pure distillation (Section 3.2).

In columns or column sections where the liquid composition is considerably different from the vapor composition, the composition of each phase does not change from one stage to the next so much because of temperature variation. Looked at from another angle, the temperature variation along the column is influenced less by phase equilibrium relations than by latent heat or enthalpy balances. Such situations occur to some extent in wide-boiling distillation columns although they are more prevalent in absorbers and strippers. In these columns, no temperature gradient is necessary to generate the compositional gradient along the column. The composition gradient is brought about by feeding streams with distinct compositions at different column locations.

The difference between the thermal behavior in predominantly distillative and predominantly absorptive processes is a consequence of the different types of mass transfer in each case. In the former, mass is transferred from the liquid to the vapor phase and vice versa at approximately the same molar rate. The net material transfer between the phases is therefore small and the ratio of liquid to vapor flow (L/V) in a column section is nearly constant. In absorption or stripping columns, there is a net mass transfer in one direction; hence, L/V is not constant.

To sum up, the temperature profile in conventional distillation columns is the result of both phase equilibrium relations and enthalpy balances. In narrow-boiling mixtures, the phase equilibrium effect is generally more pronounced, while in wide-boiling mixtures, the enthalpy balances are more significant. The importance of the distinction between the two effects is twofold. First, different mathematical solution algorithms are better suited for each situation, as will be discussed in Chapter 13. Second, the understanding and prediction of column performance is enhanced when the two effects are recognized. The following examples illustrate the two cases.

Example 7.1 Separation of a Narrow-Boiling Mixture

A mixture of methane, ethane, propane, and butane is to be split between the ethane and propane using an 11-theoretical-stage, conventional column operating at 375 psia. The feed, with the following composition, enters the column on the fourth stage from the top:

Component	Mole/hr
Methane	120
Ethane	460
Propane	27
Butane	5

The column is solved at an overhead rate of 580 mole/hr and a reflux ratio of 1. The results are given in Table 7.1.

Table 7.1   Results of Example 7.1

				Methane		
Stage	Temp °F	Tray Liq Mole/Hr	Tray Vap Mole/Hr	Y	K	L/V
COND	18	580	580	.207	3.65	
2	26	588	1160	.132	3.70	0.50
3	29	577	1168	.121	3.74	0.50
4*	32	1272	1153	.118	3.81	0.50
5	45	1259	1240	.318E-1	3.91	1.02
6	59	1189	1227	.836E-2	4.11	1.02
7	78	1120	1157	.209E-2	4.38	1.03
8	102	1092	1088	.492E-3	4.59	1.03
9	125	1097	1060	.110E-3	4.66	1.03
10	144	1103	1065	.244E-4	4.67	1.03
REB	159	32	1071	.535E-5	4.68	1.03

* Feed tray

The vapor and liquid rates for each tray are the net rates leaving that tray. In particular, the vapor rate for the condenser (or tray number 1) is the overhead product rate, and the liquid rate for the reboiler (or the bottom tray) is the bottoms product rate. The L/V ratio is the ratio of liquid to vapor rates passing each other between trays; that is, it is L_j/V_{j+1}. The methane mole fraction in the vapor and its K value are given on each tray.

Typical characteristics of a predominantly distillative process are evident in this example. The temperature rises steadily from stage to stage between the condenser and reboiler. The K values increase with temperature and the methane mole fraction decreases. The L/V ratio is very nearly constant in each section of the column.

Example 7.2 Stripping Out Light Components in a Demethanizer

The feed stream is a wide-boiling mixture with the following composition:

Component	Mole/hr
Carbon Dioxide	0.2
Nitrogen	0.5
Methane	21.7
Ethane	18.4
Propane	26.7
i-Butane	4.2
n-Butane	15.0
i-Pentane	3.3
n-Pentane	5.0
Hexane	2.5
Heptane	2.5

The feed is a saturated liquid at the column pressure of 250 psia. The purpose of the column is to remove, in the stripping section, most of the methane and lighter components from the feed. The bottoms is the purified product. There is no need to purify the overhead from heavier components; thus, no rectifying section is required. The feed is therefore introduced at the top tray. No condenser is required since the liquid feed serves as reflux.

This type of column is a special case of the conventional column and is described as a reboiled stripper. Because of the wide-boiling feed and the fact that the external feed is used as reflux, this column has mixed characteristics of distillation and absorption/stripping. The solution results are summarized in Table 7.2. In contrast

Table 7.2 Results of Example 7.2

Stage	Temp °F	Tray Liq Mole/Hr	Tray Vap Mole/Hr	Nitrogen		
				Y	K	L/V
1*	−85	103	20.0	.250E-1	26.2	5.15
2	−85	103	22.6	.433E-2	26.1	4.54
3	−84	103	23.1	.739E-3	26.1	4.46
4	−84	103	23.2	.126E-3	26.1	4.45
5	−84	103	23.2	.214E-4	26.1	4.45
6	−84	103	23.2	.365E-5	26.1	4.45
7	−82	103	23.2	.622E-6	26.3	4.45
8	−70	104	23.1	.105E-6	26.9	4.46
9	−24	111	24.1	.167E-7	27.9	4.31
REB	68	80	31.1	.195E-8	26.5	3.57

* Feed tray

to the distillation column of Example 7.1, the temperature profile in this column is fairly constant over a considerable number of trays. The K values of nitrogen on these trays are also constant, although the nitrogen mole fraction changes noticeably from tray to tray. The L/V ratio goes through significant changes at both ends of the column.

7.3 SPECIFYING COLUMN PERFORMANCE

It has been established on previous occasions (Sections 3.2.2 and 5.2.1) that a conventional two-product distillation column with a fixed feed and configuration and operating at a given pressure has 2 degrees of freedom. The configuration is defined by the number of stages, type of condenser and reboiler, and the feed location. The feed is of fixed flow rate, composition, and thermal conditions.

The description rule introduced in Section 5.2.1 may be used to check the degrees of freedom or the number of variables which must be set to define the column operation. For an existing column, the variables that can be set by construction determine the column configuration. The variables that can be controlled by external means include those variables that define the feed, the column pressure, and the condenser and reboiler duties. With a fixed feed and a pressure usually controlled to maintain an acceptable column temperature level, two variables are left for the purpose of controlling the separation — the condenser and reboiler duties. These two variables may be relaxed to allow any other two independent variables to be used for specifying the column performance.

The column may be operated to meet various performance specifications within certain ranges. The variables that can be specified in multicomponent separation include all the component compositions, rates, or recoveries in the two products as well as the product rates, properties, and temperatures; the reflux and boilup ratios; condenser and reboiler duties; and the tray temperatures and liquid and vapor rates.

The two parameters selected to define the column performance become the independent variables, and the others are calculated to satisfy the mass balances, energy balances, and equilibrium relations. Many of the parameters are interdependent to varying degrees, but the two selected as the independent variables must be independent of each other at least over certain ranges. In fact, the higher the degree of independence between the two specified variables, the wider the feasible ranges over which the column can operate and the easier for the specified values of the variables to be met.

The extent of interdependence among the different variables may be investigated by observing the surface that represents each variable as a function of the two "most independent" variables, namely, the reflux ratio and one of the product rates. Stated mathematically, any parameter p_k may be expressed as

$$p_k = p_k(R_r, B)$$

where R_r is the reflux ratio and B is the bottoms rate. The dependence of parameter p_k on either of the independent variables is measured by its partial derivatives with respect to R_r or B. The interdependence of any two parameters p_k and p_l on each other is similarly measured by the relative values of their partial derivatives with respect to the reflux ratio and bottoms rate.

It is impossible to express these derivatives in analytical form since the column equations can only be solved numerically. The interdependence of the various parameters is therefore investigated on the basis of numerical solutions for different column situations.

7.3.1 Variation of Dependent Variables with Reflux Ratio and Product Rate

The variables most often used for evaluating column performance include component fractions and recoveries in each product, the temperature of each product, and the condenser and reboiler duties. It is convenient to scale the variables in such a manner as to provide a uniform basis for comparing their variation. The reflux is expressed as a ratio and is therefore dimensionless and could vary from 0 to infinity. The product rates are expressed as fractions of the total feed, and each component recovery in each product is expressed as a fraction of the same component in the feed. These quantities could thus vary from 0 to 1. The component concentrations in the products are expressed as mole fractions and could also vary from 0 to 1. The overhead and bottoms temperatures are designated as T_D and T_B, respectively. The condenser and reboiler duties, Q_C and Q_R, are expressed on a per-mole-feed basis. The following example serves to illustrate the interdependence of the variables in a conventional column.

Example 7.3 Column Variable Interdependence

A feed stream composition is defined as follows:

Component	Mole/Hr
Methane	2
Ethane	4
Propane	17
i-Butane	9
n-Butane	26
i-Pentane	12
n-Pentane	19
n-Hexane	11
Total	100

The thermal conditions of the feed are such that it is 50% vapor at the feed tray pressure. The column has 18 theoretical stages, including a partial condenser and a reboiler. The feed is introduced at the eighth stage from the top. The column pressure is fixed at 120 psia.

Initially, the overhead product rate is fixed, and the variation of the different column variables is investigated as a function of the reflux ratio.

The flow rate of overhead or bottoms products determines roughly which components go mostly in the overhead and which ones in the bottoms. This also defines the key components where the separation takes place. In this example, an overhead rate of 50 moles/hr would include most of the methane, ethane, propane, isobutane, and n-butane. The bottoms product would include most of the hexane, n-pentane, and isopentane. The n-butane is, therefore, considered the light key component and isopentane, the heavy key component. The product compositions at a reflux ratio of 1 are given in Table 7.3.

Table 7.3   Results of Example 7.3

Reflux Ratio = 1					
Ovhd Rate:		**50 Mole/Hr**		**75 Mole/Hr**	
	Feed	**Overhead**	**Bottoms**	**Overhead**	**Bottoms**
Component	**Mole/Hr**	**Mole/Hr**	**Mole/Hr**	**Mole/Hr**	**Mole/Hr**
Methane	2.00	2.00	0.00	2.00	0.00
Ethane	4.00	4.00	0.00	4.00	0.00
Propane	17.00	16.99	0.01	17.00	0.00
i-Butane	9.00	7.65	1.35	9.00	0.00
n-Butane	26.00	17.12	8.88	25.98	0.02
i-Pentane	12.00	1.42	10.58	8.77	3.23
n-Pentane	19.00	0.82	18.18	8.22	10.78
n-Hexane	11.00	0.00	11.00	0.03	10.97
Totals	100.00	50.00	50.00	75.00	25.00

With the product rates fixed, the sharpness of separation is determined by the reflux ratio. (The column pressure and number of stages are fixed.) The reflux ratio mostly affects the product compositions of components with boiling points in the vicinity of the key components. Components that are either much lighter or much heavier than the key components are not greatly affected by variation of the reflux ratio. This may be seen by comparing the behavior of the overhead compositions $Y(C3)$ and $Y(NC4)$ at an overhead rate of 50 mole/hr (Figure 7.1).

As the overhead rate is changed, the split point shifts, and this may result in a change in the identity of the key components. In this example, at an overhead rate of 75 mole/hr, most of the isopentane and lighter components go in the overhead, while most of the n-pentane and heavier go in the bottoms (Table 7.3). Hence, isopentane is the light key component, and n-pentane is the heavy key component. At this overhead rate, since the NC4 boiling point is not close to the key components, its composition in the overhead is not sensitive to the reflux ratio. This is not the case of an overhead rate of 50 mole/hr, where the NC4 composition is quite sensitive to the reflux ratio (Figure 7.1).

Figure 7.1 shows other variables plotted vs. reflux ratio at a constant overhead rate. A graph of B_r, the bottoms rate expressed as a fraction of the feed, is, of course, constant at 0.5 when the overhead rate is 50 mole/hr. The condenser and reboiler duties, Q_c and Q_r, vary considerably with the reflux ratio as may be expected due to the variation of condensation and boilup rates as the reflux ratio varies. Since the product compositions do not vary greatly at a constant product rate, the product

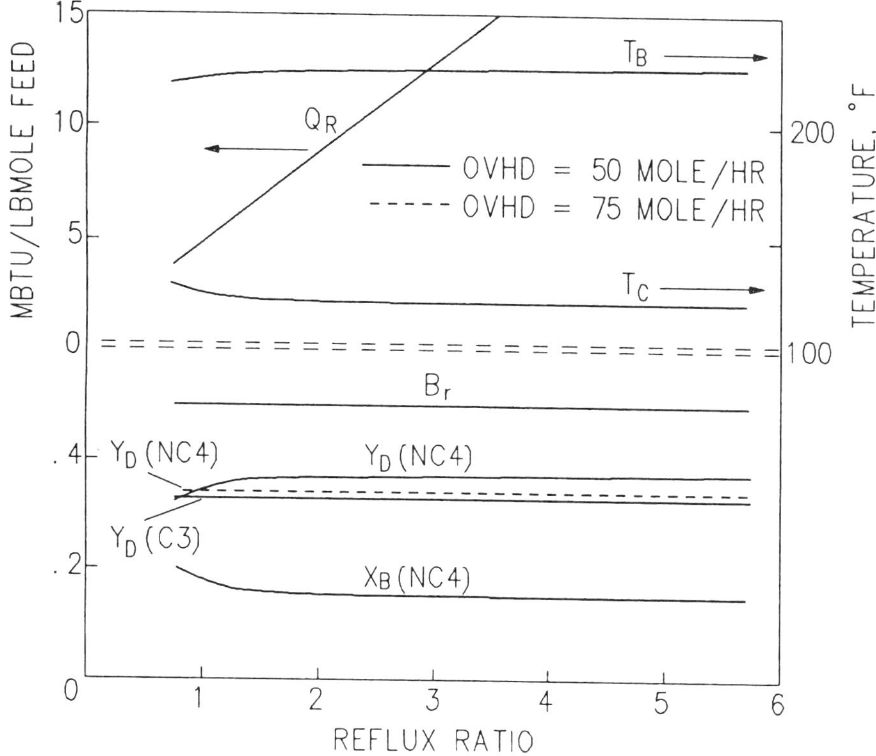

Figure 7.1 Parameter dependence on the reflux ratio.

temperatures do not vary considerably with reflux ratio. Fine-tuning of the compositions of key components and adjacent components by adjusting the reflux ratio does not affect the product temperatures considerably.

If the number of stages and reflux ratio are held constant, the fractionation attainable between the products is fixed. If the overhead rate is varied, the components where the separation takes place change. In general, most variables (except the condenser and reboiler duties) are more dependent on the product rate than on the reflux ratio. Figure 7.2 presents plots of a number of variables as a function of the overhead rate at a fixed reflux ratio of 1.

As the overhead rate is increased, the concentration of light components in the overhead goes down (Y_D for C1 in Figure 7.2). For heavy components, the concentration goes up (not shown). Intermediate components go through a maximum, which determines the overhead rate at which the column should be operated to maximize the concentration of any of these components. For example, if the objective is to maximize the concentration of NC4 in the overhead, the overhead rate should be around 65 mole/hr.

It may be desired to meet a certain component concentration, say Y_D (NC4) = 0.38, which constitutes one performance specification. To meet this concentration,

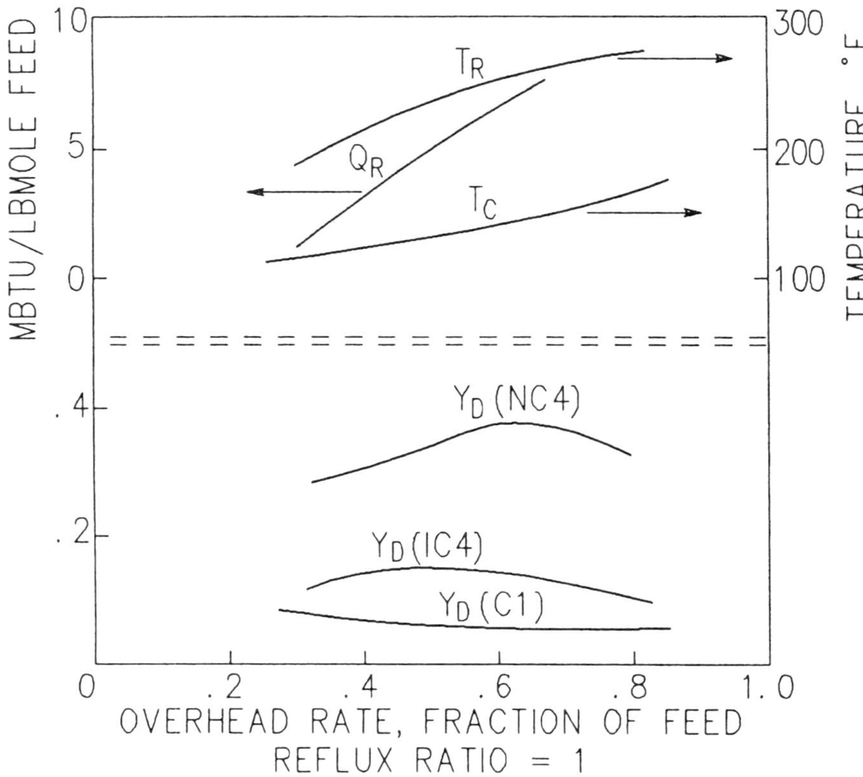

Figure 7.2 Parameter dependence on the product rate. Reflux ratio = 1.

either the reflux ratio or the overhead rate (or both) would have to be varied. Figure 7.2 indicates that, if the reflux ratio is specified at 1, the NC4 specification may be met by varying the overhead rate. It is also observed that two solutions are possible — one at an overhead rate of 58 mole/hr and another at an overhead rate of 71 mole/hr.

The variation of some of the other parameters with overhead rate at a constant reflux ratio is shown in Figure 7.2. The significance of these relationships is in determining the feasibility of specifying the column performance using a given pair of parameters. This question is explored further in the next section.

7.3.2 Parameter Feasible Ranges

Studying the interdependence between column variables is important when attempting to select parameter pairs for specifying the performance of a conventional column. The reflux ratio and product rate are generally considered the most independent pair. The other extreme may be illustrated by a situation where both product rates are specified. These are two column variables that are totally dependent since specifying one of them fixes the other by simple material balance. Once one product rate is known, no additional information about the column performance is acquired by providing the other product rate. This fact is indicated by the horizontal line for the bottoms rate shown in Figure 7.1.

For any two variables, it is possible to construct a diagram that represents the feasible area on a plane defined by the two variables as coordinates. If the two product rates are the coordinates, the area is reduced to a straight line, i.e., the feasible area is 0. The feasible area varies for different variable pairs: the larger the area, the more independent the variables. The problem discussed in Example 7.3 is examined in the following examples, each with a different pair of parameters selected as the independent variables.

Example 7.4 Independent Variables: Reflux Ratio and Overhead Rate

At a given product rate, the reflux ratio may be varied quite independently by changing the vaporization rate in the reboiler and the condensation rate in the condenser, i.e., by changing the condenser and reboiler duties. This represents internal circulation within the column and is thus independent of the product rate. Nevertheless, the two quantities are not entirely independent; if one variable is fixed, the other may vary only within a certain feasible range. Outside of this range, the vapor or liquid phase may dry up on certain column trays, or one of the products may vanish.

Figure 7.3 shows the feasible region when the independent variables are the overhead rate and the reflux ratio. The region is fairly large, underlining the high degree of independence between the two variables. The feasible range depends in part on the feed thermal conditions. At low overhead rates, a minimum boilup rate (and, hence, reflux ratio) is required to keep the vapor from drying up in the stripping section. At intermediate overhead rates, both the vapor in the stripping section and the liquid in the rectifying section start drying up below a certain reflux ratio. At higher overhead rates, the liquid in the rectifying section starts drying up below a certain reflux ratio.

Example 7.5 Independent Variables: NC4 Composition in the Overhead and Reflux Ratio

The composition of NC4 in the feed is 26%. It may be desirable either to maximize or minimize its concentration in the overhead. For a given reflux ratio, this may be achieved by varying the overhead rate. The mole fraction of NC4 in the overhead will always be less than the fraction of NC4 in the portion of the feed that includes NC4 and the lighter components or

$$Y_D(NC_4) < 26/(2 + 4 + 17 + 9 + 26)$$

$$< 0.448$$

The range between the minimum and maximum values of Y_D of NC4 increases with increasing reflux ratio as each approaches an asymptote at total reflux. At lower

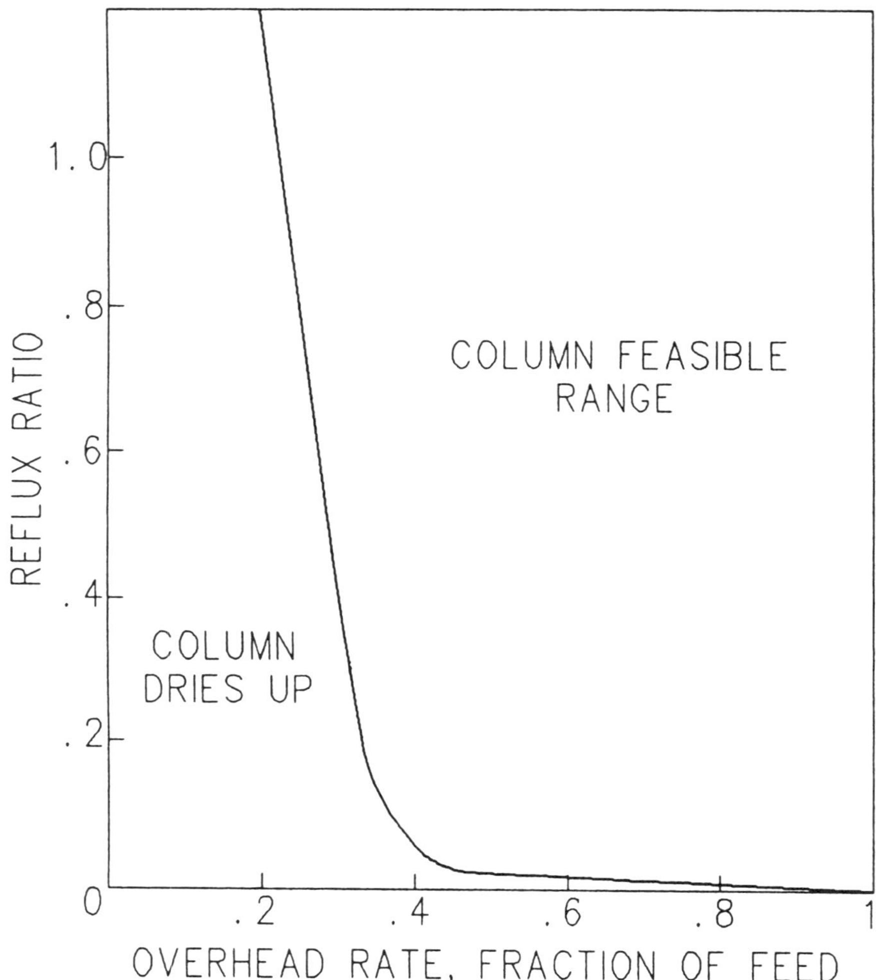

Figure 7.3   Reflux ratio — product rate feasible region.

reflux ratios, the degree of fractionation is small, and thus the range between the minimum and maximum feasible values of Y_D of NC4 is narrow. The column feasible region is plotted on a Y_D vs. reflux ratio diagram in Figure 7.4. Within this region the two variables are independent of each other.

Example 7.6 Independent Variables: NC4 Composition in the Overhead and Overhead Rate

Fixing the product rates generally determines the approximate range of compositions of the products. Thus, for a given overhead rate, the composition of NC4, for instance, can vary within a relatively small range. Conversely, if Y_D of NC4 is

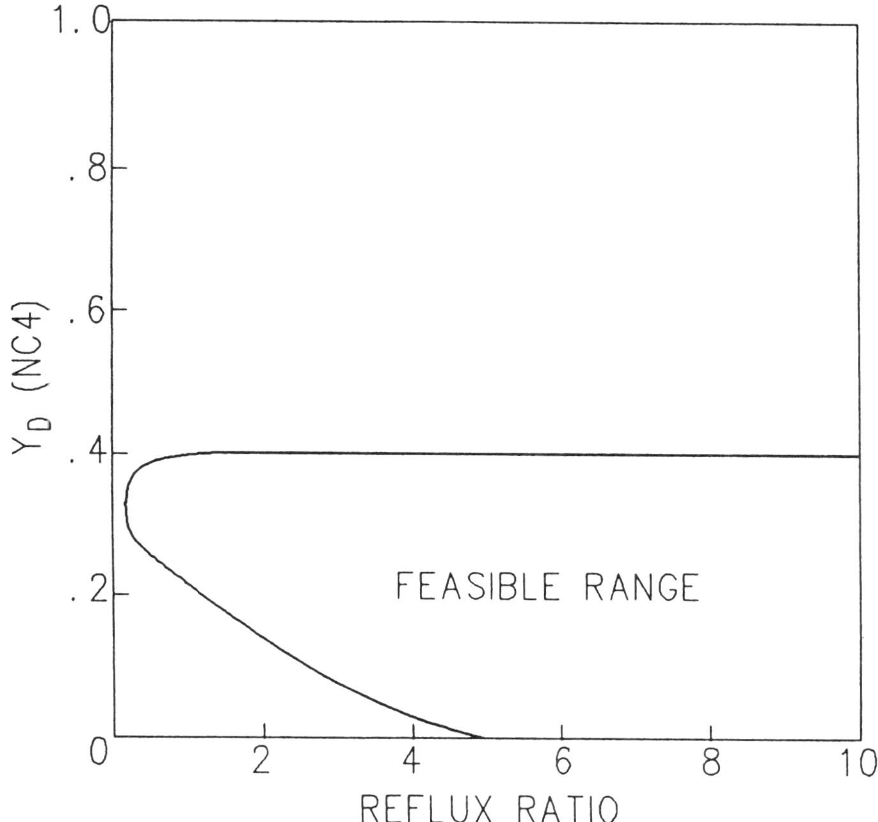

Figure 7.4 Reflux ratio — Y_D (NC4) feasible region.

specified, the overhead rate that can satisfy this specification may vary within a limited range. Therefore, the two variables, overhead rate and Y_D of NC4, are independent variables within a limited feasible region as shown in Figure 7.5.

Example 7.7 Independent Variables: NC4 Composition in the Overhead and IC5 Composition in the Bottoms

The interdependence of these two variables is best evaluated in terms of their dependence on the overhead rate and reflux ratio. Since the two components have boiling points that are not very different and both are of intermediate volatilities compared to the feed mixture, the two components are expected to be distributed between the overhead and bottoms over a considerable range of product rates. Within this range, both Y_D of NC4 and X_B of IC5 may be varied by adjusting both the overhead rate and the reflux ratio. The feasible region within which Y_D of NC4 and X_B of IC5 can vary independently is shown in Figure 7.6.

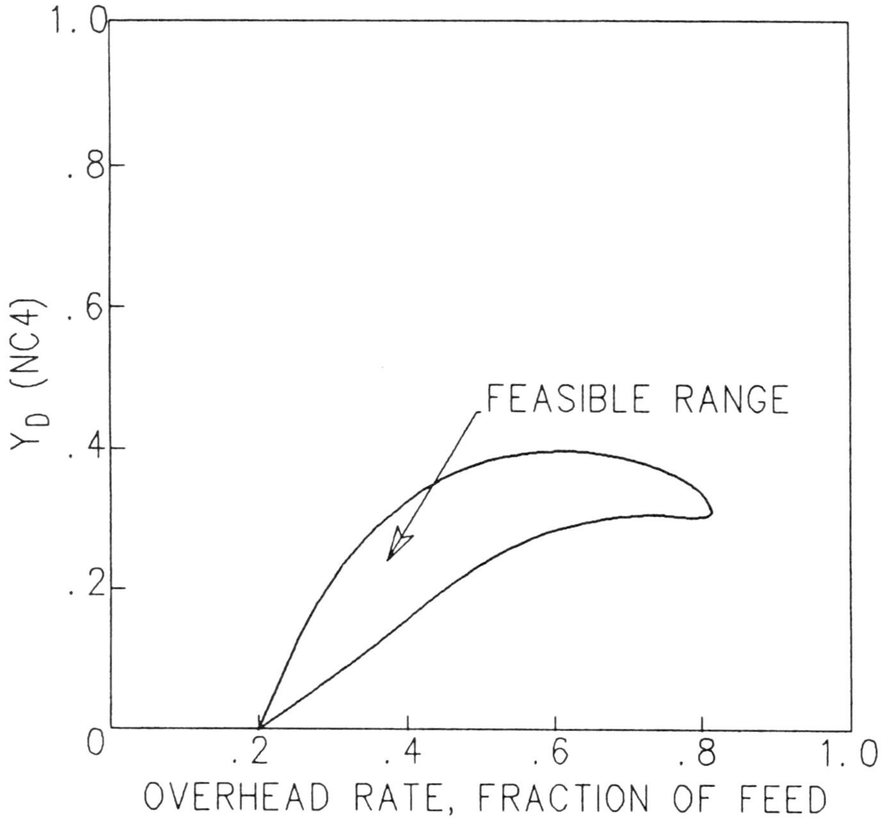

Figure 7.5   Product rate — Y_D (NC4) feasible region.

Example 7.8 Independent Variables: C1 Composition in the Overhead and NC6 Composition in the Bottoms

This is an extreme example where the designated compositions are independent only within a very narrow region. Most of the methane goes in the overhead and most of the hexane goes in the bottoms over the greater part of the range of overhead rates. The compositions of methane in the overhead and hexane in the bottoms are primarily functions of the overhead rate. The reflux ratio has a minor effect and only within certain overhead rate ranges. Since both Y_D of Cl and X_B of NC6 are primarily dependent on one variable, they are mutually independent within a small feasible range, as shown in Figure 7.7.

Product Temperature as the Independent Variable

Since a column product is at its dew point if it is vapor or at its bubble point if it is liquid, it follows that the product temperature is directly related to the product composition. The product composition was shown to be, in general, mostly dependent

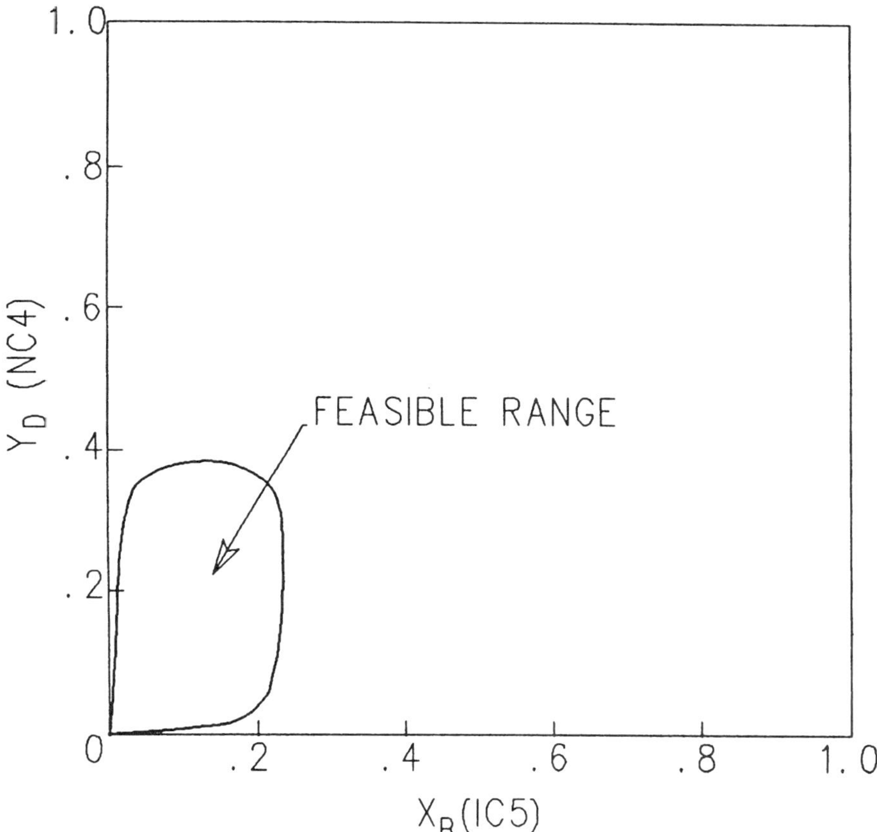

Figure 7.6 Y_D (NC4) — X_B (IC5) feasible region.

on flow rate and the composition fine-tuning is dependent on the reflux ratio. Hence, the product temperature is, for the most part, a function of its flow rate and, to a lesser degree, of the reflux ratio. Limited independence is therefore expected between a product flow rate and its temperature, while considerable independence is expected between a product temperature and the reflux ratio. The following examples, based on the problem defined in Example 7.3, are presented to illustrate the relationship between the product temperature and the column performance variables.

Example 7.9 Independent Variables: Overhead Temperature and Reflux Ratio

As the overhead product rate and reflux ratio are independent of each other over a wide region (Example 7.4), so are the overhead temperature and reflux ratio. At a given reflux ratio, the overhead temperature may be raised or lowered by increasing or decreasing the overhead rate. The practical limits are when the bottoms, overhead, or parts of the column begin drying up (vapor or liquid). The range between minimum

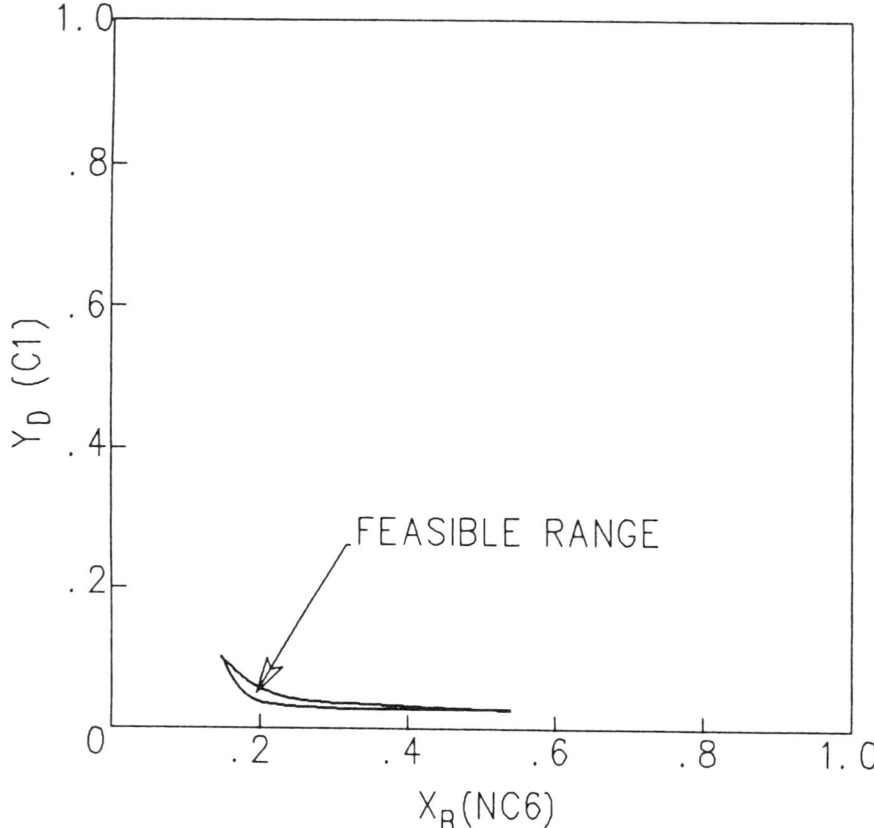

Figure 7.7 Y_D (C1) — X_B (NC6) feasible region.

and maximum possible temperatures is wider at higher reflux ratios. The feasible region where the overhead temperature and reflux ratio are mutually independent is shown in Figure 7.8.

Example 7.10 Independent Variables: Overhead Temperature and Overhead Rate

Once the overhead rate is specified, the overhead temperature may be controlled to a certain degree by varying the reflux ratio. The reflux ratio determines the fractionation and, hence, the product compositions and temperatures. However, the composition changes induced by reflux ratio variation are of a "fine-tuning" nature and do not greatly influence the temperature. Figure 7.9 shows the feasible region within which the overhead rate and temperature may be varied independently.

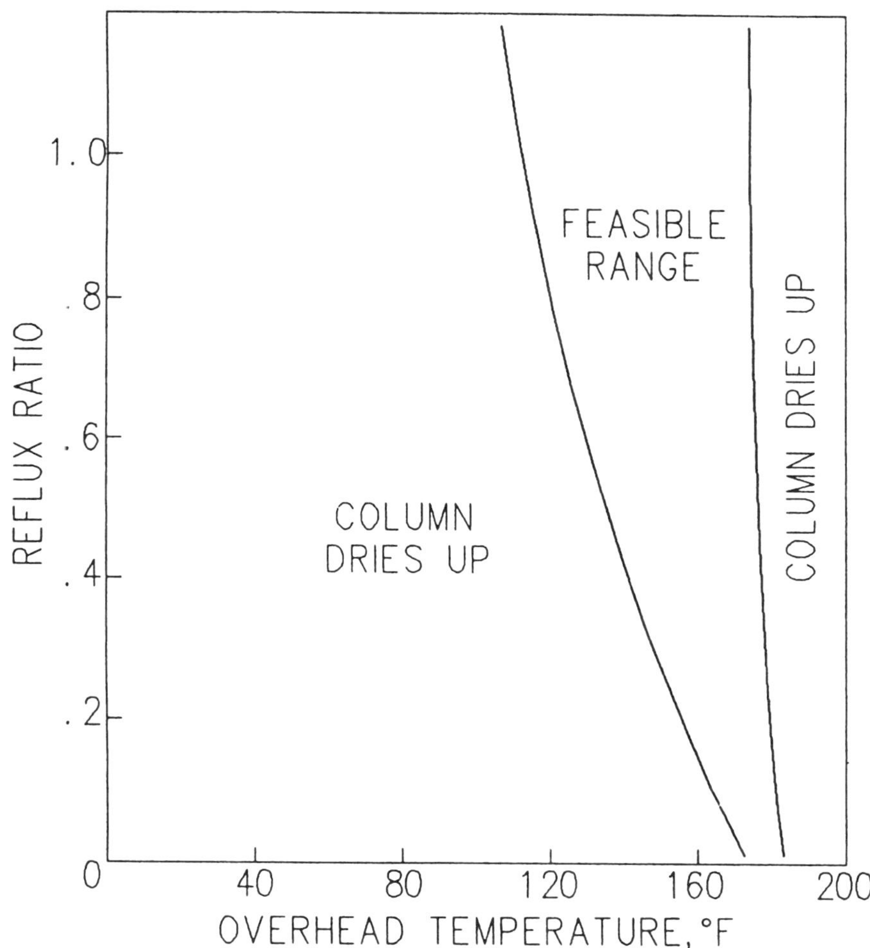

Figure 7.8   Reflux ratio–overhead temperature feasible region.

Example 7.11 Independent Variables: Overhead Temperature and Bottoms Temperature

The overhead and bottoms temperatures are determined by the product compositions, which, in turn, are controlled by the product rate and reflux ratio. A wide feasible region is expected (Figure 7.10) in which the overhead and bottoms temperatures may vary independently. The lines of constant overhead rates and constant reflux ratios may be used to determine those parameters corresponding to a given overhead and bottoms temperature pair. As the reflux ratio is increased at a constant product rate, the difference between the overhead and bottoms temperatures increases as a result of better fractionation. As the overhead rate is increased at a

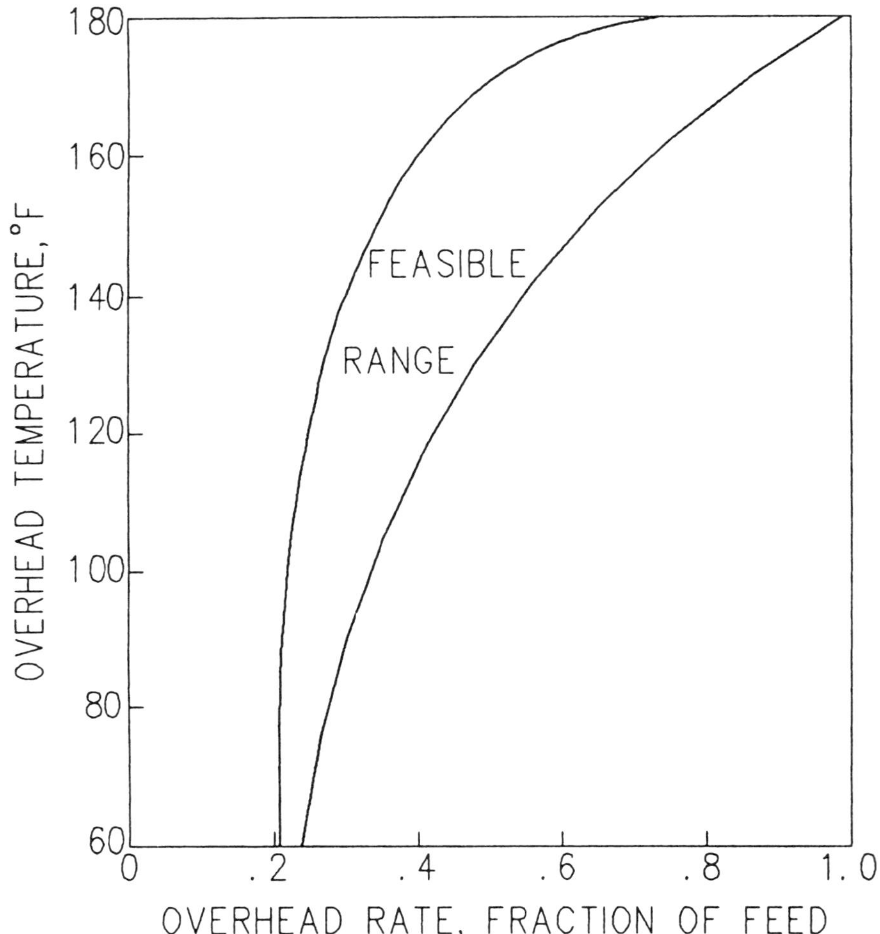

Figure 7.9   Overhead rate–overhead temperature feasible region.

constant reflux ratio, the temperatures of both products go up since the concentration of heavier components in both products goes up.

7.4 NUMBER OF TRAYS AND FEED LOCATION

The number of trays in each column section (rectifying or stripping) is determined by the required fractionation in that section at a given L/V ratio. This ratio, which is directly related to the reflux ratio, is usually limited by practical considerations such as economically acceptable reboiler and condenser duties. Also, the reflux ratio is limited by tray hydraulics considerations, discussed in Chapters 14 and 15. Column flooding can occur if the vapor or liquid velocities become excessive.

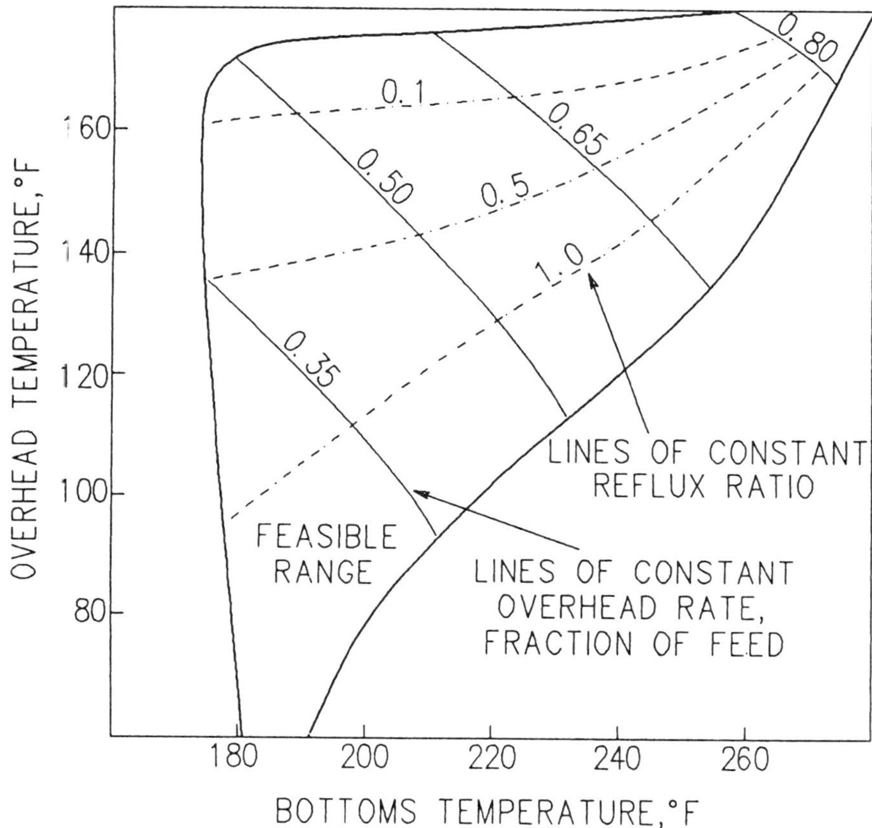

Figure 7.10 Overhead temperature–bottoms temperature feasible region.

The total column trays and feed location follow immediately once the number of trays in each section is known.

Minimum Reflux and Minimum Trays

For a given L/V ratio in a column section, better fractionation is generally achieved as the number of trays in that section is increased. Beyond a certain number of trays, however, the composition remains unchanged from tray to tray at certain points in the column. These "pinch points" may be observed on a binary Y–X diagram if the operating line touches the equilibrium curve. The same situation can occur in a multicomponent system. At a given reflux ratio, there exists a certain number of trays above which the separation cannot be improved. Therefore, a minimum reflux ratio is required to meet a specified separation. If the column is operated at that minimum, an infinite number of trays would be required to achieve that separation. Also, by analogy to binary distillation, a minimum number of trays is required for a given separation. At the minimum trays, total reflux is necessary in order to meet the specified separation.

Feed Location

For a given total number of trays, the optimum feed location is that which minimizes the reflux ratio (and therefore the reboiler and condenser duties) required to deliver a given separation. If the feed is not at its optimum location, the section with the reduced number of trays requires a higher reflux ratio to perform the same fractionation. The higher reflux ratio becomes too high for the other, overtrayed section.

In a multicomponent system, the optimum feed location depends on the light and heavy key components and their desired concentrations in the products. A feed location that is optimum for one set of specifications may be a poor selection for another. The number of rectifying trays must be sufficient to remove from the overhead as much of the components heavier than the light key as is needed to meet the required overhead composition. Similarly, the number of stripping trays must be sufficient to strip from the bottoms as much of the components lighter than the heavy key as is needed to meet the required bottoms composition.

When the feed is introduced at its optimum location, the rectifying and stripping sections are balanced with regard to the amount of fractionation taking place in each. Uniform fractionation is characterized by even variation from tray to tray of the separation parameter, defined as the logarithm of the ratio of the light to heavy key concentrations in the liquid on a given tray:

$$S = \ln(X_{lk}/X_{hk})$$

The selection of the feed tray and its effect on the column performance is examined in the following examples.

Example 7.12 Debutanizer Column

The mixture of Example 7.3 is to be separated between n-butane and isopentane in an 18-theoretical-stage column. The separation specifications are:

Mole fraction of NC4 (light key) in the bottoms: 0.04

Mole fraction of IC5 (heavy key) in the overhead: 0.018

These specifications require approximately equivalent degrees of stripping and rectification, suggesting that the feed should be placed around the middle of the column. If the feed is placed on the tenth tray from the top, the required reflux ratio calculated by simulation is 1.5.

If the feed is placed on the fourth tray from the top, there would be only four rectifying trays and, therefore, a higher reflux ratio, calculated at 2.0, would be

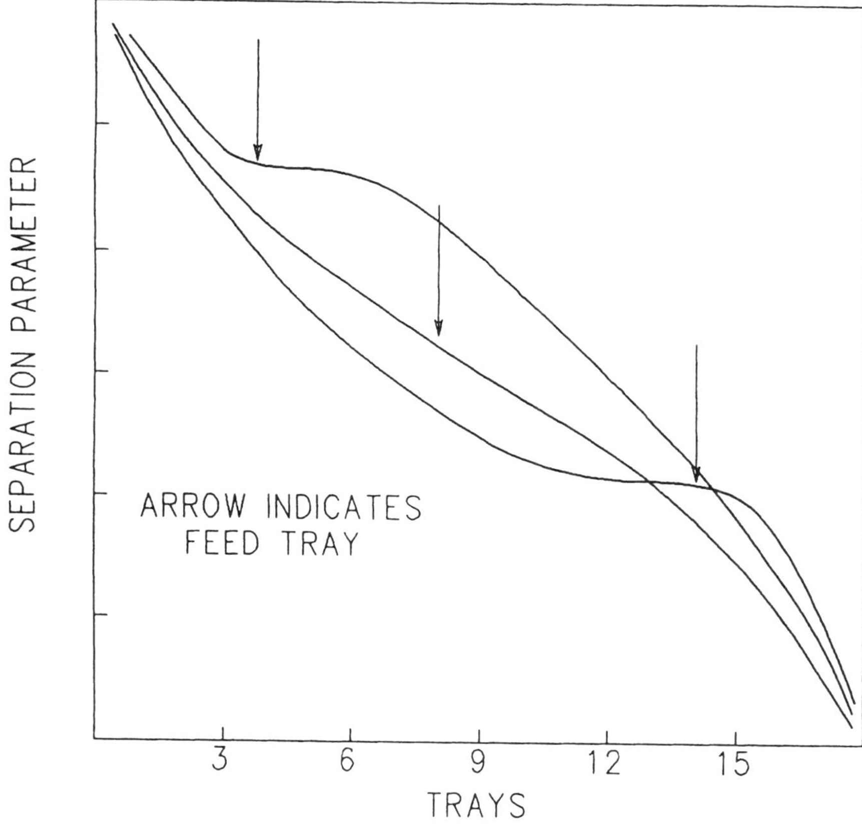

Figure 7.11 Schematic plots of the separation parameter.

required to obtain the same rectification. The stripping section becomes overtrayed, with several trays showing little change in the separation parameter. If the feed tray is too low, the rectifying section becomes overtrayed. Figure 7.11 is a schematic plot (not to scale) of the separation parameters for the different feed tray locations.

Example 7.13 Depropanizer Column

The same mixture as above is to be separated into propane and lighter components in the overhead and isobutane and heavier components in the bottoms, using an 18-stage column. The main purpose of the column is to strip as much propane and lighter components as possible from the bottoms, while tolerating some loss in the overhead of isobutane and heavier components. Hence, a large number of stripping trays is required, and a small number of rectifying trays should be adequate. The feed should therefore be placed on a tray in the upper part of the column, such as tray 5 or 6.

Effect of Feed Thermal Conditions

The feed thermal conditions affect the L/V ratio in both column sections. For a given reflux ratio, the higher the feed temperature, the lower the L/V ratio. Since better rectification is achieved at higher L/V ratios and better stripping is achieved at lower L/V ratios, it follows that a higher feed temperature requires more rectifying trays and fewer stripping trays to make the same separation. Thus, a hotter feed should be fed at a lower tray and a colder feed should be fed at a higher tray in the column.

Rectifiers and Reboiled Strippers

A rectifier is a column with no stripping section, i.e., the feed is sent to the reboiler or bottom tray. The feed could either be sent directly to the reboiler or, if sufficiently preheated, it could be sent to the bottom tray and no reboiler would be necessary. The purpose of such a column is to produce an overhead product that is enriched in lighter components from which the heavier components are refluxed in a condenser back to the column. Since no stripping section is used, a certain loss of light components in the bottoms should be expected.

A reboiled stripper is the opposite of a rectifier in that it has no rectifying section. The feed is sent either directly to the condenser or, if sufficiently cold, to the top tray, requiring no condenser. The liquid portion of the feed acts as reflux. The main purpose of this type of column is to produce a bottoms product that is enriched in heavier components and stripped of the lighter ones. With no rectifying trays, a reboiled stripper could result in some heavies being lost in the overhead.

Note that in reboiled strippers without a condenser or in rectifiers without a reboiler, there is only 1 degree of freedom or one independent variable. In a rectifier, for instance, varying the condenser duty directly affects both the reflux ratio and the overhead rate. These two parameters can no longer be varied independently and, therefore, only one performance specification may be made.

NOMENCLATURE

B	Bottoms stream designation or molar flow rate	Q	Heat duty per unit time or per mole of feed
B_r	B/F	R_r	Reflux ratio
D	Distillate or overhead stream designation or molar flow rate	S	Separation parameter
		T	Temperature
D_r	D/F	V	Tray vapor molar flow rate
F	Feed stream designation or molar flow rate	X	Component mole fraction in liquid
K	Vapor-liquid equilibrium coefficient	Y	Component mole fraction in vapor
L	Tray liquid molar flow rate	Z	Mole fraction in the feed
p_k	Column performance parameter		

SUBSCRIPTS

B	Bottoms designation	i	Component designation
C	Condenser designation	j	Tray designation
D	Distillate designation	k	Light key designation
h	Heavy key designation	R	Reboiler designation

PROBLEMS

7.1 How many variables must be specified in order to define the performance of an existing single-feed, two-product column with a partial condenser (vapor distillate only), a reboiler, and a fixed pressure profile? The feed rate, composition, and thermal conditions are also fixed.

7.2 An existing column with a partial condenser and reboiler is used to separate a given mixture of methane, ethane, propane, and butane into a distillate product containing most of the methane, and a bottoms product containing the remaining components. The column pressure profile is fixed at given values. It is required to calculate the condenser and reboiler duties. Which of the following specifications more reliably determine the required duties?

a. Distillate rate and reflux ratio.
b. Concentration of propane in the bottoms and methane in the distillate.
c. Bottoms rate and butane concentration in the bottoms.
d. Methane concentration in the bottoms and ethane concentration in the distillate.
e. Ethane recovery in the bottoms and reflux ratio.

7.3 A stream containing components A, B, and C is separated in a distillation column into distillate and bottoms products with the indicated component recoveries. (The components are listed in order of decreasing volatility.)

	Distillate Recoveries, %	Bottoms Recoveries, %
A	97	3
B	55	45
C	2	98

The bottoms rate will be set at a fixed rate. The reflux rate will be manipulated to control the recovery of either component B or C in the distillate. Which of these recoveries would you recommend as a controller input?

7.4 How many variables must be specified in order to define the performance of an existing single-feed, two-product column with a total condenser and a reboiler, and a fixed pressure profile? The feed rate, composition, and thermal conditions are fixed. How would you conceptually control the column operation?

7.5 A portion of a constant composition stream at a given temperature and pressure is used as the only feed to an existing distillation column. The column has a partial condenser and reboiler, and its only products are the bottoms and vapor distillate. What assumptions must be made to determine the degrees of freedom? Describe a conceptual control scheme to control the column pressure, product purities, and reflux rate.

7.6 A reboiled stripper column is to be designed to separate a feed stream to produce an overhead and a bottoms product. The feed is of fixed rate, composition, and thermal conditions and will be sent to the top of the column. The column will have a reboiler, but no condenser. Determine the number of degrees of freedom.

The column will be used for removing methane and nitrogen from a mixture of nitrogen and hydrocarbons ranging from methane to hexane. What specifications could be used to define the column performance once it is built?

7.7 In a vapor recompression process, heat for the reboiler of a distillation column is supplied by the vapor distillate. The temperature of the vapor distillate is raised by compressing it isentropically, and the heated vapor is used as the heat source for the reboiler. If this is the only heat source for the column, and the column feed is of fixed flow rate, composition, and thermal conditions, determine the number of degrees of freedom for an existing column. What control loops can be used to control the column performance?

CHAPTER **8**

Absorption and Stripping

The distinction was made in Chapters 3 and 7 between multistage separation processes where the composition gradient from stage to stage is primarily caused by temperature gradients, as in distillation, and those where the composition gradient is the result of external feeds with distinct compositions and which have a large difference between their average boiling points, as in absorption and stripping. In many situations both effects come into play.

This chapter is concerned with the performance of simple absorbers and strippers, defined as *countercurrent vapor–liquid multistage separation columns*, with a liquid feed at the top stage and a vapor feed at the bottom stage. A vapor product comes off the top stage and a liquid product off the bottom stage. No reboilers or condensers are used and no heat is assumed to cross the column walls (although, for the purpose of studying the thermal effects of absorption and stripping, heat sources or sinks may be included in a hypothetical model).

In absorption, net mass is transferred from the vapor to the liquid, resulting in the recovery of certain components from the vapor feed by the liquid. In stripping, net mass is transferred from the liquid to the vapor, resulting in the recovery of certain components from the liquid feed by the vapor. The end effect is that the main feed (vapor in absorbers, liquid in strippers) is separated, part of it going with the vapor product and the other part with the liquid product. As in distillation, the heaviest component going mostly with the vapor is the light key, and the lightest component going mostly with the liquid is the heavy key. In absorbers or strippers, however, it is mainly the heavy key or the light key, respectively, that is of interest. Thus, in these columns, reference is usually made to one key component only.

The factors that affect absorption or stripping column performance include the feed compositions and thermal conditions, the ratio of liquid to vapor feed flow rates, the column temperature and pressure, and the number of stages.

8.1 THERMAL EFFECTS

The basics of absorption and stripping were discussed in Section 3.3 in terms of ternary single- or two-stage systems. One characteristic feature of these processes is the large difference between the average boiling points of the liquid and vapor in the column. A key component, or group of components that include the key component, having intermediate boiling points are transferred either from the vapor to the liquid (absorption) or vice versa (stripping). On each stage the intermediate components distribute themselves between the vapor and liquid until their fugacities in both phases are equal or until $Y_i = K_i X_i$, where K_i is the vapor–liquid equilibrium coefficient and Y_i and X_i are the mole fractions in the vapor and liquid of component i. In ideal solutions this relationship simplifies to Raoult's law (Chapter 1), $PY_i = p_i^0 X_i$, where P is the stage pressure and p_i^0 is the vapor pressure of component i at the stage temperature.

The mass transfer that takes place to achieve phase equilibrium involves primarily the intermediate components. Because of the large difference between the boiling points of the main components in the liquid and vapor, these components experience little mass transfer between the phases. Referring to Raoult's law, the vapor pressure of the light components is too high to allow them to move in any appreciable amounts to the liquid. (If p_i^0 is large, X_i must be small to maintain their product equal to the partial pressure of i, PY_i.) Conversely, the vapor pressure of the heavy components is too low to allow it to be transferred significantly to the vapor. Hence, in absorption or stripping, net mass is transferred from one phase to the other. This is in contrast to distillation, where mass transfer between the phases takes place in both directions.

In absorption, the transfer of molecules from the vapor to the liquid is a condensation process that is accompanied by the release of an amount of heat equivalent to the latent heat of condensation of the components being absorbed. If the process is adiabatic, where no heat crosses the system boundaries, the heat released by absorption is converted to sensible heat, resulting in a temperature rise. This thermal effect is reversed in stripping since the stripped components are transferred from the liquid state to the vapor state. The latent heat of vaporization is responsible for a temperature drop in adiabatic stripping processes.

The temperature variation in an absorption or stripping column is a side effect and is not a required contributor to separation, as in distillation. In fact, in situations where the temperature variation is excessive, the column may have to be cooled or heated in order to counter the absorption/stripping thermal effects. The column may be operated isothermally, with all stages maintained at the same temperature, by applying the right amounts of heater/cooler duties.

The thermal effects of absorption and stripping are demonstrated in the following example using a single stage, i.e., an adiabatic flash.

Example 8.1 Single-Stage Absorption and Stripping of Propane

Propane (the key component) is recovered in one case from a nitrogen–propane gas mixture by absorption with a liquid consisting mostly of decane. In another case propane is recovered from a decane–propane solution by stripping it with a gas that is mostly nitrogen. All the feed streams are maintained at a temperature of 75°F and a pressure of 250 psia. The absorption or stripping takes place in an insulated vessel where the vapor and liquid feeds mix, then separate into vapor and liquid products. This single-stage column, or adiabatic flash, is held at a pressure of 75 psia. The feed and product compositions and thermal conditions are given in Table 8.1.

Table 8.1   Results of Example 8.1

Component	Gas Feed Mole/Hr	Liquid Feed Mole/Hr	Gas Product Mole/Hr	Liq Prod Mole/Hr
A. Adiabatic Absorption				
N2	100.0	0.0	97.6	2.4
C3	14.9	0.1	5.3	9.7
NC10	0.0	100.0	0.0	100.0
Totals	114.9	100.1	102.9	112.1
Temperature, °F	75.0	75.0	82.8	82.8
Pressure, psia	250.0	250.0	250.0	250.0
B. Adiabatic Stripping				
N2	100.0	0.0	97.6	2.4
C3	0.1	14.9	4.8	10.2
NC10	0.0	100.0	0.0	100.0
Totals	100.1	114.9	102.4	112.6
Temperature, °F	75.0	75.0	70.9	70.9
Pressure, psia	250.0	250.0	250.0	250.0
C. Isothermal Absorption				
N2	100.0	0.0	97.6	2.4
C3	14.9	0.1	5.0	10.0
NC10	0.0	100.0	0.0	100.0
Totals	114.9	100.1	102.6	112.4
Temperature, °F	75.0	75.0	75.0	75.0
Pressure, psia	250.0	250.0	250.0	250.0
Duty = −67,000 Btu/Hr				
D. Isothermal Stripping				
N2	100.0	0.0	97.6	2.4
C3	0.1	14.9	5.0	10.0
NC10	0.0	100.0	0.0	100.0
Totals	100.1	114.9	102.6	112.4
Temperature, °F	75.0	75.0	75.0	75.0
Pressure, psia	250.0	250.0	250.0	250.0
Duty = 35,000 Btu/Hr				

The process involves no pressure change; hence, any temperature change is solely the result of mixing and separation of liquid and vapor in an adiabatic environment. Table 8.1A shows the liquid product to be richer in propane than the liquid feed. Propane absorbed from the gas was transferred from the vapor phase to the liquid phase. A small amount of nitrogen was also absorbed, and a small amount of decane was vaporized in the process. The vaporized decane is negligible compared to the propane that was condensed. Overall, the liquid product flow rate is larger than the liquid feed rate. Net condensation took place, which resulted in a temperature rise of about 8°F.

Table 8.1B summarizes the results of the stripping process. A certain amount of propane was transferred from the liquid to the vapor and a smaller amount of nitrogen went in the opposite direction. There was net vaporization since the vapor product rate is larger than the vapor feed rate. A temperature drop of about 4°F resulted.

The processes are carried out again under isothermal conditions. A heat source or sink is assumed which maintains the vessel temperature at 75°F. Tables 8.1C and 8.1D indicate that better absorption and stripping are achieved isothermally. The cost, however, is a cooling duty of 67,000 BTU/hr for absorption and a heating duty of 35,000 BTU/hr for stripping.

8.2 LIQUID-TO-VAPOR RATIOS

In distillation columns a higher reflux ratio means higher liquid-to-vapor traffic, or L/V ratio, in the rectifying section and a lower L/V ratio (higher V/L ratio) in the stripping section. For a given number of stages the separation is enhanced by raising the reflux ratio, with maximum separation obtained at total reflux. In absorbers the separation of the intermediate components from the gas feed is also improved with higher L/V ratios. Similarly, in strippers the separation of intermediate components from the liquid is improved with higher V/L ratios. Normally, in absorbers, the gas feed is at a constant rate, and the L/V ratio is varied by varying the liquid feed or solvent rate. In strippers the liquid feed is usually fixed, and the V/L ratio is varied by varying the stripping gas rate.

The limiting condition of total reflux in distillation is approached when most of the overhead is refluxed back to the top of the column and most of the bottoms is reboiled and sent back to the bottom of the column. In absorbers and strippers the concept of total reflux does not apply since the liquid reflux (solvent) or stripping gas are external feeds. The rates of these feeds may be increased indefinitely, whereupon the product compositions reach asymptotic values. The required feed rate for a given number of stages is determined to satisfy certain product composition specifications. Obviously, the column size sets practical limitations on the allowable flow rate ranges. Moreover, as the liquid solvent or stripping gas rates are increased, larger quantities of the gas or liquid feeds get carried along with the intermediate components to be recovered. This translates into poor separation with the existing number of stages, and a larger number may be required.

Since absorption and stripping processes involve mostly one-way mass transfer from one phase to the other, the L/V ratio in absorbers and strippers varies considerably from stage to stage. In absorbers the L/V ratio usually varies most appreciably in the bottom stages where most of the absorption takes place. Any thermal effects causing a temperature rise are also most noticeable in these stages. In the upper stages the final purification takes place, which involves somewhat less mass transfer and temperature change. In strippers most of the L/V and temperature variations occur in the upper stages where most of the stripping takes place.

The following example examines the variation of the L/V ratio, temperature, and key component composition across the column and the effect of the relative feed rates on the performance of an absorber.

Example 8.2 Absorption of Propane by NC12

Dodecane (NC12) is the solvent or lean oil used to recover propane and heavier components from a rich gas mixture by absorption. The initial conditions of both feed streams are fixed at 100°F and 400 psia. Under these conditions the lean oil is a subcooled liquid, and the rich gas is a superheated vapor. The column has ten theoretical stages and is maintained at 400 psia.

The lean oil enters the column at the top, and the rich gas at the bottom. The product streams are the rich oil leaving the bottom of the column and the lean gas leaving the top. The flow rates, compositions, and thermal conditions of all four streams are given in Table 8.2. Most of the nitrogen, methane, and ethane, which are the components lighter than the key component, leave the column with the vapor product. Most of the propane (the key component) and heavier leave with the liquid product. The separation of the gas feed thus takes place between the ethane and propane. By far, most of the mass transfer is from the vapor to the liquid, as only 0.1 mole of lean oil (dodecane) is vaporized and becomes part of the vapor product.

Table 8.2 Stream Data for Example 8.2

Component	Rich Gas Mole/Hr	Lean Oil Mole/Hr	Lean Gas Mole/Hr	Rich Oil Mole/Hr
N2	20.0	0.0	19.4	0.6
C1	800.0	0.0	722.8	77.2
C2	80.0	0.0	48.5	31.5
C3	70.0	0.0	3.3	66.7
NC4	20.0	0.0	0.0	20.0
NC5	10.0	0.0	0.0	10.0
NC12	0.0	500.0	0.1	499.9
Totals	1000.0	500.0	794.1	705.9
Temperature, °F	100.0	100.0	106.1	117.8
Pressure, psia	400.0	400.0	400.0	400.0

Table 8.3 shows the column profiles of temperature, L/V ratio, and K-value and vapor concentration of propane. The profiles are also plotted in Figure 8.1. The temperature rise due to absorption is highest at the bottom stage where the liquid leaves at about 18°F above the inlet temperature of both feeds. The column temperature rise tapers off at the upper stages where the absorption rates are lower. The L/V ratio goes up progressively from the top of the column down as more material is transferred from the vapor to the liquid.

Going up the column, the vapor is at equilibrium with leaner oil (dodecane with lower concentrations of propane and other dissolved gases). Hence, the propane concentration in the vapor is lower in the upper column stages. The K-values of propane are higher in the lower column stages because the temperature is higher. If

Table 8.3　Column Profiles for Example 8.2

Stage	Temperature, °F	L/V	Y(C3)	K(C3)
1 (TOP)	106.1		0.0042	0.574
2	107.4	0.671	0.0086	0.584
3	108.5	0.675	0.0137	0.592
4	109.6	0.678	0.0193	0.599
5	110.8	0.680	0.0255	0.607
6	112.0	0.682	0.0323	0.615
7	113.4	0.685	0.0396	0.624
8	115.1	0.687	0.0471	0.636
9	117.2	0.690	0.0547	0.649
10 (BOTTOM)	117.8	0.694	0.0620	0.656

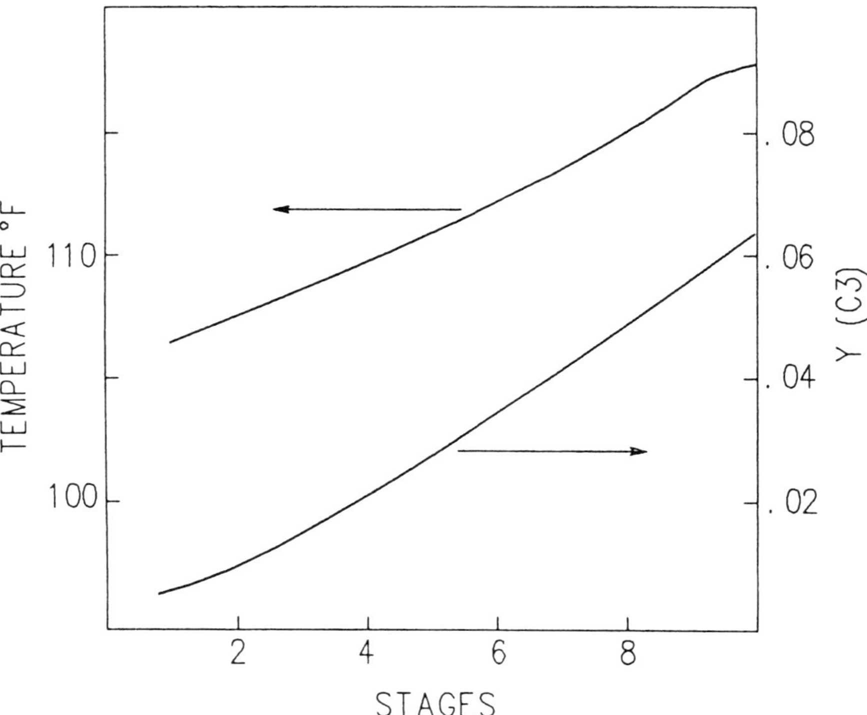

Figure 8.1　Absorber column profiles.

external cooling were used to keep the temperature from rising and thereby keep the K-values down, better absorption would be expected.

In order to investigate the effect of the L/V ratio on the separation, the lean oil rate is varied from 25 to 5,000 mole/hr at a fixed rich gas feed rate of 1,000 mole/hr. Table 8.4 lists the recoveries in the rich oil and concentrations in the lean gas of several components. The recovery is the percentage of a component in the rich gas

Table 8.4   Effect of Lean Oil Rate on Column Performance

Feed Gas Rate = 1,000 Mole/Hr (Example 8.2)

Lean Oil Rate Mole/Hr	Percent Recovery in Rich Oil				
	C1	C2	C3	NC4	NC5
25	0.73	3.16	9.19	26.50	69.50
100	2.24	9.19	25.11	63.60	99.40
300	5.99	24.20	65.76	99.90	100.00
600	11.47	47.37	98.88	100.00	100.00
750	14.19	59.40	99.86	100.00	100.00
1,000	18.71	78.14	99.99	100.00	100.00
5,000	94.83	100.00	100.00	100.00	100.00

	Mole Fraction in Overhead Vapor				
25	0.816	0.080	0.065	0.015	0.003
100	0.837	0.078	0.056	0.008	0.0
300	0.879	0.071	0.028	0.0	0.0
600	0.919	0.055	0.001	0.0	0.0
750	0.930	0.044	0.0	0.0	0.0
1,000	0.947	0.025	0.0	0.0	0.0
5,000	0.738	0.0	0.0	0.0	0.0

feed that is recovered in the rich oil. As implied by the definition of key components in absorbers, the key component is that which has the lowest recovery above 50%. At low lean oil rates, only the heavier components' recoveries are over 50%. As the lean oil rate is increased, the recovery of all components goes up and the key component changes. Table 8.4 indicates that, at a lean oil rate of 25 mole/hr, the key component is pentane. At 100 mole/hr, the key component is butane. At 300 and 600 mole/hr, it is propane, and so on. The required lean oil rate is thus determined in part by where the separation of the rich gas should take place. A closer look at the effect of the liquid solvent feed rate (or stripping gas rate) on the column performance is deferred to Section 8.4.

8.3 NUMBER OF STAGES

The component separation in absorbers or strippers depends both on the number of stages in the column and on the ratio of liquid-to-vapor feed rates. This ratio is bracketed on the basis of key component selection (Section 8.2). Most of the mass transfer of different components from one phase to the other usually takes place at different stages in the column. The transfer of the key component generally takes place over a larger number of stages than the other components. This implies that the number of stages affects mostly the key component recovery.

Consider, for instance, an absorber with a small number of stages operating with the feeds at a fixed ratio and resulting in certain component recoveries. If more stages are added to the column, the recovery of most components will go up. Beyond a certain number of stages, the heavier components are recovered in their entirety, while the liquid becomes saturated with the lightest components. As the number of stages is further increased, the recovery of the lightest components is not affected, although more of the heavier components (still lighter than the key) are recovered. If more stages are added, these components will, in turn, reach saturation, at which point additional stages do not affect their recovery. Depending on the ratio of feeds and on the vapor–liquid equilibrium data, further increases in the number of stages will either saturate the liquid with the key component or will result in 100% recovery of this component. In either case, additional stages are not warranted.

With a fixed number of stages, the liquid-to-vapor feed ratio is adjusted for fine-tuning of one of the component's (usually the key) recovery or concentration in the bottoms or overhead. Note that any of these parameters determines the rest since they are all related by the column equations.

Example 8.3 Effect of the Number of Stages on the Absorption of Light Hydrocarbons

The process described in Example 8.2 is used to study the effect of the number of stages on the absorption characteristics. The rich gas feed rate is fixed at 1,000 mole/hr, and the lean oil rate (NC12) is selected so that the key component is propane. Table 8.4 reveals that this condition can be met with lean oil rates between 300 to 600 mole/hr.

Tables 8.5 and 8.6 present the recoveries of ethane, propane, and n-butane for a variable number of stages. Propane is the key component, ethane is the next lightest (light key), and n-butane is the next heaviest component. Table 8.5 gives the results for a lean oil rate of 300 mole/hr, and Table 8.6 for 600 mole/hr. The recoveries are plotted in Figures 8.2 and 8.3.

Table 8.5   Effect of Number of Stages on Absorber Performance

Lean Oil Rate = 300 Mole/Hr (Example 8.3)

Number of Stages	Recovery in Bottoms, Percent of Feed		
	Ethane	Propane	n-Butane
2	23.58	54.39	85.64
5	23.98	63.01	97.88
10	23.95	64.74	99.89
20	23.95	64.94	100.00
30	23.95	64.94	100.00
40	23.95	64.94	100.00

At each lean oil rate, the recoveries reach a constant value after a certain number of stages. At a lean oil rate of 300 mole/hr, propane reaches a constant recovery of 64.94% after 20 stages and, at a lean oil rate of 600 mole/hr, it reaches 100% recovery with 30 stages. The ethane recovery reaches a constant value after five stages at a lean oil rate of 300 mole/hr and after eight stages at a lean oil rate of 600 mole/hr.

Table 8.6 Effect of Number of Stages on Absorber Performance

Lean Oil Rate = 600 Mole/Hr (Example 8.3)

Number of Stages	Recovery in Bottoms, Percent of Feed		
	Ethane	Propane	n-Butane
2	42.94	78.21	95.69
5	47.07	93.68	99.93
10	47.29	98.88	100.00
20	47.29	99.96	100.00
30	47.29	100.00	100.00
40	47.29	100.00	100.00

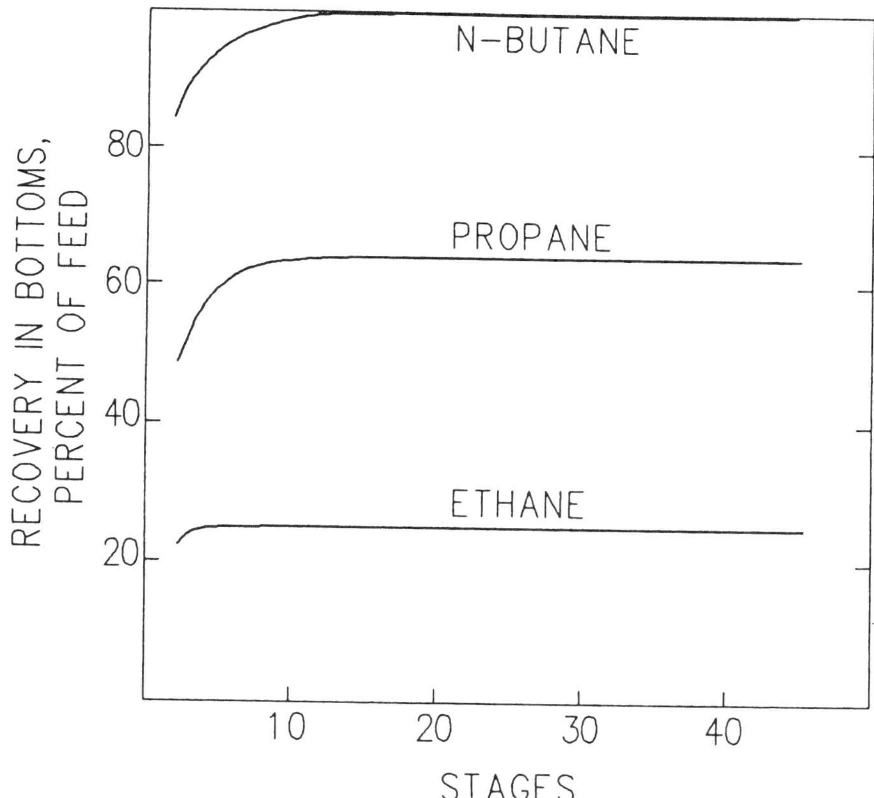

Figure 8.2 Effect of number of stages on absorber performance. Lean oil rate = 300 mole/hr (Example 8.3).

N-butane is totally absorbed after 20 stages at a lean oil rate of 300 mole/hr and after 10 stages at a lean oil rate of 600 mole/hr.

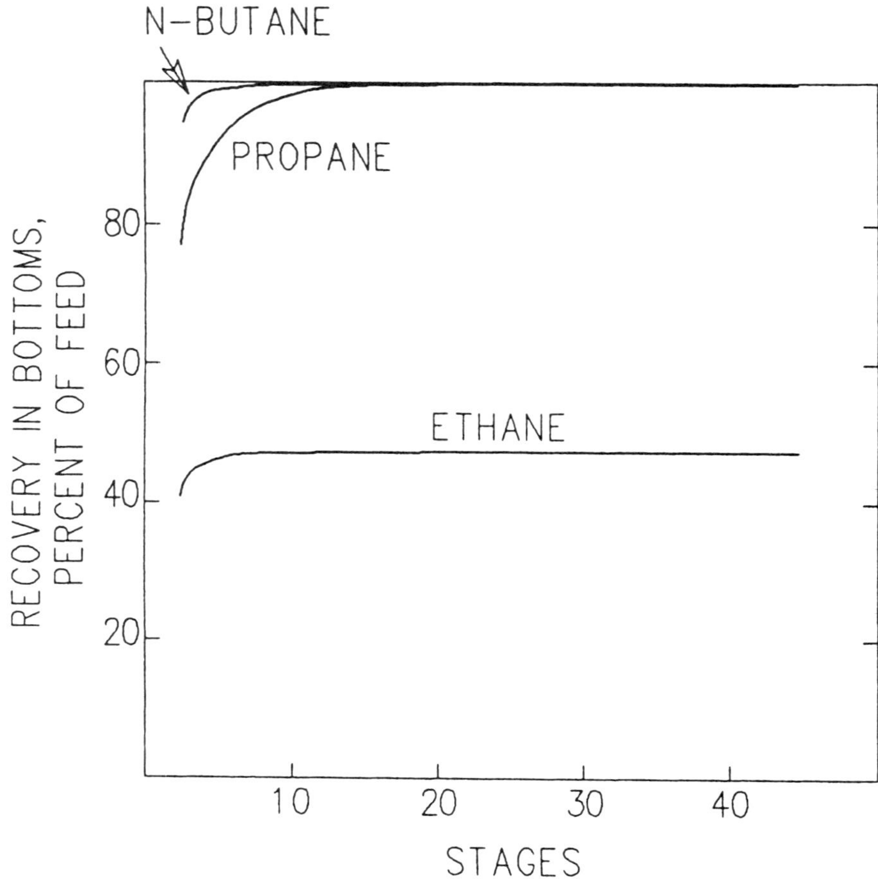

Figure 8.3   Effect of number of stages on absorber performance. Lean oil rate = 600 mole/hr (Example 8.3).

8.4 PERFORMANCE SPECIFICATIONS

An absorber or stripper is designed to meet certain process requirements, such as a specified recovery of desirable components or a specified maximum concentration of an impurity in a given product. In the design phase, the engineer has at his or her disposal a number of parameters that may be manipulated to meet the process specifications. The engineer can decide on such parameters as column size, number of stages, feed conditions, operating pressure, whether or not heaters or coolers will be used, and so on. Once a column is built, the ability to control its performance is greatly restricted. In fact, the degrees of freedom are reduced to 0 for an absorber or stripper that has a fixed number of stages, is operated adiabatically at a fixed pressure, and whose feeds are of fixed flow rates, compositions, and thermal conditions (Section

3.3.3). The product rates and compositions in such a column are determined by the heat and material balances and the vapor–liquid equilibrium relations. Under these circumstances the operator has no control over the column performance.

The number of independent variables required to define the operation of an absorber or stripper may also be determined by applying the description rule, stated in Section 5.2.1. The number of trays or the column height is set by construction and may, in the design phase, be used as a design variable. Since by definition the feeds are introduced at the top and bottom of the column, the feed locations are not variables. The feed compositions and thermal conditions are set outside the column region and are therefore beyond the operator's control. The operator can, however, control the valves on the two feeds and the two products. One of these four valves, usually the bottoms product valve, cannot be controlled independently since it must be set at steady state so as to maintain the required liquid level in the bottom of the column. The overhead valve is usually used to control the column pressure. The two feed valves may be controlled independently; one controls the main process stream rate and the other controls the solvent or stripping gas flow rate.

The parameter used as the control variable to achieve performance specifications is the absorbent rate in absorbers or the stripping gas rate in strippers. With one feed rate allowed to vary, this type of column acquires 1 degree of freedom; hence, one specification would be required to define its operation.

If the variable absorbent or stripping gas feed rate in an absorber or stripper is considered as the independent variable, the dependent variables would include product rates and compositions, component recoveries, product temperatures, and so on. When one of these parameters is specified, the variable feed rate must be calculated to satisfy the specification. The ability of the feed rate to satisfy a given parameter specification depends on the relationship between that parameter and the feed rate. Each parameter may be specified within a certain feasible range, determined by the dependence of the parameter on the feed rate.

As discussed in Section 8.2, the identity of the key component in an absorber- or stripper-type separation depends on the ratio of the liquid-to-gas feed rates. Within any given range of liquid-to-gas feed ratios, the component whose concentration in the product stream is most strongly dependent on the feed rates ratio is the key component. The other component concentrations are less dependent on this ratio, and the farther their boiling points are from the key component boiling point, the less the dependence. The following example illustrates the different ways of specifying an absorber and also looks at the feasible ranges of various performance specifications.

Example 8.4 Absorber Performance Specifications and Feasible Ranges

This example is an expansion on Example 8.2, the dodecane absorber. The recovery in the rich oil and the mole fractions in the overhead in Table 8.4 are plotted vs. the lean oil rate for a number of components in Figure 8.4.

If one variable is specified, the lean oil rate and the column performance are determined. For instance, if the mole fraction of methane in the overhead is specified

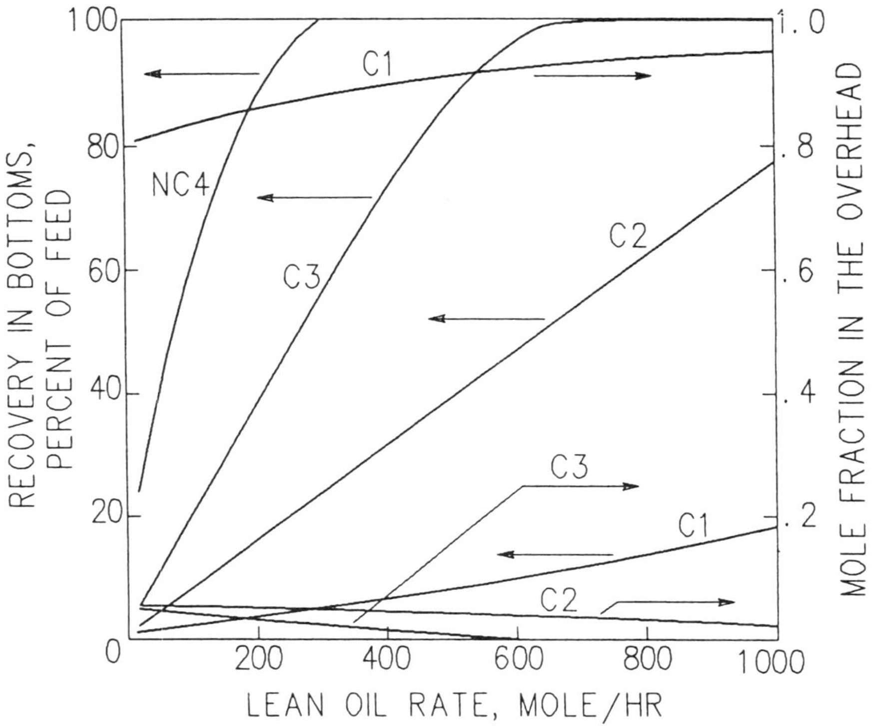

Figure 8.4   Effect of lean oil rate on column performance. Feed gas rate = 1,000 mole/hr (Example 8.4).

at 0.9, the lean oil rate, required to meet this specification is 440 moles/hr. At this lean oil rate, the mole fraction of ethane in the overhead is 0.06, the recovery of methane in the liquid is 8%, that of ethane 35%, propane 84%, n-butane 100%, and so on (Figure 8.4).

Each variable may be specified within a certain feasible range. The methane mole fraction in the overhead may be specified at any value between 0.80 and 0.96. Outside this range, the specification cannot be met regardless of the lean oil rate. Mathematically, no solution to the problem exists.

A glance at Figure 8.4 shows that the feasible ranges for some variables are larger than others. For instance, the recovery of propane in the bottoms may be specified at practically any point within the entire range from 0 to 100%. It is essential, however, for defining the column performance, not only for the specification to be feasible, but for a unique solution to exist. If, for instance, the propane recovery in the bottoms is specified at 100%, no unique solution exists since any lean oil rate over about 800 mole/hr satisfies the specification.

Within different lean oil rate regions, different specifications provide "more unique" definitions of the column performance. For instance, at a lean oil rate of around 700 mole/hr, the key component is ethane. Therefore, in this region the column performance is better defined by specifying the ethane recovery or concentration than, say, the methane concentration in the overhead. This is because, at about this lean oil rate, the dependence of the ethane recovery or concentration on the lean oil rate is stronger than that of the methane concentration in the overhead. At lower lean oil rates, propane becomes the key component; hence, its concentration or recovery would be a better variable for defining the column performance.

8.5 GRAPHICAL REPRESENTATION

The absorption or stripping process may be described graphically, if certain assumptions are made, in a manner similar to the McCabe–Thiele representation of binary distillation. As in binary distillation, it is emphasized here that the graphical representation is not a substitute for a rigorous solution but rather a tool that can be used in conjunction with it to obtain a visual appreciation of the effect of the various parameters on the performance of the column.

The graphical method assumes an idealized absorption/stripping model that is defined in terms of essentially three components or groups of components: a liquid, a gas, and a distributed component. The liquid is assumed not to vaporize (i.e., it has a very low K-value), and the gas is assumed not to dissolve in the liquid (i.e., it has a very high K-value). The distributed component is the key component to be absorbed by the liquid or stripped by the gas. It distributes itself between the two phases to satisfy vapor–liquid equilibrium criteria (Chapter 1).

Another simplification in the idealized model is the exclusion of enthalpy balances and the heat of absorption or stripping. The tray temperatures must be assumed or determined independently. An isothermal column may be assumed with the implication that heat sources or sinks are available for maintaining a constant temperature.

Since the liquid and gas components do not change phases, their flow rates remain constant in the column. It follows that, if the distributed component compositions are expressed as mole ratios to the liquid or gas instead of mole fractions, the column operating line becomes a straight line.

The operating line equation is derived from a material balance on the distributed component around trays 1 through j, counting from the top:

$$LX'_0 + VY'_{j+1} = LX'_j + VY'_1$$

or

$$Y'_{j+1} = (L/V)(X'_j - X'_0) + Y'_1$$

Since the liquid and vapor flows, L and V, are constant, this equation represents a straight operating line. The mole ratios of the distributed component to the liquid or gas, X' or Y', are related to the mole fractions as follows:

$$X' = \frac{X}{1-X}$$

$$Y' = \frac{Y}{1-Y}$$

The component subscript is dropped since reference is made exclusively to the distributed component. The subscripts in the operating line equation refer to the tray number, with X_0' referring to the external liquid feed composition and Y_1' to the overhead vapor composition.

Since the operating line equation must be plotted on a $Y'-X'$ diagram to produce a straight line, the equilibrium data must be converted to Y' vs. X' points to generate the equilibrium curve on the same diagram. The operating line relates compositions (expressed as mole ratios) of vapor and liquid streams counterflowing between trays j and $j+1$. In absorbers, Y_{j+1}' is greater than Y_j' and, therefore, the operating line is above the equilibrium curve. The opposite is true for strippers.

The following example illustrates the graphical solution of an absorber problem.

Example 8.5 Absorption of SO$_2$ from Nitrogen by Water

A countercurrent absorber is used to remove part of the sulfur dioxide contained in a nitrogen–SO$_2$ gas mixture by contact with water as the absorbent. The inlet gas is 3% mole SO$_2$ and 97% mole nitrogen. The concentration of SO$_2$ in the treated gas must be brought down to 2% mole. The column is maintained at 65°F and 15 psia.

It is required to calculate the number of stages that would achieve the desired separation at different water-to-gas ratios.

Solution. As in Example 6.5 (the deethanizer column), this example is intended to provide graphical visualization based on a rigorous computer column simulation. The equilibrium curve shown in Figure 8.5 may be constructed based on equilibrium vapor and liquid compositions from the simulation, converted from mole fractions to mole ratios.

The lower point of the operating line corresponds to the conditions at the top of the column where the inlet liquid is pure water (SO$_2$ mole ratio $X_0' = 0$), and the overhead gas has an SO$_2$ mole ratio $Y_1' = 0.02$. (At these low concentrations the mole fractions and mole ratios are almost identical.) The top of the operating line corresponds to the column bottom, where the concentration of the inlet gas is given as $Y_{N+1}' = 0.03$. Thus, the intersection of the operating line with a horizontal line drawn through $Y_{N+1}' = 0.03$ marks the upper end of the operating line. Its slope, according to the equation above, is equal to L/V, the liquid-to-gas ratio. The X' coordinate of the top of the operating line corresponds to the SO$_2$ concentration in the water leaving the column. The number of stages required at a given liquid-to-gas ratio is determined by stepping off equilibrium stages between the operating line and the equilibrium curve (Figure 8.5).

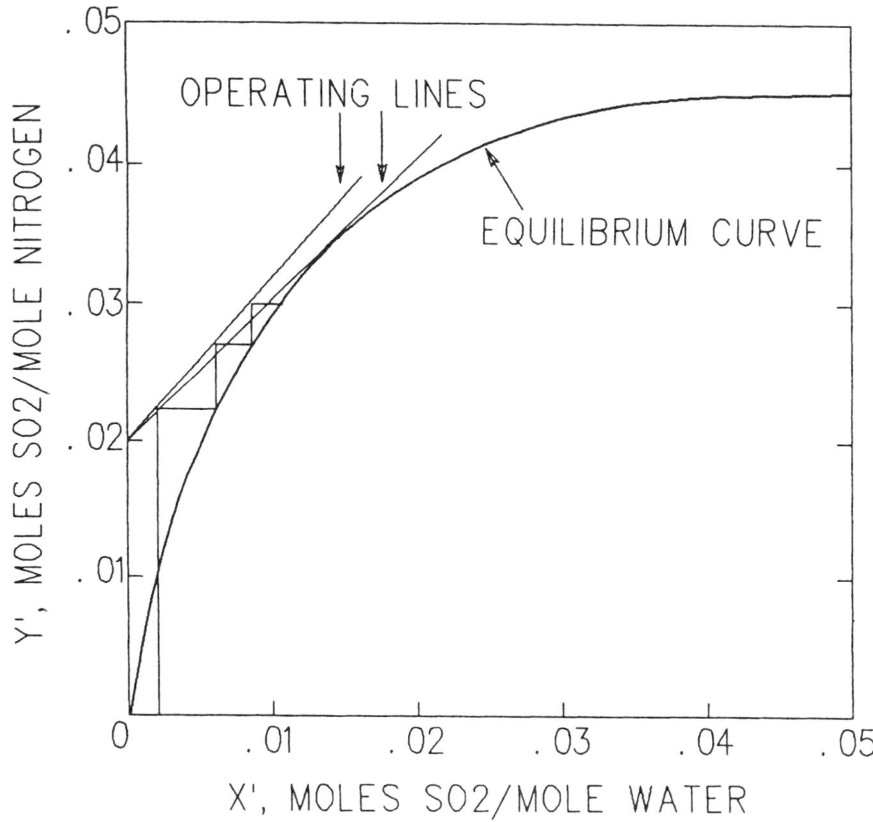

Figure 8.5 Absorption of SO_2 from nitrogen by water.

It can be seen from the diagram that the higher the liquid-to-gas ratio, the fewer the number of trays required for the separation. The concentration of SO_2 in the water effluent also drops as the water-to-gas ratio increases. As the L/V ratio is lowered, the operating line moves closer to the equilibrium curve, and the number of stages increases. The minimum L/V ratio required for the separation is determined by the operating line that touches the equilibrium curve, at which conditions the number of stages becomes infinite. If L/V is lowered below the minimum ratio, the operating line would cross the equilibrium curve, and the SO_2 would start moving from the liquid to the gas, as in a stripping process.

As in distillation, the column design should avoid pinch conditions; that is, the column should be designed to handle adequate water flow rates so the operating line would not get too close to the equilibrium curve. The decision on the number of trays vs. the L/V ratio is an economical one, where the capital investment in column construction is weighed against operating (as well as capital) costs involved in pumping the required amounts of water, etc.

NOMENCLATURE

K	Vapor–liquid equilibrium coefficient	X′	Mole ratio of the key component to the total liquid
L	Tray liquid molar flow rate	Y	Component mole fraction in the vapor
P	Stage or column pressure		
P^0	Component vapor pressure	Y′	Mole ratio of the key component to the total gas
V	Tray vapor molar flow rate		
X	Component mole fraction in the liquid		

SUBSCRIPTS

i   Component designation
j   Tray designation

PROBLEMS

8.1 How many variables must be specified in order to define the performance of an existing absorber with a liquid feed at the top and a vapor feed at the bottom? The column pressure and the flow rates, compositions, and thermal conditions of both feed streams are fixed.

8.2 A column is designed to remove lighter components from a liquid feed stream by stripping with steam. The feed stream is sent to the top tray, and is at constant rate, composition, and pressure, but its temperature may be controlled. The stripping steam, going to the bottom tray, is at constant temperature and pressure, but its rate may be controlled. How many degrees of freedom exist during the design phase? If the separation is a function of the ratio of overhead rate to bottoms rate, the column bottom temperature, and the column pressure, what control loops would you use to control the separation?

8.3 An absorber column is used to remove most of the propane and heavier components from a gas stream containing methane, ethane, propane, butane, and pentane by treating with a lean oil. The gas rate is constant and the lean oil rate is to be calculated to satisfy a performance specification. Which of the following specifications better defines the lean oil rate?

a. The butane recovery in the bottoms.
b. The propane concentration in the overhead.
c. The overhead rate.
d. The bottoms rate.
e. The column temperature.
f. The column pressure.

8.4 The acetone concentration in an effluent air stream is to be reduced from 1% mole to 0.05% mole by scrubbing with water in a countercurrent absorber. Assuming Henry's law holds at these concentrations, and that it may be expressed as $Y = 2.45X$, where X and Y are equilibrium acetone mole fractions in the liquid and gas, calculate the minimum water rate required for an air–acetone mixture rate of 100 lbmole/hr. How many equilibrium stages would be required if the water rate is twice the minimum? What is the concentration of acetone in the water leaving the absorber? Assume water vapor in the gas and air dissolved in the water are negligible.

8.5 A countercurrent stripper column with three equilibrium stages is to be used for removing 90% of a component dissolved in a liquid stream by stripping with an inert gas. The liquid feed rate is 100 lbmole/hr, and the mole fraction of the dissolved component is 0.01. This component is also present in the stripping gas at a mole fraction of 0.0001. Calculate the required stripping gas rate. Assume Henry's law, given in the form $Y = 2X$, holds for the dissolved component. The stripping gas may be assumed immiscible in the liquid and the liquid essentially nonvolatile. What is the composition of the stripping gas as it leaves the column?

CHAPTER 9

Complex Distillation and Multiple Column Processes

Complex distillation may be defined as a multistage vapor–liquid separation process that includes one or more of the following features: multiple feeds, side-draws, pumparounds, and side heaters or coolers.

Absorbers and strippers are examples of multifeed columns. In general, in multiple feed columns, the feeds are of different compositions and/or thermal conditions and are fed to different column trays. In certain processes, multiple products are taken from the column, each from a different tray. Each product has a distinct composition, corresponding to the draw tray fluid composition.

Side coolers or heaters are used for redistributing the heat load along the column in order to control liquid and vapor traffic or L/V ratios in the various sections of the column. Pumparounds may be used as the means of transferring heat between the side heaters or coolers and the column.

Complex columns often perform the functions of a number of columns combined into one. In analyzing the performance of complex columns, it is helpful to consider the column as made up of sections, where a section consists of a number of adjacent trays having no feeds, products, or heaters/coolers except at the top and bottom trays. The separation effected in a given column section depends on the number of stages and the L/V ratio in that section. The L/V ratio is a function of the feeds and product rates, heater and cooler duties, and pumparounds associated with the section. The fractionation may be of both a distillative or an absorptive or stripping nature (Chapters 7 and 8), although one effect or the other may be predominant in different situations. The various features of complex distillation are described in this chapter.

Separations that cannot be achieved in a single column are carried out in multiple columns, frequently interconnected in various complex configurations aimed at optimizing the process while meeting the separation specifications. Multiple column processes are also discussed in this chapter.

9.1 MULTIPLE FEEDS

When more than one stream is fed to a column, it is possible to distinguish between two types of feed: one as a stream to be separated and the other as a stream intended to bring about or enhance the separation of another stream. The gas feed to an absorber and the liquid feed to a stripper are of the former type, while the absorbent and the stripping gas are of the latter. Multiple feed columns may be grouped in the following categories: columns with no condenser or reboiler, columns with a reboiler only, columns with a condenser only, and columns with both a condenser and a reboiler. The first category includes simple absorbers and strippers, covered in Chapter 8. The other categories are discussed below. Different feed types are considered and the factors that determine the optimum feed tray are examined.

9.1.1   Columns with a Reboiler and No Condenser

Since these columns have no condenser, one of the feeds provides external reflux and must be fed at the top of the column. The external reflux is either a high-boiling liquid, such as an oil absorbent, or a stream that is sufficiently cold to maintain a liquid reflux down the column. Obviously, no rectification takes place on the top feed and a certain fraction of its components that are volatile enough at the top tray temperature will exit the column with the overhead. The heavier components in the lower feed are recovered by the external reflux in the rectifying trays between the feed tray and the top of the column. The lighter components are stripped by the reboiler vapors in the stripping trays below the lower feed.

Two-feed columns with a reboiler usually have 1 or 2 degrees of freedom, depending on whether only the reboiler duty or both the reboiler duty and one of the feed rates are variable. Other column parameters, namely the number of trays, the feed tray locations, the feed compositions and thermal conditions, and the column pressure, are assumed fixed as in an existing column. If one or two of the parameters are not fixed, the column operation must be defined by setting one or two performance specifications, such as product rates and compositions. The number of variables required to define the operation of this type of column may be verified by applying the description rule (Section 5.2.1). For an existing column with fixed-feed compositions and thermal conditions, the variables that can be controlled independently by external means include the two feed rates, one product rate, and the reboiler duty. One feed rate is usually determined based on design throughput, and the product rate is used to control the column pressure, leaving potentially two variables to control the separation.

Depending on the difference between the average boiling points of the feeds, the column and mainly its rectifying section may have stronger either distillative or absorption characteristics. The following are examples of these two cases.

Example 9.1 Demethanizer Column

The separation of light hydrocarbon gases into methane and lighter components as one product and ethane and heavier components as another is typically carried out in expander plants. In a simplified version of this process, the high-pressure feed gas is cooled, then flashed and separated in a high-pressure separator into vapor and liquid streams. The vapor is expanded in a turbo-expander, a polytropic process that drops its temperature and partially liquefies it. The liquid from the flash drum is throttled through a valve down to about the same pressure as the expander discharge. The adiabatic pressure drop across the valve lowers the stream temperature, although this temperature drop is smaller than that across the expander. The expander discharge is thus at a lower temperature than the valve effluent.

The liquid from the expander is fed to the top of the demethanizer column as external reflux, and the valve outlet stream is fed to an intermediate tray. In the present example, the column has 12 theoretical trays and a reboiler (a total of 13 theoretical stages). As shown in Figure 9.1, one stream goes to tray 1 and the other to tray 5, counting trays from the top of the column down. Streams 3 and 4 are the overhead and bottoms products, respectively. The column and the reboiler operate at a constant pressure of 425 psia. The compositions, flow rates, and thermal conditions of the streams are given in Table 9.1. The process requirement is that the bottoms product should have a maximum of 0.001 mole fraction methane.

Since this column has a reboiler but no condenser, and if the feeds are held at constant rates, 1 degree of freedom remains if the reboiler duty is variable. One specification is therefore required to define the column performance: the methane concentration in the bottoms in this example. The reboiler duty required to meet this specification is to be determined.

Design Considerations

Since no condenser is used, the colder feed, serving as external reflux, must be fed to the top tray. With no rectifying trays available above this feed, some of its heavy key component is "lost" in the overhead. In this example ethane is the heavy key (methane is the light key), and Table 9.1 shows that, of the 250 moles ethane in the upper feed, over 230 moles go overhead.

The distribution of components in the main feed (stream 2) between the overhead and bottoms is consistent with the desired heavy key/light key separation. Most of the ethane in this feed goes with the bottoms and most of the methane with the overhead. Overall, this column may be considered not highly efficient from the standpoint of recovering ethane in the bottoms product but is very efficient in purifying it from the methane (hence, its designation as a demethanizer).

In selecting the feed tray for the main feed, it is desirable to maximize the number of stripping trays in order to minimize the reboiler duty required to achieve the specified methane stripping. However, the feed tray should not be too high. If it is, an excessive amount of ethane is lost in the overhead because of an insufficient number of rectifying trays. Moreover, if the feed is too high, the ethane lost in the overhead reduces the reflux, thereby requiring additional boilup to meet the specified bottoms purity. Table 9.2 shows reboiler duty requirements and ethane recoveries in the bottoms for different feed tray locations. The methane mole fraction in the bottoms is fixed in all cases at 0.001. The results indicate that the optimum feed tray is the third tray from the top.

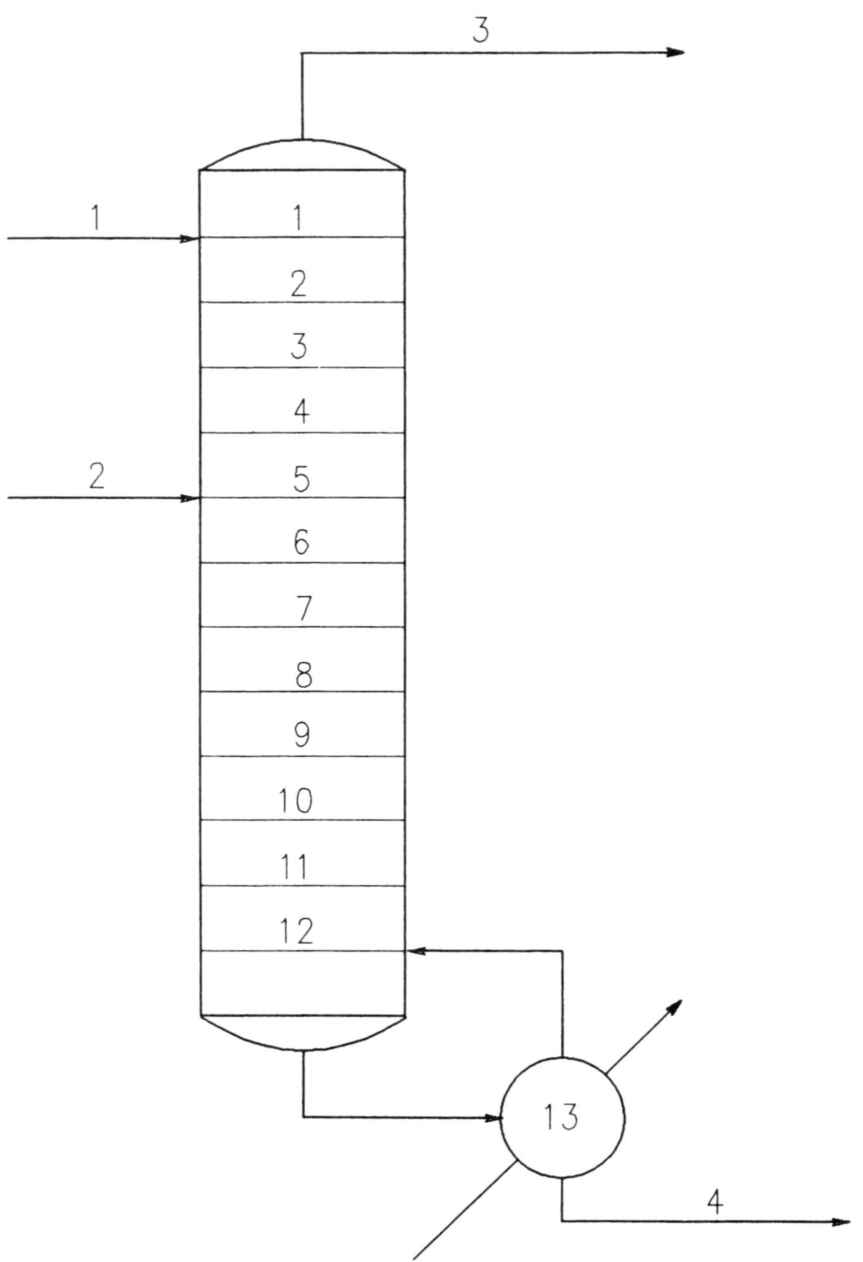

Figure 9.1   Demethanizer column (Example 9.1).

Table 9.1 Stream Data for Example 9.1

| | Mole Fraction | | | |
Component	Stream 1 Upper Feed	Stream 2 Lower Feed	Stream 3 Overhead	Stream 4 Bottoms
N_2	0.002	0.004	0.0049	0.0000
CO_2	0.050	0.030	0.0376	0.0285
Methane	0.580	0.640	0.8708	0.0010
Ethane	0.250	0.170	0.0824	0.4630
Propane	0.080	0.085	0.0039	0.2865
i-Butane	0.020	0.015	0.0003	0.0568
n-Butane	0.010	0.020	0.0001	0.0617
i-Pentane	0.005	0.010	0.0000	0.0309
n-Pentane	0.002	0.010	0.0000	0.0283
Heptane	0.001	0.016	0.0000	0.0433
Flow Rate, lbmole/hr	1,000	3,000	2,870	1,130
Temperature, °F	−100	−50	−81	105
Pressure, psia	425	425	425	425

Table 9.2 Effect of Feed Tray Location on Reboiler Duty and Ethane Recovery (Example 9.1)

Feet Tray for Stream 2	Reboiler Duty MMBTU/hr	Ethane Recovery in Bottoms lbmole/hr
1	7.08	513
2	6.52	528
3	6.44	534
4	6.49	533
5	6.64	523
7	7.36	453

Column Characteristics

Even though the column has no condenser and the reflux is from an external feed, the column, including its rectifying section, behaves more like a distillation column than an absorber. The reason is that the two feeds — the external reflux and the main feed — have similar compositions and average boiling points. Also, the vapor and liquid compositions on each tray are not much different from each other so that mass transfer takes place both ways between the phases. As a result, the L/V ratio variation from tray to tray is fairly small in each column section, and the temperature rises steadily from tray to tray going down the column. The column profiles are shown in Table 9.3.

Example 9.2 Reboiled Absorber

The purpose of this column is to recover propane from a light gas mixture by absorption with a lean oil. In addition, the stream containing the recovered propane

Table 9.3   Column Profiles for Example 9.1

Tray	Temperature °F	Liquid Flow lbmole/hr	Vapor Flow lbmole/hr	L/V
1*	−81	850	2,870	
2	−69	821	2,720	0.313
3	−63	791	2,691	0.305
4	−59	698	2,660	0.297
5*	−41	1,844	686	1.017
6	−19	1,932	714	2.584
7	7	2,076	801	2.411
8	29	2,233	946	2.195
9	44	2,357	1,102	2.026
10	55	2,442	1,227	1.921
11	64	2,484	1,312	1.862
12	76	2,433	1,354	1.835
13	105	1,130	1,303	1.868

* Feed Trays

(the rich oil stream) is to be stripped of some of the lighter gases using a reboiler. The column has thus two functions and its full designation would be a reboiled stripper absorber. A condenser could be used instead of an external reflux, making it a distillation column. The benefit gained by using absorption is that the absorber operates at a moderate temperature, whereas a condenser would have to be held at a considerably lower temperature and/or higher pressure in order to condense the propane. The drawback of using an absorber is the additional downstream processing required to separate the propane from the oil.

The column has two feeds, as in Example 9.1, but in this application the feeds are of very different average boiling points and they feed the column at about the same temperature. The oil feed has the higher boiling point and is fed at the top of the column as external reflux, while the gas mixture is fed at an intermediate tray. The top section of the column between the two feeds is the absorption section where most of the propane recovery takes place. The lower section, between the gas feed and the reboiler, is the reboiled stripping section where ethane and lighter components are stripped off as in conventional columns.

A schematic of the column is shown in Figure 9.2 and the stream data are given in Table 9.4. The column has nine theoretical trays and a reboiler and operates at 150 psia. The rich gas is fed on tray 7, allowing for seven absorption trays and leaving trays 8 and 9 and the reboiler for stripping. The oil feed is a relatively high-boiling stream, and by far most of it flows down the column as reflux with very little being lost in the overhead. The main feed is separated between ethane and propane, the light and heavy key components.

The column profiles are given in Table 9.5, showing the temperature, liquid and vapor flow, L/V ratio, and ethane and propane mole fractions in the liquid on each tray. The profiles indicate that the upper six trays act as typical absorption trays with a rise in temperature, peaking on tray 6, due to heat of absorption and also due to hot vapors from the reboiler. The L/V ratio varies steadily from tray to tray in the absorption section, underscoring the one-way mass transfer of propane from the

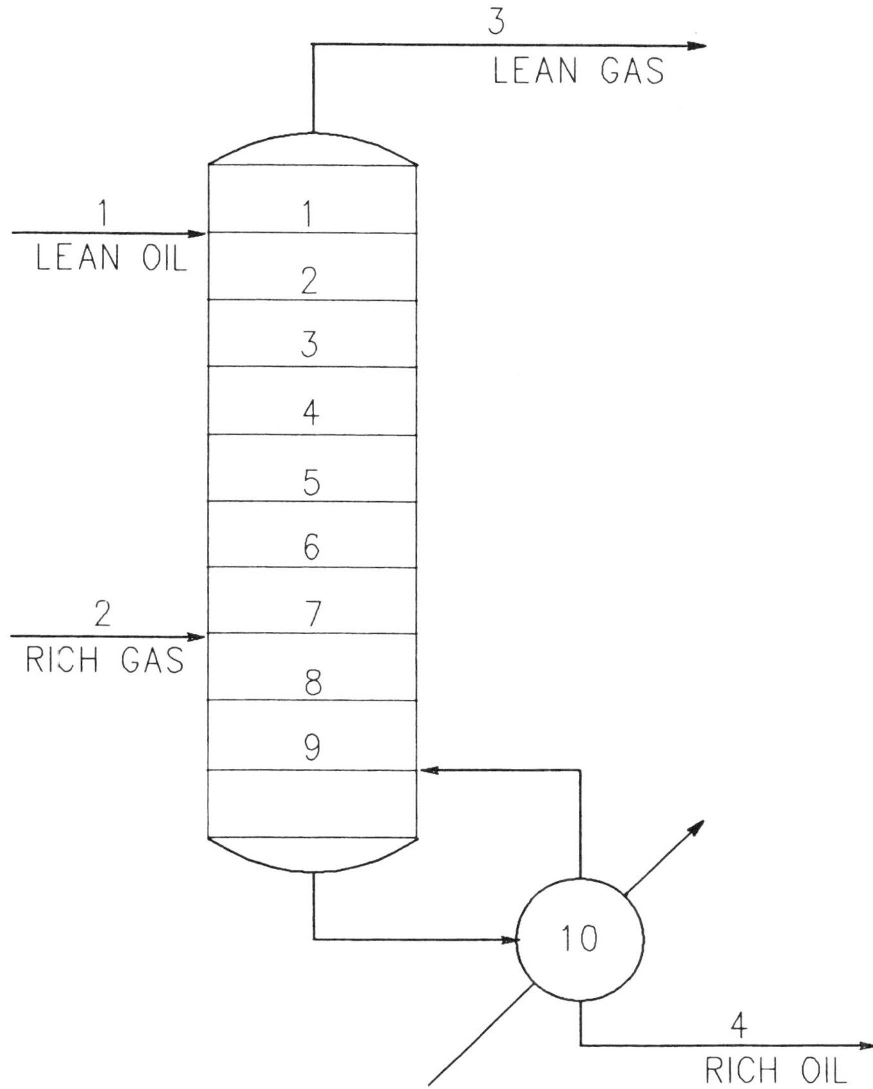

Figure 9.2 Reboiled absorber (Example 9.2).

vapor to the liquid. In the stripping section, the temperature changes quite steeply, while L/V starts leveling off at the bottom trays. This is because the stripping is accomplished with a reboiler rather than an external stripping medium. The column may thus be viewed as a hybrid of an absorber column and the stripping section of a distillation column. The ethane and propane concentration profiles in Table 9.5 demonstrate the absorption of propane in the upper column section and the stripping of ethane in the lower section.

Table 9.4   Stream Data for Example 9.2

	Mole Percent			
Component	Stream 1 Lean Oil	Stream 2 Rich Gas	Stream 3 Lean Gas	Stream 4 Rich Oil
N_2	0	2.50	6.67	0
CO_2	0	10.00	26.04	0.15
Methane	0	20.00	53.26	0.03
Ethane	0	2.50	5.35	0.30
Propane	0	65.00	8.68	37.23
Dodecane	100.00	0	0.004	62.29
Flow Rate, lbmole/hr	1,033	1,000	374	1,659
Temperature, °F	75	75	87	183
Pressure, psia	150	150	150	150

Table 9.5   Column Profiles for Example 9.2

Tray	Temperature °F	Liquid Flow lbmole/hr	Vapor Flow lbmole/hr	L/V
1*	87	1,234	374	
2	101	1,401	576	2.14
3	114	1,572	742	1.89
4	122	1,693	914	1.72
5	126	1,758	1,034	1.64
6	127	1,793	1,100	1.60
7*	124	1,856	1,134	1.58
8	132	1,985	197	9.42
9	140	2,074	327	6.07
10	183	1,659	415	5.00

	Mole Fraction in Liquid	
Tray	Ethane	Propane
1*	0.0156	0.0838
2	0.0178	0.1937
3	0.0144	0.2910
4	0.0106	0.3505
5	0.0081	0.3801
6	0.0068	0.3946
7*	0.0064	0.4152
8	0.0074	0.4578
9	0.0061	0.4890
10	0.0030	0.3723

* Feed Trays

Assuming a fixed-configuration, fixed-pressure column, there could still be two independent operating variables: the reboiler duty and the oil feed rate. These two parameters may be varied independently to meet specified propane recovery and ethane rejection in the bottoms. Although all the parameters are interdependent to varying degrees, it is logical to visualize the lean oil rate as primarily affecting the propane recovery, and the reboiler duty mainly determining the ethane rejection. The process requirements for this example are to recover in the bottoms 95% of the propane in the feed and to bring the ethane mole fraction in the bottoms down to 0.003. To meet these specifications, a reboiler duty of 7.82 MMBTU/hr and a lean oil rate of 1,033 lbmole/hr are required.

Feed Tray

The reboiler duty and oil feed required to meet the process specifications correspond to a fixed gas feed location on tray 7. If the feed tray were to be altered to meet the same specifications, the reboiler duty and oil rate requirements would, in general, be different. A higher feed tray implies fewer absorption trays and, therefore, a higher oil rate would be required to meet the recovery specification. A higher feed tray also means more stripping trays are available. Two conflicting factors could affect the reboiler duty: (1) A lower duty is needed since more stripping trays are available, or (2) a higher duty is needed since the reboiler must handle a higher liquid flow caused by the higher oil rate. The opposite is true if the feed tray is lowered. If the second factor is controlling, an optimum feed location would exist that minimizes both the oil rate and reboiler duty. If the first factor is controlling, i.e., a higher feed tray requires a smaller reboiler duty, then the feed location may be optimized to minimize either the oil rate or the reboiler duty or some function (such as cost) associated with both parameters.

In the present example, both the reboiler duty and oil rate are minimized when the feed is on the seventh tray. The reboiler duty and oil rate requirements are summarized for several feed trays in Table 9.6.

Table 9.6 Effect of Feed Tray Location (Example 9.2)

Specifications: Propane Recovery in Bottoms: 95%
Ethane Mole Fraction in Bottoms: 0.003

Feed Tray	Reboiler Duty MMBTU/hr	Lean Oil Rate lbmole/hr
5	8.88	1071
6	8.15	1040
7	7.82	1033
8	7.94	1058
9	8.77	1143

9.1.2 Columns with a Condenser and No Reboiler

With no reboiler, an external feed, at least partially vapor, is necessary at the bottom tray to act as an external stripping medium. This feed could be the only one, in which case the column may be designated as a rectifier (Section 7.4). Such a column would perform the function of separating the feed into an overhead and a bottoms product, where the heavies are removed from the overhead product to a specified level. No particular concern is placed on the amount of light components lost in the bottoms.

In this section, consideration is given to columns where the bottom feed serves primarily as a stripping medium and where the main feed to be separated is placed on an intermediate tray. Although the stripping stream must be at least partially vapor, the main feed could be at any thermal condition. The stripping stream should be sufficiently hot or of low enough boiling point to maintain vapor traffic throughout the column. Steam is commonly used as the stripping medium because of its relatively low cost and ease of its downstream separation from the stripped gases by condensation and water decantation.

In some situations external stripping is favored over reboiled stripping, just as in certain situations external reflux is favored over a condenser. Stripping with an external stream can take place at lower temperatures than would exist in a reboiler. Reboiled stripping requires high temperatures in order to drive up the K-values, to cause lighter components to cross from the liquid to the vapor. In external gas stripping, the K-values and temperature need not be raised since the transfer of light components from the liquid to the vapor takes place as a result of lowering their partial pressure or vapor mole fraction by introducing a lighter component (the stripping medium). As the lighter components' vapor mole fraction is lowered, their liquid mole fraction must also come down to maintain the vapor–liquid equilibrium constant. The result is that some of the light and intermediate components transfer from the liquid to the vapor.

Examples of steam stripping include stripping of light components from crude oil or regenerating lean oil from rich oil. Using a reboiler in such applications could require unacceptably high temperatures. Example 9.3 compares two columns that perform the same function, one using a reboiler and the other using steam stripping.

Columns with a condenser and an external stripping stream include the usual rectifying section between the condenser and the main feed as well as the stripping section between the two feeds. A fixed-configuration, fixed-pressure column of this kind has two independent operating parameters or 2 degrees of freedom: the condenser duty and the stripping stream flow rate. If the two parameters are variable, two process specifications are required to define the column performance, such as the recovery of one component and the rejection of another in one of the products.

Example 9.3 Rich Oil Stripper

An oil stream contains heavy oil components decane (n-$C_{10}H_{22}$), undecane (n-$C_{11}H_{24}$), dodecane (n-$C_{12}H_{26}$), and tridecane (n-$C_{13}H_{28}$), plus absorbed components pentane and hexane. The stream is to be separated to recover the pentane and hexane and also to regenerate the lean oil. A 10-stage column operating at 25 psia is to be used for this purpose. The column is equipped with a partial condenser (first stage) and may either be stripped with steam entering the tenth stage or with a reboiler as the tenth stage. The rich oil enters the column on the fifth stage from the top as a slightly subcooled liquid at 250°F and 25 psia. Superheated steam at 400°F and 25 psia is used for stripping. The schematic of the columns is shown in Figure 9.3, and the stream data are given in Table 9.7.

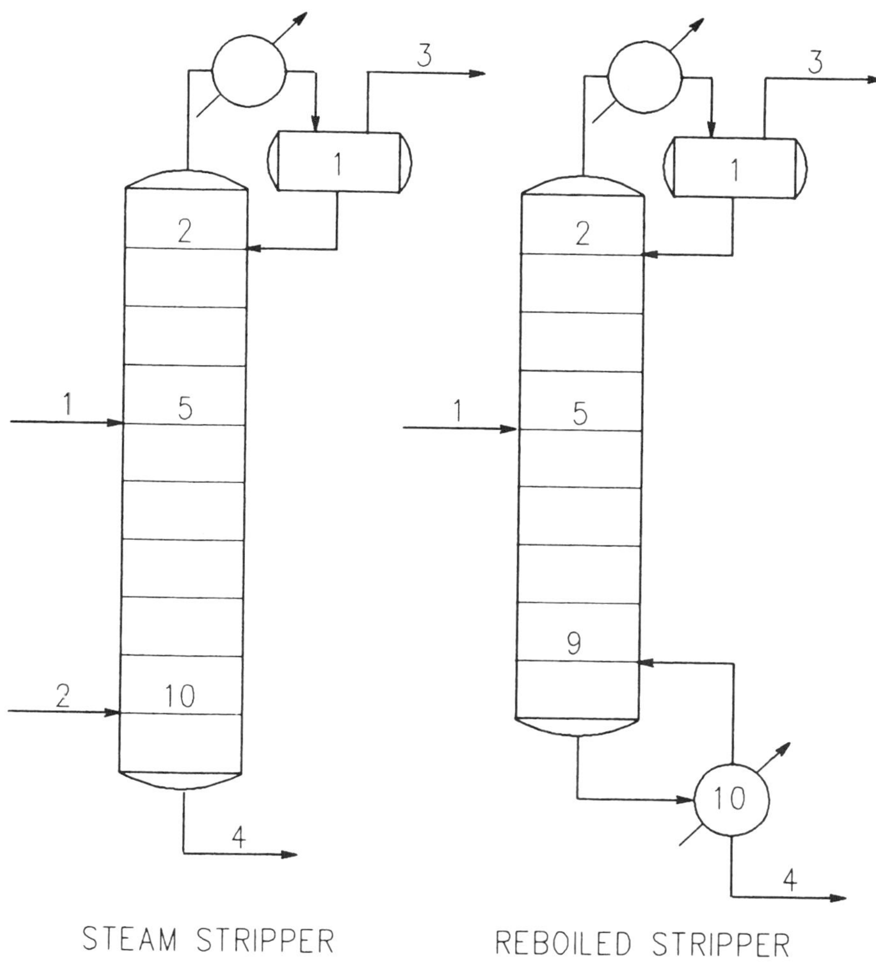

Figure 9.3   Steam stripper and reboiled stripper.

The process specifications require a minimum recovery of 99% for both the stripped hydrocarbons and the regenerated oil. At a reflux rate of 50 lbmole/hr and a bottoms rate of 78 lbmole/hr, both columns surpass the two specifications. The column profiles, however, are different in the two cases, as shown in Table 9.8. Much of the stripping steam condenses in the column and is withdrawn as decanted water. The decanted water is not shown in Table 9.8, and this accounts for the apparent material imbalance on some of the trays. A significant difference between the two columns is the temperature profile, which is substantially lower in the steam-stripped column. Also, the temperature range in the reboiled column is wider than that of the stripped column, which is consistent with the distillative nature of the former and the external stripping characteristics of the latter. The steam-stripping column requires a much lower L/V ratio (higher V/L ratio) than the reboiled column because

Table 9.7 Stream Data for Example 9.3

| | | | Stream 3 Overhead Vapor | | Stream 3 Stripped Oil | |
	Stream 1 Rich Oil	Stream 2 Steam	Steam	Reboiled	Steam	Reboiled
H_2O	0.00	330	3.88	0.00	1.24	0.00
Pentane	13.20		13.20	13.20	0.00	0.00
Hexane	8.58		8.58	8.58	0.00	0.00
Decane	8.25		0.01	0.20	8.24	8.05
Undecane	34.00		0.00	0.02	34.00	33.98
Dodecane	33.00		0.00	0.00	33.00	33.00
Tridecane	2.97		0.00	0.00	2.97	2.97
Totals	100.00	330	25.67	22.00	79.45	78.00
Temperature, °F	250	400	151	185	251	435
Pressure, psia	25	25	25	25	25	25

The column above "Lbmole/Hr" spans the six numeric columns.

Table 9.8 Column Profiles for Example 9.3

Tray	Temperature °F	Liquid Flow lbmole/hr	Vapor Flow lbmole/hr	L/V
		Steam Stripping		
1	151	50.0	25.7	
2	175	16.6	400.6	0.125
3	202	14.4	367.2	0.045
4	221	13.2	365.0	0.039
5*	230	95.4	363.8	0.036
6	234	95.1	345.9	0.276
7	239	96.4	345.7	0.275
8	244	97.5	347.0	0.278
9	247	98.1	348.1	0.280
10*	251	79.4	348.6	0.281
		Reboiled Stripping		
1	185	50.0	22.0	
2	301	44.3	72.0	0.694
3	360	49.8	66.3	0.668
4	378	50.1	71.8	0.694
5*	389	202.2	72.1	0.694
6	416	228.0	124.2	1.628
7	422	232.3	150.0	1.520
8	425	232.4	154.3	1.506
9	430	231.6	154.4	1.505
10	436	78.0	153.6	1.508

*Feed Trays

the vapor flow in the steam column includes the stripping steam. This is a factor that must be taken into account in selecting the column type for the process. The vapor flow affects column hydraulics, such as flooding conditions and pressure drop, which are discussed in Chapters 14 and 15.

9.1.3   Columns with a Condenser and a Reboiler

In many applications, different process streams that contain essentially the same components must undergo similar separation specifications. The process streams may have different flow rates, compositions, and/or thermal conditions. Each stream could be separated in a different column or the different streams could be first mixed together then separated into the desired products in one column. Alternatively, the streams could be fed directly to the multistage column without prior mixing. Since the optimum feed tray depends on the feed's composition and thermal conditions, each stream is, in general, fed to a different tray.

Excluding cases where some of the feed streams serve as absorption or stripping media, a multifeed column requires a condenser and a reboiler to generate internal reflux and boilup for fractionation, just as in single-feed conventional columns. For a fixed total number of trays, the feed locations should be optimized for a specified separation to minimize the reflux and boilup ratios or the condenser and reboiler duties. The column section below the lowest feed is the stripping section, and that above the highest feed is the rectifying section. In the column sections between the feeds, the L/V ratio, and hence the fractionation, is under the combined effect of the reflux and boilup and the external feeds. The feeds should thus be located to take full advantage of their differential compositions and thermal conditions to enhance separation, while at the same time sufficient stripping trays and rectifying trays should be allowed. All these factors are interrelated and highly complex and must be incorporated in the overall column solution. The toluene–xylene column described below provides a qualitative insight into the performance of a multifeed column.

Example 9.4 Toluene–Xylene Column

Two mixtures of toluene and xylene are separated into toluene and xylene using one column consisting of 36 theoretical stages, including a partial condenser and a reboiler and operating at a pressure of 20 psia. The two feed streams are at the same temperature and pressure but are of different flow rates, compositions, and phase (Table 9.9). The process requirements are a toluene recovery of 99.9% in the overhead and a xylene recovery of 96.2% in the bottoms. Recoveries are based on total feed.

Since the system is binary, the two recovery specifications determine the total product rates by overall material balance. With the total number of stages fixed, the reflux ratio required to achieve the desired separation depends on the feed locations. It is desirable to operate the column at the lowest possible reflux ratio in order to minimize the condenser and reboiler duties.

Table 9.9 Stream Data for Example 9.4

| | lbmole/hr | | | |
Component	Feed 1 Vapor	Feed 2 Liquid	Overhead	Bottoms
Toluene	640.00	80.00	719.27	0.73
Xylene	160.00	120.00	10.53	269.47
Totals	800.00	200.00	729.80	270.20
Temperature, °F	275	275	252	304
Pressure, psia	20	20	20	20

The optimum feed location for a single-feed binary stream may be estimated graphically using a Y–X diagram (Chapters 5 and 6). The complexity of graphical estimation increases steeply with the number of feeds and is infeasible for multi-component mixtures. Taken separately, the optimum tray for feed 1 would, on the one hand, be higher than that of feed 2 because feed 1 is richer in the lighter component. On the other hand, feed 1 optimum tray would be lower than feed 2 optimum tray because feed 1 is a vapor and feed 2 is a liquid. Generalizing the graphical binary concepts to multicomponent systems, streams are fed at upper trays if they are lighter in composition and/or if they have a higher liquid fraction or are subcooled to a lower temperature, and vice versa. In the present example, if the streams are fed separately, the phase condition appears to be a more predominant factor than the composition, and feed 1 optimum tray is lower than feed 2 optimum tray. Table 9.10 shows the variation of the reboiler duty with the feed tray for each of streams 1 and 2 when fed separately to the column. The optimum for feed 1 is tray 18 and that for feed 2 is tray 11.

Table 9.10 Reboiler Duty for a Single-Feed Column (Example 9.4)

| | MMBTU/hr | |
Feed Tray	Feed 1	Feed 2
8		4.09
10	12.98	4.00
11	11.55	3.96
13	10.84	4.00
14	10.42	
15		4.16
16	10.15	
18	10.05	
19	10.07	
20	10.11	

When both streams are fed to the same column, the above considerations still apply, but their relative importance may vary. If one of the streams is predicted to have a higher optimum feed tray than the other on account of both its composition and thermal conditions, then the optimum feed trays in the two-feed column would correspond to the same relative locations as with the streams fed separately. However, if the two factors are contradictory, as in this example, the optimum feed locations in the two-feed column may or may not correspond to the relative single-feed optimum locations. In the two-feed toluene-xylene column, the reboiler duty is minimized when feed 1 is higher than feed 2, which is contrary to the single-feed optimum feed trays. With the two-feed column, the compositional effect on the feed location outweighs the phase effect. Table 9.11 lists reboiler duties for several combinations of feed trays. The optimum corresponds to feed 1 on tray 15 and feed 2 on tray 22.

Table 9.11 Reboiler Duty for a Multiple-Feed Column (Example 9.4)

Feed 1 Tray	Feed 2 Tray	Reboiler Duty, MMBTU/hr
10	20	14.43
11	22	13.45
15	22	12.92
16	22	12.98
16	24	13.00

9.2 MULTIPLE PRODUCTS

Column products other than the overhead and bottoms may be taken out as liquid or vapor side draws from any tray. Multiple product columns are used in certain applications as a one-step process for separating multicomponent streams into a number of products with different compositions. A multiproduct column could, therefore, perform the equivalent task of several columns. In many situations the separation realized in the multiproduct column is only a crude separation to be followed by further downstream processing.

The compositions of the side products depend on their location in the column, in accordance with the composition profile that exists along the column. A small liquid product taken from a column tray has the same composition as the undisturbed liquid on that tray. As the product rate varies, the column L/V ratio and other column parameters also vary. New steady-state conditions are reestablished with a new column composition profile. The side product composition is therefore also dependent on its flow rate.

9.2.1 Column Sections

In a multicomponent column, the concentration of each component peaks at a certain tray, as shown schematically in Figure 9.4. The lighter components reach their peak at the upper trays, and the heavier components at the lower trays. Side products may be drawn at the concentration peak trays or at other trays to obtain products with different compositions. The set of trays between two adjacent products constitutes a column section, and the fractionation between the two products depends

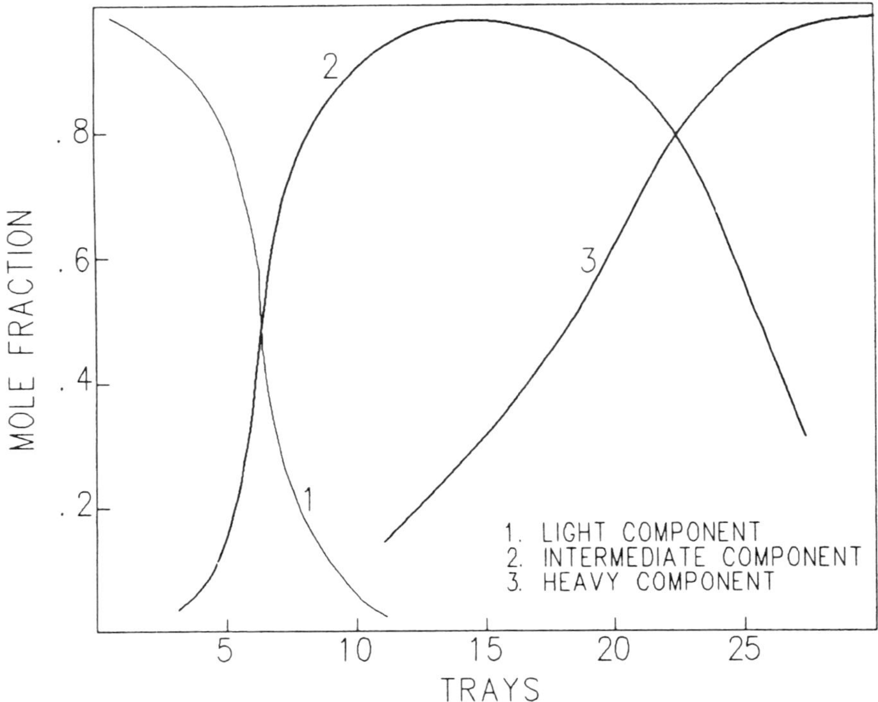

Figure 9.4 Composition profiles in a multicomponent column.

on the number of trays and the L/V ratio in that section. The L/V ratio is a function of the column reflux ratio and the side product flow rates.

In analyzing the performance of multiproduct columns, several types of column sections are recognized, among which are the following: a section with a condenser, an overhead, and a side product; a section with a reboiler, a bottoms, and a side product; a feed section with two side products; and sections with just two side products. While the separation pattern, or key components in each section, is determined by the product rates and relative locations of the feeds and products, the separation sharpness is determined by the L/V or reflux ratio.

The separation of components between two products identifies a light key and a heavy key in each section. The feed split coincides with the light and heavy key components of the section where the feed is introduced. In very general terms, the feed section or feed tray location should be such that products lighter than the feed are drawn above it and products heavier than the feed are drawn below it. The optimum feed location can only be determined on the basis of rigorous calculations. In multiproduct columns the optimum feed location could be highly constrained by the separation requirements in the various products.

9.2.2 Degrees of Freedom

Each side product constitutes one additional independent column parameter. To define the column performance, the flow rate of each side product must be known. Alternatively, a performance specification may be defined for any side product such as the concentration of one of its components. The side product flow rate becomes a dependent variable which must be calculated to satisfy the performance specification. It has been concluded earlier (Chapter 7) that a fixed-feed, fixed-configuration, fixed-pressure column with a condenser and a reboiler has 2 degrees of freedom; that is, two parameters, such as the condenser and reboiler duties, may be varied independently. Each side product adds to the column 1 degree of freedom. Hence, a column as defined above, having S side products, has $2 + S$ degrees of freedom. The duties and side product flow rates can each be varied independently, allowing $2 + S$ performance specifications to be achieved. This conclusion can also be reached by applying the description rule since each additional product rate can be controlled independently by external means.

Once a column has been designed and built, the total number of trays and the feed and product locations are fixed. The column pressure may be varied within certain limits but is usually dictated by other process considerations, such as condenser cooling medium temperature and reboiler temperature, etc. Assuming a fixed column pressure, the operating variables that are available for achieving product specifications, such as purities or recoveries, are the reflux ratio and the product rates. Higher reflux ratios generally improve fractionation throughout the column. If the fractionation between two products appears adequate (i.e., sufficient number of trays and L/V in the column section between the two products), but the products are not split at the desired point, then the product rates must be varied to meet their composition specifications. The separation in each section and the properties of each product may be selectively controlled by varying each individual product rate independently.

9.2.3 Partial and Total Condensers

The condenser is the stage where overhead vapors are condensed and returned to the top of the column as reflux. The condenser is *partial* if only part of the vapor is condensed and refluxed and the remainder leaves the condenser as vapor distillate. This type of condenser adds one equilibrium stage to the column trays because it holds a vapor phase and a liquid phase at equilibrium with each other. A *total condenser* is one where the entire overhead vapor is condensed (cooled to the bubble point temperature or subcooled to a lower temperature), part of the condensate returned as reflux, and the remaining part taken as liquid distillate. This type of condenser does not count as an equilibrium stage because no vapor–liquid separation takes place in it. The liquid distillate composition is identical to the composition of the vapor leaving the column top tray.

A third type of condenser is often used, one which is a partial condenser but where only part of the condensate is refluxed and the remaining part is drawn as liquid distillate alongside the vapor distillate. This type of condenser constitutes an equilibrium stage. The two overhead products have distinct compositions, being the equilibrium vapor and liquid phases. A column with vapor and liquid overhead products may be regarded as a special case of a multiproduct column since the liquid overhead is essentially a side product from the condenser, or top stage. Consequently, such a column has 3 degrees of freedom, whereas, with the other types of condensers, the column has 2 degrees of freedom.

The factors influencing the performance of multiproduct columns are best understood by analyzing an example column as described below.

Example 9.5 Multiproduct Column

A stream containing ethane, propane, butane, and pentane is to be separated into four products, each containing mainly one of the four components. A single column with 30 theoretical stages, including a partial condenser and a reboiler, is to be used for this purpose. The column pressure is set at 200 psia. The feed composition and thermal conditions are given in Table 9.12.

Table 9.12 Stream Data for Example 9.5

Component	Feed	Overhead	Side Draw 1	Side Draw 2	Bottoms
			lbmole/hr		
			Reflux Ratio = 8		
C2	12.00	8.75	2.34	0.91	0.00
C3	48.00	3.25	35.60	8.82	0.33
NC4	25.00	0.00	10.05	9.52	5.43
NC5	15.00	0.00	0.01	5.75	9.24
Totals	100.00	12.00	48.00	25.00	15.00
Temp, °F	150.00	40.00	109.00	155.00	244.20
P, psia	200.00	200.00	200.00	200.00	200.00
			Reflux Ratio = 16		
C2	12.00	9.82	1.66	0.53	0.00
C3	48.00	2.18	39.77	6.01	0.02
NC4	25.00	0.00	6.57	16.93	1.50
NC5	15.00	0.00	0.00	1.53	13.48
Totals	100.00	12.00	48.00	25.00	15.00
Temp, °F	150.00	27.90	106.60	165.80	274.80
P, psia	200.00	200.00	200.00	200.00	200.00

The relative locations of the products along the column may be estimated on the basis of total reflux shortcut calculations, to be discussed in Chapter 12. The side draw trays are fixed in this example at the locations indicated in Figure 9.5. The effect of the feed location is discussed in the example. Initially it is fixed at tray 25 from the top.

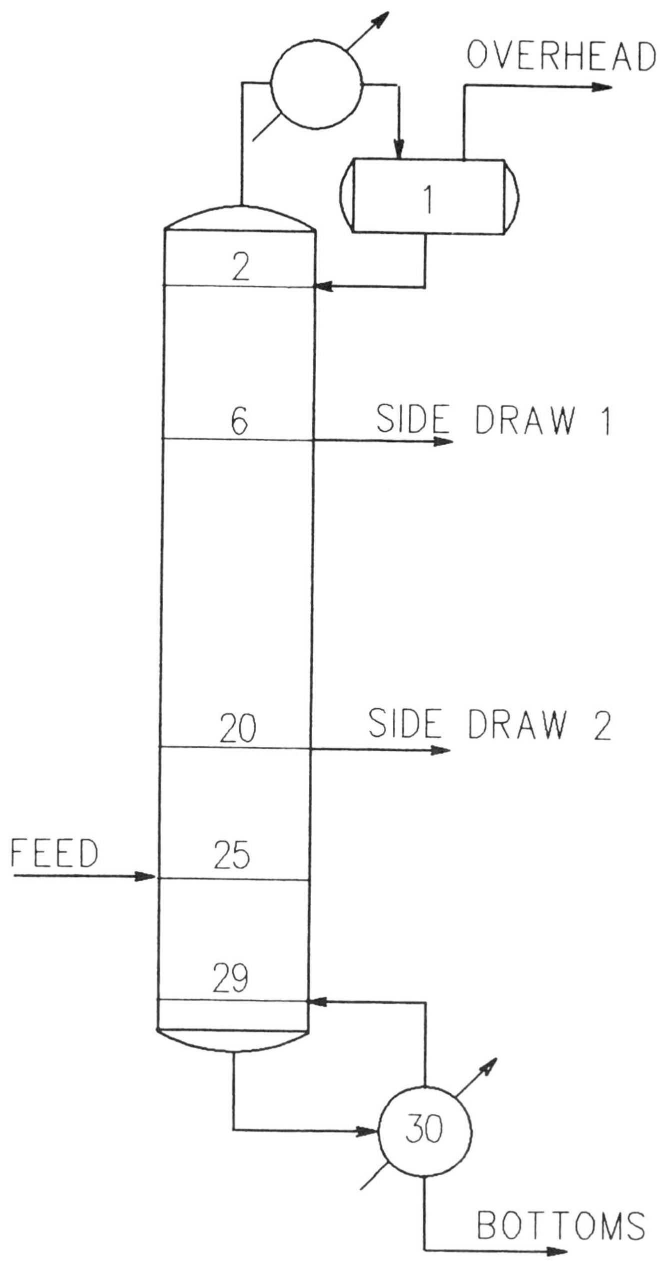

Figure 9.5 Multiproduct column (Example 9.5).

Column Specifications

The column has four products, two of which are side draws. Since it also has a condenser and a reboiler, it has a total of 2 + S, or 4, degrees of freedom and, therefore, four performance specifications are required to define its operation.

One possible set of specifications would include the flow rates of three of the products plus the column reflux ratio. The way the components are split among the products and the identity of the key components in each section are determined by the product rates and locations of the feed and products. In this example the products are specified at flow rates that correspond to the component flow rates in the feed. Thus, the overhead is specified at 12 lbmoles/hr, the upper side draw at 48 lbmoles/hr, and the lower side draw at 25 lbmoles/hr. By overall material balance, the resulting bottoms flow rate is 15 lbmoles/hr.

The overall objective of the column specifications is to recover most of the ethane in the overhead, the propane in the upper side draw, the butane in the lower side draw, and the pentane in the bottoms. Ethane is the light key in the upper section, propane is the heavy key in the upper section and the light key in the middle section, butane is the heavy key in the middle section and the light key in the lower section, and pentane is the heavy key in the lower section. The distinction between rectification and stripping is relative in a multiproduct situation. In the upper section, for instance, ethane is stripped from the upper side draw and propane is absorbed by reflux from the overhead. Similarly, propane is stripped from the lower side draw, butane is separated from the upper side draw by liquid reflux, butane is stripped from the bottoms, and pentane is separated from the lower side draw by reflux.

The fourth column specification is the reflux ratio, which controls the quality of separation between the products. The reflux ratio sets the L/V ratios in the column. These ratios tend to be fairly constant within each section, depending on the distillation characteristics of the mixture. They are different, though, in each section because the side draws alter the fluid flow in the column. If, for instance, a side draw is a liquid, the liquid flow on the trays below the draw tray is smaller than that on the trays above it. In the feed section, the L/V ratio is generally different above and below the feed tray, depending on the feed thermal conditions.

If all other parameters are held constant, a higher reflux ratio results in better fractionation throughout the column and better separation between the products. Table 9.12 summarizes the stream data at reflux ratios of 8 and 16.

Feed Location

If the feed were placed in the column top section, it would be separated in that section into ethane and the rest of the components, which would be separated into the other products in the lower sections. If the feed were placed in the middle section, it would be separated between propane and butane in that section. The upper and lower sections would then make the separation between ethane and propane, and butane and pentane, respectively. The feed could be placed in the lower section, where the separation would take place between butane and pentane. Further separation into the lighter products would take place in the upper sections.

If the feed is introduced in the upper section, at a given reflux ratio some of the heavier components escape in the upper side draw due to inadequate rectification. If, at the same reflux ratio, the feed is introduced in the middle section, an appreciable degree of rectification takes place between the feed and the upper side draw, resulting in better separation in the lower products at the expense of slightly lower purity of the overhead. Putting the feed in the lower section enhances the purity of the bottoms although less stripping of the light components takes place, causing part of them to escape in the lower side draw. The product compositions at three different feed locations are summarized in Table 9.13.

Table 9.13 Effect of Feed Location (Example 9.5, Reflux Ratio = 16)

Component	lbmole/hr			
	Overhead	Side Draw 1	Side Draw 2	Bottoms
Feed on Tray 4 (Upper Section)				
C2	10.62	1.37	0.00	0.00
C3	1.37	34.78	11.84	0.01
NC4	0.01	7.93	10.61	6.46
NC5	0.00	3.92	2.55	8.53
Feed on Tray 10 (Middle Section)				
C2	10.13	1.87	0.00	0.00
C3	1.87	44.74	1.38	0.00
NC4	0.00	1.36	19.66	3.98
NC5	0.00	0.03	3.95	11.02
Feed on Tray 25 (Lower Section)				
C2	9.82	1.66	0.53	0.00
C3	2.18	39.77	6.01	0.02
NC4	0.00	6.57	16.93	1.50
NC5	0.00	0.00	1.53	13.48

Effect of Product Rates

With a fixed configuration and reflux ratio, each product composition may be controlled independently by varying its draw rate. The overhead contains primarily ethane, while heavier components are refluxed. Ethane is the lightest component; therefore, any impurities in this product are heavier components. Lowering the overhead rate should improve its purity by decreasing the concentration of heavier components. The obvious penalty is a lower ethane recovery in this product. Since the fractionation is fixed, the overhead rate must be determined to achieve some balance between purity and recovery. Table 9.14 indicates an ethane recovery in the overhead of $12 \times 0.818 = 9.82$ mole/hr at a purity of 0.818 mole fraction or a recovery of $10 \times 0.912 = 9.12$ mole/hr at a purity of 0.912 mole fraction.

An intermediate product such as the upper side draw, which is mostly propane, may contain impurities from components both lighter and heavier than the main component. The upper side draw is, at the same time, the bottom product of the top

Table 9.14 Effect of Product Rates
(Example 9.5, Reflux Ratio = 16)

	Mole/Hr	C2	C3	NC4	NC5
Overhead	12	0.818	0.182	0.000	0.000
Side Draw 1	48	0.035	0.828	0.137	0.000
Side Draw 2	25	0.021	0.241	0.676	0.062
Bottoms	15	0.000	0.002	0.101	0.897
Overhead	10	0.912	0.088	0.000	0.000
Side Draw 1	48	0.046	0.824	0.130	0.000
Side Draw 2	25	0.027	0.300	0.610	0.063
Bottoms	17	0.000	0.005	0.206	0.789
Overhead	13	0.773	0.227	0.000	0.000
Side Draw 1	47	0.031	0.840	0.129	0.000
Side Draw 2	25	0.019	0.222	0.704	0.055
Bottoms	15	0.000	0.001	0.090	0.909
Overhead	12	0.819	0.181	0.000	0.000
Side Draw 1	45	0.035	0.868	0.097	0.000
Side Draw 2	28	0.021	0.241	0.682	0.056
Bottoms	15	0.000	0.002	0.102	0.896

The column header spanning C2, C3, NC4, NC5 is **Mole Fraction**.

column section and the top product of the second section. The fractionation in the upper section determines to what extent ethane is stripped off from the propane product, and the fractionation in the second section determines to what extent butane is removed from the propane product. Again, in this situation since the number of stages in each section and the reflux ratio are all fixed, the fractionation is fixed. The propane recovery and purity depend mostly on its flow rate and on the flow rates of the adjacent products above and below it. If the propane product contains too much ethane, its flow rate should be cut back and the overhead rate increased. If the propane product contains too much butane, its flow rate should be cut back and the lower side draw rate increased. Table 9.14 summarizes the purities of components in different products at different flow rates. The recoveries can also be calculated from this table. The dependence of the other products' compositions on their rates may be analyzed in a similar manner.

Example 9.6 Ammonia–Acetone–Water Column

A distillation column operating at 5 atm is used to recover pure ammonia in the overhead from a water–acetone solution. The column has 18 theoretical trays, a total condenser, and a reboiler; the feed is introduced on the tenth tray from the top. In the same column, crude separation is to be made between the acetone and the water by taking a liquid side stream. The feed, which is superheated vapor at 200°C and 5 atm before entering the column, has the following component rates:

Component	Feed Rate, kgmole/hr
Ammonia	4,000
Acetone	200
Water	500
Total	4,700

It is required to determine the best draw tray for the side stream and the effect of the draw rate on the separation.

The column is solved by computer simulation with vapor–liquid equilibrium calculations based on the van Laar liquid activity coefficient equation (Chapter 1). Initially the column is solved without a side draw to determine the composition profiles in the column. The column is solved at a reflux ratio of 1 and a bottoms rate of 800 kgmoles/hr. The composition profiles with no side draw are shown in Figure 9.6.

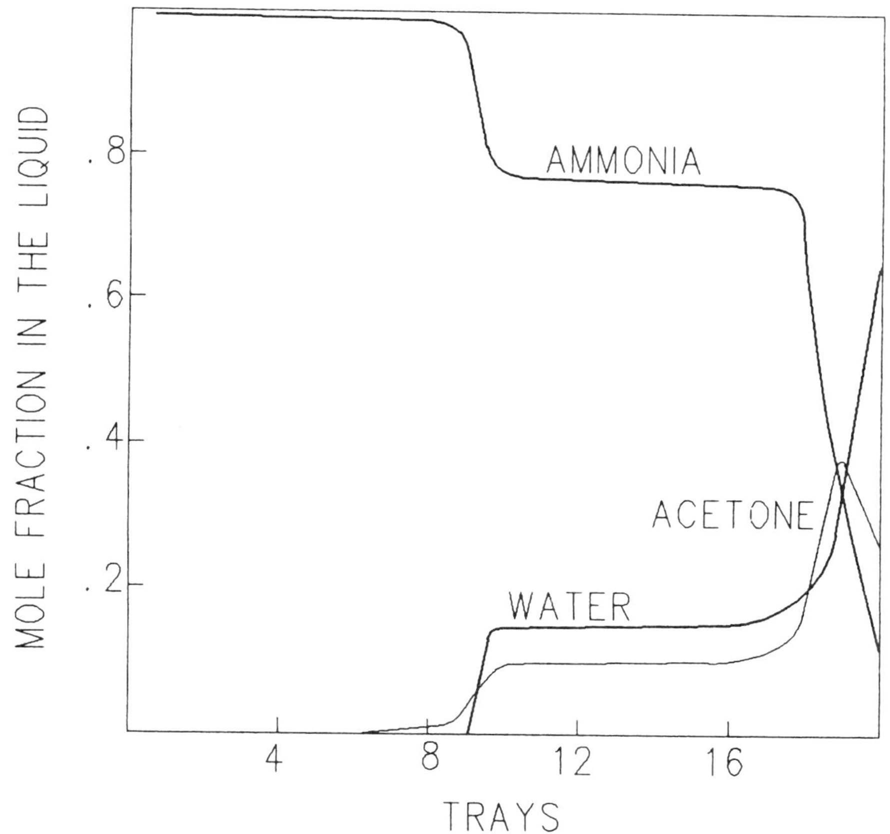

Figure 9.6 Composition profiles in the ammonia–acetone–water column (Example 9.6).

The results indicate that the column is overtrayed, especially in the rectifying section, since almost pure ammonia is obtained only two or three trays above the feed. Although the bottoms rate of 800 kgmole/hr ensures a high-purity ammonia in the overhead, it does that at the expense of ammonia recovery since much of it is forced down the column and leaves with the bottoms.

The acetone concentration peaks at tray 19, a logical place to put the side draw. As the side draw rate is increased, its acetone concentration goes down since more and more water gets drawn with the acetone. The variation of the acetone concentration with the side draw rate is shown in Figure 9.7.

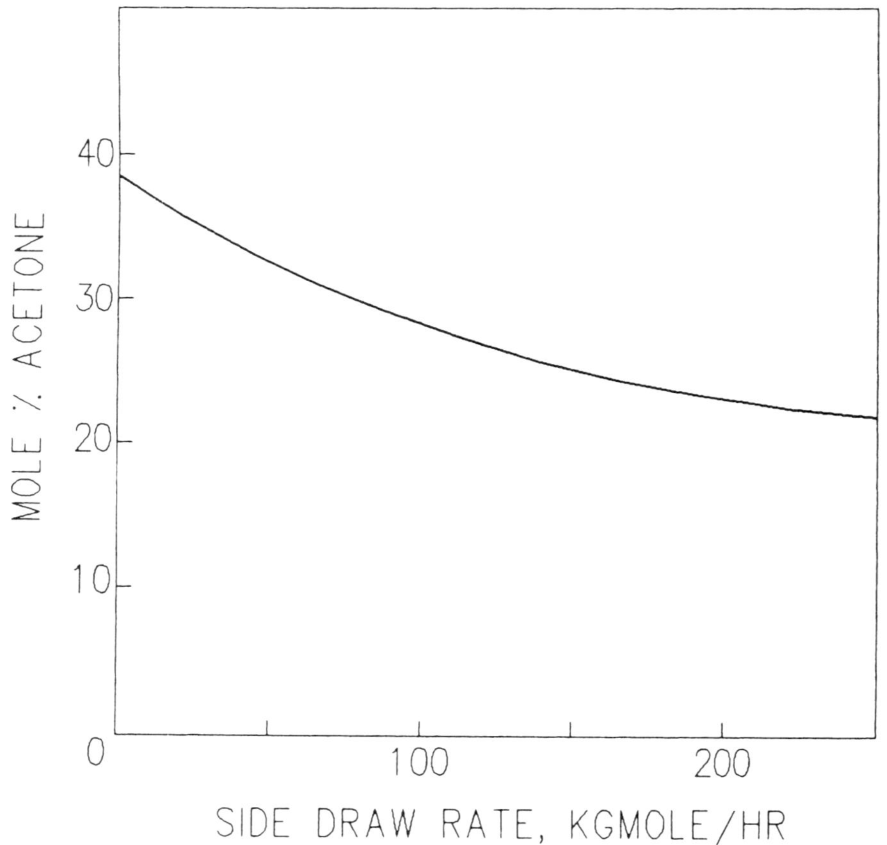

Figure 9.7   Effect of side draw rate on acetone concentration (Example 9.6).

9.3 SIDE HEATERS/COOLERS AND PUMPAROUNDS

Distillation columns and other multistage columns generally require heat addition and/or removal, which normally takes place at the reboiler and/or condenser. For a variety of reasons, it may be desirable in certain types of columns to add or remove heat at intermediate trays aside from the condenser and reboiler. Examples of such columns include demethanizers that utilize a side reboiler alongside the bottom reboiler, multiproduct columns where intermediate condensers are associated with some of the side products, and absorber intercoolers used for partial removal of the heat of absorption. The method most commonly employed for exchanging heat between a column tray and a heat source or sink is the pumparound, where a fluid is drawn from the tray, sent to a heat exchanger, then pumped back to the column. The following sections pertain to the various applications of side heaters and coolers and the different types of pumparounds.

9.3.1 Applications

The purposes of side heaters and coolers may be categorized as follows: utilization of heat sources or sinks at different temperature levels, removal of heat of absorption, and control of vapor and liquid flows in the column.

Temperature Levels

Whenever feasible, processes are designed for maximum energy integration aimed at recovering heat from process streams to the extent that is technically and economically advantageous. For instance, a hot process stream may be a suitable heat source for a column reboiler, while a relatively cold stream may be used as the heat sink for the condenser. The net outcome is that the hot and cold streams are brought closer to their respective target temperatures, thereby resulting in utility savings in both the column and process streams. It happens quite frequently, however, that the temperature levels of the process streams may not be right for heat exchange at the reboiler or condenser temperatures. For instance, consider the removal of most of the heat in a column to be done at the condenser, using cooling water at 85°F. It may be desired to supplement the heat removal in the condenser with a process stream that is to be heated, say from 120 to 140°F. This stream cannot be used at the condenser if the condenser temperature must be maintained at 100°F. The process stream could be used for heat removal at a lower tray where the temperature is high enough to allow heat exchange with the 120°F process stream. A side cooler or condenser could be used for this purpose.

The use of process streams to supplement the reboiler duty using side reboilers is similar in principle. In demethanizers, for example, one process stream may be used as a heat source for the bottom reboilers, while another stream, at a lower temperature, may be used as a side reboiler a few trays above the bottom reboiler, where the temperature level is low enough to allow heat exchange with the process stream.

The implications of side heaters or coolers on the column performance are discussed later in this section.

Heat of Absorption

The effect of heat of absorption on the performance of absorbers was discussed in Chapter 8. Under certain circumstances, it may be necessary to improve the efficiency of absorbers by compensating for the heat of absorption with heat removal using side coolers on certain trays along the column.

Column Vapor and Liquid Flows

Side coolers or condensers are commonly used in multiproduct columns to prevent excessive variations in the vapor and liquid flow profiles along the column. The multiproduct column described in Section 9.2 had, for simplicity, only one condenser and a reboiler. If the feed in this column was introduced at its lower part, the vapor and liquid rates would continuously increase above the feed to the top of the column. A similar situation would exist in crude columns if no side coolers were used.

The variations in the column flow profiles are caused by the liquid side draws. The liquid flow at the top of the column is the reflux rate. Each liquid draw removes a portion of the column liquid flow. With a reduced liquid flow, the vapor flow also decreases since the L/V ratio remains unchanged, assuming other factors are constant. The column flow profiles can be manipulated by redistributing the condenser duty among side coolers along the column.

If a condenser or side cooler is placed on a draw tray or right below it, a portion of the rising vapor condenses. This causes the liquid flow, and consequently the vapor flow, below the side draw to rise, which tends to equalize the profiles above and below the draw tray. The side cooler duty may be adjusted to obtain the desired profiles. In this manner, maintaining the desired liquid and vapor flows in the lower sections of the column does not require unnecessarily high flows in the upper sections.

It is evident, therefore, that each side cooler or heater placed on the column adds 1 degree of freedom to the column performance. By varying the side cooler or heater duties, it is possible to meet certain performance specifications such as liquid or vapor profiles or side product purities. The side condensers enhance the side product purities since better fractionation is achieved with the increased liquid and vapor flows below the side draw.

The following example examines the effect of side coolers on the column performance.

Example 9.7 Column Side Coolers

The liquid and vapor profiles in the multiproduct column of Example 9.5 are shown in Table 9.15 for the case of a single condenser at the top stage. The feed is on theoretical tray 25 counting from the condenser, and the reflux ratio is 16. As the table indicates, the liquid and vapor profiles increase from the feed tray to the condenser with discontinuities at each draw tray.

Keeping the reflux ratio at 16, additional condensers are placed as heat sinks on trays 7 and 21, just below the draw trays, to maintain fairly uniform liquid and vapor profiles along the column. The new profiles are given in Table 9.16.

The extra side coolers result in better fractionation for the column products even though the reflux ratio is kept the same. The new product compositions are shown

Table 9.15 Column Profiles: No Side Coolers (Example 9.7, Reflux Ratio = 16)

Tray	Temperature °F	Liquid Flow lbmole/hr	Vapor Flow lbmole/hr	Product lbmole/hr	Duty MMBTU/hr
1	28	192	12	12	−1.15
2	58	185	204		
3	80	187	197		
4	92	188	199		
5	99	186	200		
6	107	130	198	48	
7	119	123	190		
8	132	118	183		
9	143	116	178		
10	151	115	176		
11	156	115	175		
12	159	115	175		
13	160	115	175		
14	161	115	175		
15	162	115	175		
16	162	115	175		
17	162	114	175		
18	163	114	174		
19	164	113	174		
20	166	85	173	25	
21	169	83	170		
22	174	80	168		
23	179	77	165		
24	184	76	162		
25 Feed	205	120	102		
26	229	122	105		
27	247	125	107		
28	259	127	110		
29	268	129	112		
30	275	15	114	15	0.88

in Table 9.17 and should be compared with the compositions in the upper part of Table 9.14. With the side coolers, a lower reflux ratio is required to obtain a separation equivalent to the case with a single condenser. Table 9.17 also summarizes the product compositions at a reflux ratio of 10 with side coolers. The compositions at this reflux ratio are comparable to those at the higher reflux ratio of 16 (Table 9.14). The result of adding side coolers is that the condenser duty is redistributed among the top and side condensers. The variations in the column profiles are diminished, and the actual vapor and liquid flows are reduced in the top column section since the reflux ratio is lower. It is highly preferable to design a column with a mostly uniform diameter throughout than to have to design the upper sections with larger diameters to handle the higher fluid flows.

9.3.2 Pumparounds

The addition or removal of heat in side heaters or coolers is commonly carried out by drawing a fluid from a column tray, pumping it through a heat exchanger,

Table 9.16 Column Profiles with Side Coolers (Example 9.7, Reflux Ratio = 16)

Tray	Temperature °F	Liquid Flow lbmole/hr	Vapor Flow lbmole/hr	Product lbmole/hr	Duty MMBTU/hr
1	27	192	12	12	−1.15
2	57	185	204		
3	79	187	197		
4	91	188	199		
5	98	186	200		
6	105	132	198	48	
7	115	169	192		−0.30
8	129	161	229		
9	143	158	221		
10	153	156	218		
11	160	156	216		
12	164	156	216		
13	166	157	216		
14	167	157	217		
15	168	157	217		
16	168	157	217		
17	169	157	217		
18	169	156	217		
19	170	155	216		
20	171	128	215	25	
21	174	165	213		−0.30
22	181	165	250		
23	188	155	245		
24	196	150	240		
25 Feed	218	195	177		
26	240	200	180		
27	255	203	184		
28	266	206	188		
29	273	209	191		
30	278	15	194	15	1.48

then returning it to the column. Different pumparound configurations may be envisioned for circulating the liquid or the vapor. In general, it is the liquid that is pumped for exchanging heat, by either subcooling it or partially vaporizing it. The pumparound action itself, even with no heat exchange, can affect the column performance. Types of pumparounds with no associated heat exchange are considered next.

Either the entire amount of the liquid on a tray or part of it may be taken out and returned to the tray directly below it. The combined liquid flow (the pumparound and the liquid flowing directly to the tray below) would be the same as the liquid flow down to the tray below had there been no pumparound. From the standpoint of equilibrium stages, the column would perform exactly as though the pumparound did not exist although the tray efficiency may be affected due to the different flow pattern. If the liquid is returned several trays below the draw tray, the trays between the draw tray and the return tray are bypassed by the pumparound liquid. The amount of fractionation in that column section is therefore reduced. This pumparound thus tends to lower the overall number of effective trays in the column.

Table 9.17 Product Compositions with Side Coolers (Example 9.7)

	Mole/Hr	MMBTU/Hr	Mole Fraction			
			C2	C3	NC4	NC5
Reflux Ratio = 16						
Overhead	12		0.823	0.177	0.000	0.000
Condenser		−1.15				
Side Draw 1	48		0.035	0.852	0.113	0.000
Side Cooler 1		−0.30				
Side Draw 2	25		0.017	0.199	0.741	0.043
Side Cooler 2		−0.30				
Bottoms	15		0.000	0.001	0.070	0.929
Reboiler		1.48				
Reflux Ratio = 10						
Overhead	12		0.770	0.230	0.000	0.000
Condenser		−0.73				
Side Draw 1	48		0.045	0.804	0.151	0.000
Side Cooler 1		−0.30				
Side Draw 2	25		0.024	0.266	0.658	0.052
Side Cooler 2		−0.30				
Bottoms	15		0.000	0.001	0.086	0.913
Reboiler		1.05				

In another type of pumparound, the liquid drawn from the column is returned at a point a number of trays above the draw tray. This process also tends to lessen the effective number of trays and fractionation in that column section because of backmixing higher-boiling liquid from the lower tray with lower-boiling liquid in the upper tray.

The above indicates that, while heat removal in the upper sections of multiproduct columns improves fractionation, the means for heat removal could undermine it. The net outcome depends on the particular situation, but, in general, pumparounds should be designed to minimize any negative effect they might have on fractionation. For instance, the number of trays between the draw tray and return tray should be kept at a minimum. If a downward pumparound is used and no trays are bypassed between the draw tray and the return tray, no loss in fractionation is incurred.

The amount of heat exchanged in pumparounds depends on the pumparound rate and its temperature change across the heat exchanger. In downward pumparounds, the pumparound rate is limited by the liquid flow in the column. Since the temperature change is also limited by the temperature level of the available cooling or heating medium, the heat exchange rate in downward pumparounds is subject to practical limitations that could be restrictive. In upward pumparounds, the pumparound rate could, in principle, be increased indefinitely by increasing liquid circulation in the column between the draw tray and the return tray. This, of course, also has its practical limits imposed by column hydraulics considerations, discussed in Chapters 14 and 15. In general, higher heat exchange rates are achievable with upward pumparounds than with downward pumparounds.

9.4 MULTIPLE COLUMN PROCESSES

While multiproduct columns could be used to separate mixtures into products with distinct compositions, sharper separations usually require processing with multiple columns to achieve the desired splits and purities. A possible scheme would be to break down the separation process into a forward flowing series of single-feed, two-product distillation columns, each of which makes the separation between two key components. In a more complex multicolumn approach, the process may include recycle streams, external separation-enhancing streams, stream splits, recombination of streams, energy integration, etc., all aimed at achieving an economic optimum.

It is evident that a host of alternative arrangements of columns could be used for making the same set of separation specifications, with the number of arrangements escalating steeply as the number of components to be separated increases. Finding the optimum thus becomes a truly challenging task. Although computer simulation is the most efficient method for final evaluation and comparison of different plausible options, narrowing down the field on the basis of some systematic, logical approach is highly desirable for saving both computer time and human hours. Unfortunately, no systematic method for evaluating complex, multicolumn separation processes seems to exist, although workable heuristic rules are available for selecting the better sequences among the simple, forward-flowing processes.

For processes belonging to this category and involving only single-feed, two-product distillation columns, a single sequence exists for separating a binary:

$$AB \rightarrow A + B$$

That is, mixture AB is fed to a distillation column; the lighter component, A, is recovered in the overhead; and the heavier component, B, is recovered in the bottoms. (Components are arranged in the order of decreasing volatility.) Two sequences are possible for separating a ternary:

$$ABC \rightarrow A + BC \rightarrow A + B + C$$

$$ABC \rightarrow AB + C \rightarrow A + B + C$$

Two columns are needed in either of these sequences: the first to recover one of the components in the ternary and the second to separate the remaining two. For a four-component mixture, five alternative sequences could be used, each requiring three distillation columns:

$$ABCD \rightarrow A + BCD \rightarrow A + B + CD \rightarrow A + B + C + D$$

$$ABCD \rightarrow A + BCD \rightarrow A + BC + D \rightarrow A + B + C + D$$

$$ABCD \rightarrow AB + CD \rightarrow A + B + C + D$$

$$ABCD \rightarrow ABC + D \rightarrow A + BC + D \rightarrow A + B + C + D$$

$$ABCD \rightarrow ABC + D \rightarrow AB + C + D \rightarrow A + B + C + D$$

Sequences could be listed in this manner for any number of components.

For these simple processes, certain rules can be used to select a number of preferred sequences from which an optimum could be chosen. The rules are not absolute and will generally generate several options. Many factors that influence the final decision are specific to the particular process and its economics, etc., and are not considered in the selection rules. The final selection should be based on rigorous simulation models of the chosen sequences.

According to the rules proposed by Seader and Westerberg (1977), the components in the feed are arranged by decreasing volatility and two sets of parameters are checked: the relative volatilities of adjacent pairs and the molar percentages of the components in the feed. The relative volatility, α_{ij}, of component i with respect to component j is used as an indicator of how readily they could be separated and is defined as the ratio of their vapor–liquid equilibrium coefficients.

$$\alpha_{ij} = \frac{K_i}{K_j}$$

The separation sequence is determined according to the following rules:

1. Where the relative volatilities vary widely, splits are sequenced in the order of decreasing relative volatility.
2. Where the relative volatilities do not vary widely, the removal of components is sequenced in the order of their molar percentages in the feed, if these percentages vary widely.
3. If neither relative volatilities nor molar percentages vary widely, the components are separated in a direct sequence of columns; that is, the lightest component is separated as the overhead product of the first column, the bottoms is sent to the next column where the next heavier component is separated as overhead product, and so forth.

In the following example, separation strategy is considered for producing pentanes from a naphtha stream. The above heuristic rules are taken into account as well as other factors that are specific to the problem.

Example 9.8 Separating Pentanes from Naphtha

A mixture of primarily C5 components is to be separated from a naphtha stream containing hydrocarbons ranging from C3 to C10. It is proposed to utilize two existing columns to carry out the separation, one having 18 theoretical trays and the other 5. It is required to investigate alternative configurations for the process.

The feed composition, along with approximate average relative volatilities for the adjacent components over the expected range of conditions, is given in Table 9.13. The required separation is not between individual components but rather between three groups of components: those lighter than the pentanes (C3 through NC4), the pentanes (IC5 and NC5), and those heavier than the pentanes (NC6 through NC10). Consequently, the splits that matter are the NC4/IC5 and the

Table 9.18 Naphtha Feed and Component Relative Volatilities (Example 9.8)

	lbmole/hr	α_{ij}
C3	148.2	
		1.75
IC4	91.0	
		1.18
NC4	239.2	
		1.79
IC5	135.2	
		1.19
NC5	174.2	
		2.29
NC6	218.4	
		1.25
Benzene	65.0	
		1.12
Cyclohexane	101.4	
		1.45
NC7	288.6	
		1.21
Toluene	426.4	
		2.16
NC10	712.4	
Total	2600.0	

NC5/NC6 splits. Table 9.18 indicates that the relative volatility of the NC5/NC6 pair is somewhat higher than that of the NC4/IC5 pair. That is, the NC5/NC6 split is the easier one.

Although the relative volatilities do not vary too widely, the first heuristic rule suggests that the first split should be between NC5 and NC6. Thus, the NC6 and heavier components would be removed in the bottoms of the first column. The overhead would then be separated in the second column into NC4 and lighter components in the overhead and the C5s in the bottoms (Figure 9.8A).

Another factor that should be considered in this application is the fact that the separation is to be carried out in two existing columns of different sizes. Since the easier separation requires fewer trays, the logical choice would be to use the smaller column for the initial split and the larger column for recovering the C5s. Figure 9.8 shows some of the possible configurations for the separation. Figure 9.8D shows a sidestripper arrangement which differs from the rest in that it separates the pentanes as a side draw from the main column. This stream is then sent to the smaller column, which is used as a stripper to remove some of the lighter components in the overhead. The overhead stream is recycled back to the main column to recover some of the pentanes carried over.

Comparing the different options requires an economic analysis, although a good approach would be to compare the total reboiler duties required in each configuration

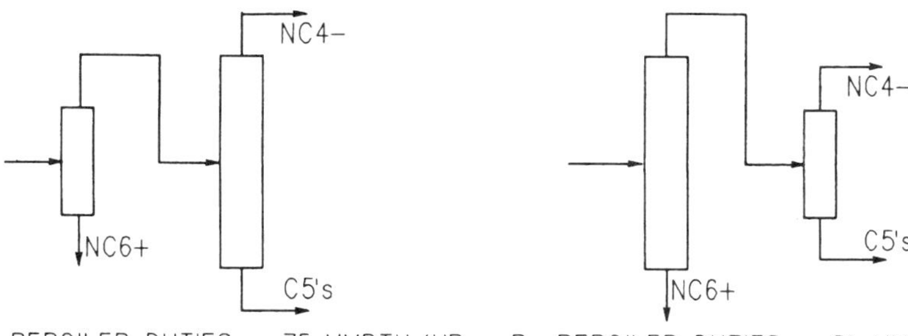

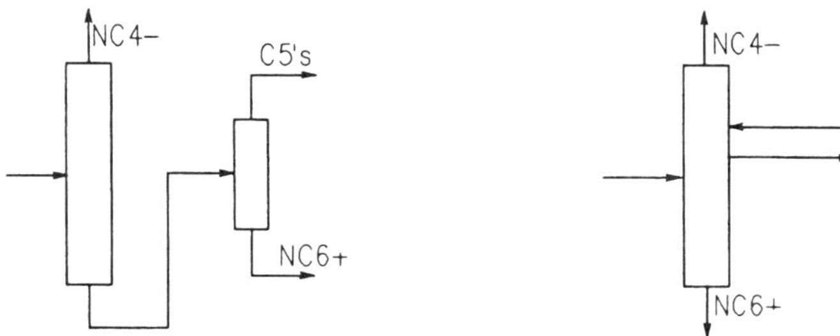

Figure 9.8   Alternative configurations for naphtha splitting (Example 9.8).

to achieve similar separations and recoveries. The duty requirements for the different configurations are obtained from simulation and are shown in Figure 9.8. Based only on this criterion, the results appear to support the heuristic rules, which favor option A. This is followed closely by the recycle scheme, option D, which is not considered in the heuristic rules.

NOMENCLATURE

K_i Vapor–liquid equilibrium coefficient

L Liquid molar flow rate in the column

S Side products

V Vapor molar flow rate in the column

α_{ij} Relative volatility of component i with respect to j

SUBSCRIPTS

i Component designation

j Component designation

REFERENCE

Seader, J. D. and A. W. Westerberg, *AIChE J.*, 23, 951, 1977.

PROBLEMS

9.1 An existing column has a reboiler, no condenser, and two feeds, one of which is at the top of the column. The top feed is of fixed temperature, pressure, and composition, but its rate is variable. The other feed is of fixed rate, composition, and pressure and is flowed through a heat exchanger where its temperature may be controlled. How many independent variables are required to define this process? If the key components are ethane and propane, describe conceptual control loops to control the separation and performance of this process.

9.2 Two light hydrocarbon streams are to be separated into propane-rich and butane-rich products using one distillation column. Assuming the streams are of fixed rates, compositions, and thermal conditions, and that the column pressure is predetermined, how many degrees of freedom are available at the design phase if

a. The streams are mixed and fed at the same tray.
b. Each stream is fed separately to the column.

Given the following feed stream definitions, what would be the recommended relative feed locations if they are fed separately?

	Feed 1	Feed 2
	lbmole/hr	
Ethane	15	25
Propane	65	82
Butane	37	77
Pentane	13	16
Temperature, °F	50	75
Pressure, psia	100	100

9.3 A column has two feeds, a partial condenser, a partial reboiler, a vapor distillate, a bottoms product, and a side draw. The feeds are of fixed flow rates, compositions, and thermal conditions.

a. How many degrees of freedom exist during the design phase?
b. How many degrees of freedom exist once the column is built?

9.4 A distillation column is to be designed to separate a stream into two products, and to remove light impurities. A partial condenser is proposed, where the impurities would be removed in the vapor distillate, and the liquid distillate and bottoms would be the main column products. Preliminary simulation calculations indicate that too much of the distillate product would be lost with the impurities, and the distillate product would contain too many impurities. How would you modify the column design to improve its performance?

9.5 A distillation column with a total condenser and a reboiler separates a stream into a distillate product, heavy residuals in the bottoms, and a main intermediate product as a side draw several trays above the reboiler. For an existing column, with a feed of fixed rate, composition, and thermal conditions, determine the number of variables required to define the column operation. Assume the column pressure is set, and no heat leaks occur except in the condenser and reboiler. How would you control the separation in the column?

9.6 It is proposed to utilize a hot process stream as a heat source to supplement the reboiler duty of an existing distillation column. The heat from the process stream would be added to the column via a pumparound a few trays above the reboiler. How many additional degrees of freedom would result from the proposed modification?

9.7 A column is designed for f feeds, a partial condenser with only a vapor product, a reboiler, a bottoms product, s side draws, and h pumparound heaters/coolers. Assuming the feeds are of fixed rates, compositions, and thermal conditions, determine the number of independent variables required to uniquely define the column during design, and its operation after construction.

9.8 A distillation column side draw constitutes the main product of the column, where lighter impurities are removed in the distillate and heavier impurities are removed in the bottoms. Because of anticipated variations in the feed composition the column is equipped with a side cooler just below, and a side heater just above the side draw to improve operating flexibility. What potential operating problems could the side cooler and heater help alleviate? What specifications would you use to define the column operation?

9.9 A stream containing components A, B, C, D, E, and F is to be separated into these six components using five two-product ordinary distillation columns. The composition of the feed and the relative volatilities of the components are listed below, with the components arranged by decreasing volatility. What separation sequence would you recommend for this process?

	Mole Percent in Feed	Relative Volatility
A	5	
		1.6
B	17	
		1.5
C	35	
		2.4
D	14	
		1.7
E	16	
		1.3
F	13	

9.10 How many operating degrees of freedom exist for the column configuration shown in Figure 9.8D? The main column has a reboiler and a partial condenser with only a vapor distillate. The sidestripper has a reboiler but no condenser. The external feed is of fixed rate, composition, and thermal conditions.

CHAPTER 10

Azeotropic and Extractive Distillation Three-Phase Distillation

The formation of azeotropes due to deviations from Raoult's law was discussed in Section 1.3. An azeotrope is a mixture that, at a given pressure (the azeotropic pressure), boils at a constant temperature (the azeotropic temperature), and has the same composition (the azeotropic composition) in the equilibrium vapor and liquid phases. Homogeneous azeotropes are those that form one liquid phase at equilibrium with the vapor; heterogeneous azeotropes are those that form two liquid phases at equilibrium with each other and the vapor.

Homogeneous azeotropes can be either minimum boiling, where the boiling point of the azeotrope is lower than any of those of the constituents, or maximum boiling, where the boiling point of the azeotrope is higher than any of those of the constituents. Heterogeneous azeotropes are invariably minimum boiling. Azeotropes can be multicomponent although those most commonly encountered in practice are either binary or ternary.

The absence of an interphase compositional differential makes the separation of azeotropes into their constituent components impossible by conventional vapor–liquid separation processes. The azeotropic pattern, however, may be altered by adding an external component — an entrainer — which breaks the original azeotrope while forming new ones with the feed components. The process may be designed so that the new azeotropes may be separated from individual components or from other azeotropes by what is known as *azeotropic distillation*. Azeotropic distillation is not limited to the separation of azeotropes but is also used for separating close boilers that are difficult to separate by conventional distillation.

Extractive distillation is another process used for separating azeotropes or close boilers. In this process, a solvent is added that alters the relative volatilities of the feed components through its preferential affinity for one or more of the components over the others.

In analyzing the complex processes of azeotropic and extractive distillation, it is convenient to represent the system as a ternary that includes the two key components to be separated and the entrainer or solvent. Real mixtures are, of course, normally multicomponents, but focusing on the key components and the separating agent helps in obtaining a basic grasp of the problem. A qualitative understanding of the operation is important for preliminary evaluations aimed at selecting the entrainer or solvent; estimating the amount of entrainer or solvent, the approximate number of trays required, and the feed location; and considering any associated ancillary operations that may be essential for the process. With this background, the use of computer programs for accurate simulation of the process becomes highly effective.

10.1 THE FUNCTION OF ENTRAINERS AND SOLVENTS

The terms *entrainer* and *solvent* are commonly used interchangeably to refer to the separating agent used to enhance the separation of close boilers or azeotropes by azeotropic or extractive distillation. For consistency, the term "entrainer" will be used to designate the azeotropic distillation agent and "solvent" the extractive distillation agent.

In azeotropic distillation, when two components are difficult to separate due to the proximity of their boiling points or because they form an azeotrope, an entrainer is added which forms an azeotrope with at least one of the components to be separated. The entrainer must be selected so that the azeotropes that are formed have sufficiently spaced boiling points to facilitate their separation by distillation. Moreover, it must be verified that the azeotropes that are formed can themselves be separated by relatively straightforward and economical means. The number of suitable entrainers for a given situation is, therefore, quite limited, especially when other factors in the selection process are taken into account such as cost, stability, and safety.

The entrainer is usually introduced to the column with the feed or may be added in the column as reflux.

In extractive distillation the solvent attracts one of the components to be separated more strongly than the other on account of their different chemical structures. The solvent has a markedly higher boiling point than the components in the feed and tends to lower the volatility of one of the components preferentially. This component leaves the column at the bottoms with the solvent. Recovering the solvent from the bottoms is relatively easy because of its low volatility compared to the component to be separated.

Hydrogen bonding is one contributor to the preferential attraction between the solvent and certain components. Similarity of the chemical structure is another. For instance, the separation of benzene and cyclohexane, two close boilers that also form an azeotrope, can be achieved by extractive distillation using phenol as the solvent. Benzene is more strongly attracted to phenol because both contain an aromatic ring, whereas cyclohexane does not. Therefore, in this process, benzene is taken in the column bottoms with the phenol, and cyclohexane is recovered in the column overhead.

The choice of a selective solvent is easier the more the components to be separated differ in their chemical structure. It would be difficult or impossible, for instance, to find a selective solvent for the separation of stereoisomers. Nevertheless, the restrictions on extractive distillation solvents are less severe than those on azeotropic distillation entrainers since the boiling point limitation on solvents is virtually nonexistent on the high end.

Due to its low volatility, the solvent in extractive distillation is introduced near the top of the column, several trays above the main feed tray. This is to ensure adequate concentration of the solvent on most trays, thereby maintaining the desired relative volatility throughout the column.

10.2 SEPARATION PROCESSES

Various processes are used for separating components that are difficult or impossible to separate by conventional distillation. Whether the difficulty of separation arises from the components' close boiling points or their tendency to form azeotropes, the separation processes must take into account the complex vapor–liquid equilibrium relationships of the system. The system to be considered involves both the components to be separated and the separating agent that, in one way or another, enhances the desired separation. The vapor–liquid equilibria of such mixtures is highly nonideal, and it is precisely this nonideality that is capitalized on to effect the separation.

In view of the complexity of these processes, they are typically examined on a case-by-case basis with regard to selection of the separating agent, configuration of the column(s), process conditions, etc. Nevertheless, certain general principles apply that allow categorizing the methods as described below. The processes are discussed generically, and actual applications are cited as examples for each process.

10.2.1 Separating Azeotropes with Pressure-Sensitive Composition

This is one case in which the separation of azeotropes may be carried out without the use of a separating agent. The method applies to minimum- or maximum-boiling homogeneous azeotropes that are characterized by an appreciable variation of the azeotropic composition with pressure. The effect of pressure on the azeotropic composition may be estimated for a binary, provided the activity coefficients can be adequately represented by one of the activity coefficient equations (Section 1.3.4).

The Y–X and T–X diagrams of a typical minimum-boiling, homogeneous, binary azeotrope are shown in Figure 10.1, where compositions refer to component A, the lighter (lower boiling) component to be separated from component B. The diagrams are given at two pressure levels, showing the directional effect of the pressure on the azeotropic composition.

The separation process employs two columns at two different pressures to take advantage of the effect of pressure. The details of the process may vary from case

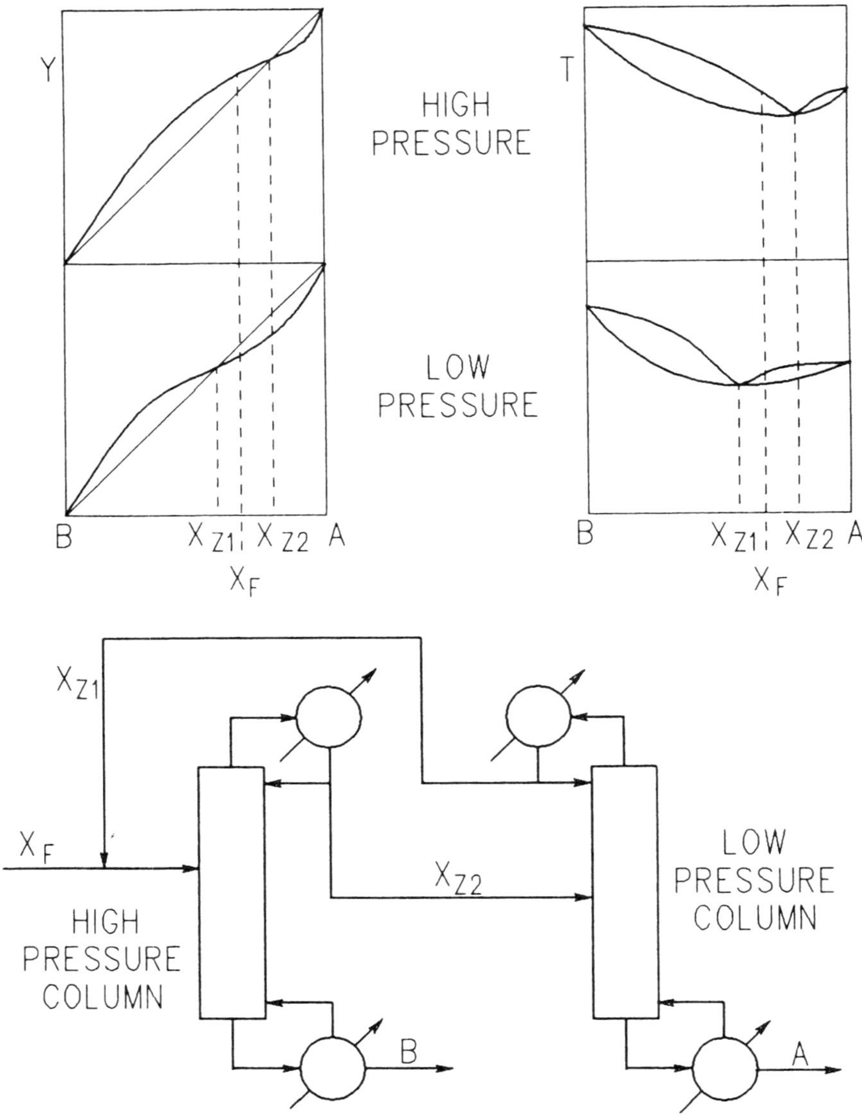

Figure 10.1 Separating azeotropes with pressure-sensitive composition.

to case, depending in part on the feed composition relative to the azeotropic compositions at the two column pressures. Assume the column pressures are such that the feed composition, X_F, is between the azeotropic compositions, X_{Z1} and X_{Z2}, at the lower and higher column pressures, respectively. Sending the feed to the lower pressure column would produce relatively pure component A in the bottoms and the azeotrope in the overhead. Sending the feed to the higher pressure column would produce relatively pure component B in the bottoms and the azeotrope in the overhead.

Figure 10.1 shows a schematic for one possible process. The feed is mixed with the overhead from the low-pressure column then fed to the high-pressure column, where component B is separated in the bottoms and the higher pressure azeotrope is removed in the overhead. The overhead is sent to the lower pressure column, where the azeotropic composition is leaner in component A than the feed to this column. This column thus separates component A in the bottoms and the lower pressure azeotrope in the overhead. To complete the cycle, this overhead stream is mixed with fresh feed and sent to the high pressure column. The combined composition is leaner in component A than the fresh feed, thus ensuring that the combined feed is to the left of the higher pressure azeotropic composition.

The column sequence could be reversed and other configurations could be laid out for different feed compositions. For instance, if $X_F < X_{z1}$, the feed must go first to the high-pressure column so that the resulting azeotropic composition will be on the other side of the lower-pressure azeotropic composition. If $X_F > X_{z2}$, the feed must first be sent to the lower-pressure column. The final selection of a particular configuration among possible alternatives is best made on the basis of computer simulation of the various options to determine the economic optimum.

The concept is the same for separating maximum-boiling, homogeneous, binary azeotropes by using two columns at different pressures. In this case, however, the pure components are recovered as overhead products since they are lighter than the azeotrope.

Examples of pressure-sensitive, homogeneous, binary azeotropes that are amenable to separation by processes as described above are given in Table 10.1, extracted from Othmer (1963).

Graphical Representation

Vapor–liquid equilibrium diagrams of homogeneous binary azeotropes may be viewed as consisting of two parts, each resembling a regular binary diagram. For instance, the Y–X diagrams in Figure 10.1 consist of a curve between the heavy component and the azeotrope and another one between the azeotrope and the light component. Each of these curves resembles a regular binary Y–X curve, although the second one lies below the diagonal. This is because the azeotrope, with its lower boiling point, is to the left of the higher boiling component A.

The significance of this analogy to regular binary systems is that the graphical techniques of binary distillation discussed in Chapters 5 and 6 may be applied to distillation columns used for separating a pure component from an azeotrope. The effect of such parameters as number of trays, reflux ratio, and feed location may be studied, at least qualitatively, using these graphical techniques for preliminary column design and for determining performance trends. Once a fundamental understanding of the process is achieved, a computer model can be set up to carry out the detailed rigorous calculations.

Table 10.1 Homogeneous Azeotropes with Pressure-Sensitive Compositions

	Pressure (mmHg)	Temperature (°C)	Composition (1st comp)
Minimum-Boiling Binaries			
n-Propyl alcohol/water	740	87	71.7 wt.%
	5,930	151	73.3 wt.%
Acetonitrile/water	150	34	72.5 mol%
	760	77	84.5 mol%
Ethyl alcohol/water	100	34.2	89.0 mol%
	1,450	95.3	99.6 mol%
Methyl alcohol/benzene	200	26	33.1 wt.%
	11,000	149	63.0 wt.%
Methyl alcohol/MEK	100	18.4	52 wt.%
	2,040	92.3	85 wt.%
Acetone/water	2,570	95.8	78.0 mol%
	10,320	155.8	96.5 mol%
Ethyl acetate/ethyl alcohol	25	−1.4	61.13 wt.%
	1,475.5	91.4	87.15 wt.%
Maximum-Boiling Binaries			
HCl/water	50	48.7	19.3 wt.%
	1,210	123	23.5 wt.%
HBr/water	100	74.12	47.03 wt.%
	1,200	137.34	49.80 wt.%
MEK/water	200	39.9	65.4 mol%
	760	73.3	72.2 mol%

Equilibrium curve sections on the Y–X diagram to the left and right of the azeotrope are used each for a different column. Operating lines drawn on one section cannot be extended across the azeotropic point to the other section. The section of the equilibrium curve that lies below the diagonal may be redrawn above the diagonal if the azeotrope and the heavy pure component places are switched by expressing the compositions in terms of the heavier rather than the lighter component.

Corresponding to each equilibrium curve section, a column may be designed with the required number of trays and reflux ratio to achieve the desired component purity at one end of the column and the azeotropic composition at the other.

10.2.2   Separating Heterogeneous Minimum-Boiling Azeotropes

This is another case where azeotropes can be separated without the addition of a separating agent. The process applies to heterogeneous minimum-boiling azeotropes: those that form two liquid phases in certain regions of temperature, pressure, and composition. Figure 10.2 shows a schematic T–X diagram of a typical heterogeneous binary minimum-boiling azeotrope. The compositions are expressed as mole fractions of the lighter component.

A fresh feed with a composition in region L1 (single liquid with predominantly component 1) may be distilled in a conventional column to produce essentially pure component 1 in the bottoms and an overhead at or close to the azeotropic composition X_{az}. If the feed composition is in region L2 (single liquid with predominantly

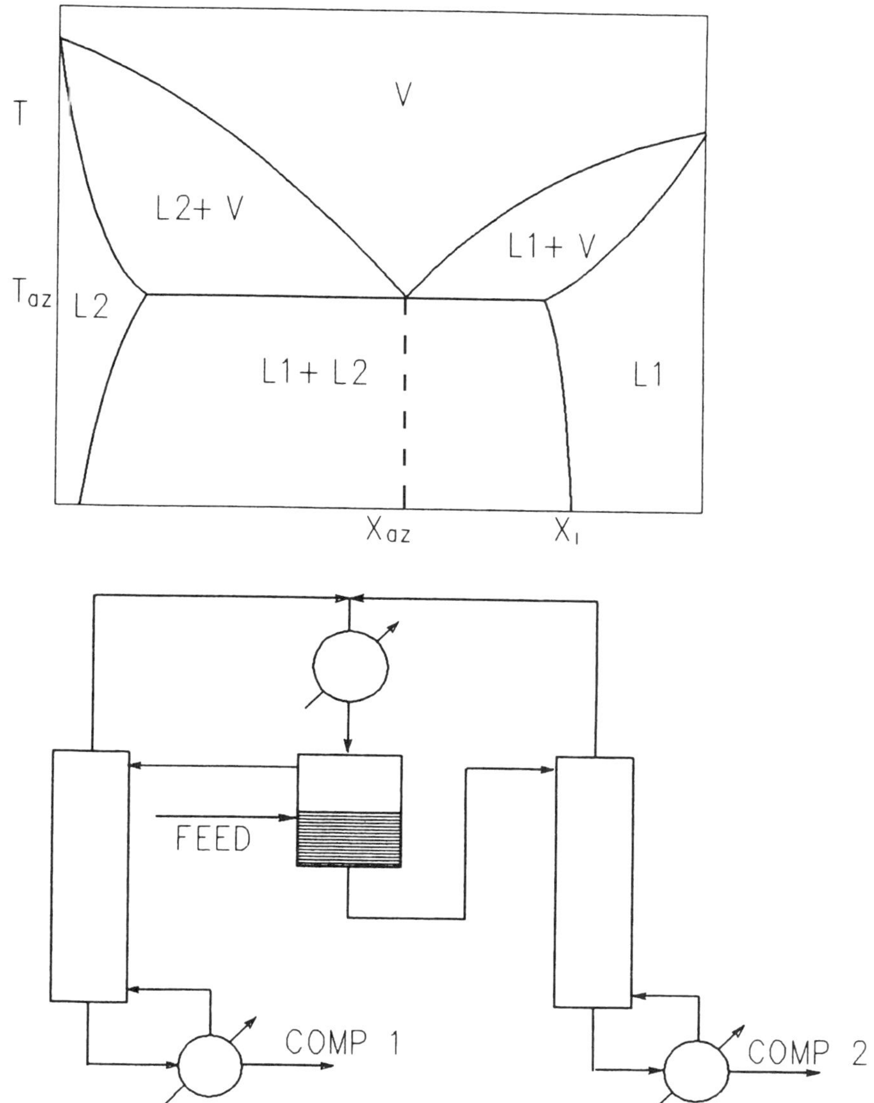

Figure 10.2 Separating heterogeneous minimum-boiling azeotropes.

component 2), it may be separated by conventional distillation into a bottoms product that is relatively pure component 2 and an overhead that is around the azeotropic composition. The overhead from either column may be cooled below the azeotropic temperature, T_{az}, to the two-liquid phase region where the two phases, one rich in component 1 and the other rich in component 2, are separated by decantation.

Figure 10.2 shows a schematic of a separation process for a feed with a composition in the two-liquid phase region. The liquid phases are separated and each is distilled in a separate column as described above. The overhead streams from both columns are recycled to the two-phase separator, where they are combined with the fresh feed.

Examples of the separation of heterogeneous azeotropes using such a process include n-butanol–water, ethyl acetate–water, and ethyl ether–water (Othmer, 1963).

Graphical Representation

Each of the distillation columns used in the heterogeneous azeotropic distillation is essentially a conventional column separating a binary: one column separates component 1 and the azeotrope, and the other separates component 2 and the azeotrope. As such, the corresponding Y–X diagrams can be used for preliminary evaluation of the process using graphical methods as described in Chapters 5 and 6. On this foundation rigorous computer simulation provides accurate process computations for design and optimization.

10.2.3 Separation by Forming an Azeotrope with One Component

This process consists of adding to the feed a separating agent or entrainer, E, which forms an azeotrope with only one of the components to be separated, A or B. By adding the entrainer, the composition is shifted in favor of forming an azeotrope, AE, between the entrainer and component A while breaking the original azeotrope AB. If the purpose is to produce pure component B, then B is separated from the azeotrope AE by distillation. The problem is obviously transformed from separating AB to separating AE. For the process to be viable, there should be an economical means for separating AE. If AE has a pressure-sensitive azeotropic composition or is a heterogeneous azeotrope, methods 10.2.1 or 10.2.2 could be used to round out the process.

The separation process depends on the nature of the vapor–liquid equilibrium relationships of the system, which can be represented on a ternary diagram. Figure 10.3a shows a ternary diagram at some fixed system pressure. Components A and B are close boilers, and A forms an azeotrope with the entrainer E. The curves in the triangle represent liquid isotherms. A corresponding vapor isotherm (not shown) could be drawn to represent the vapor at equilibrium with each liquid curve with tie lines joining vapor and liquid compositions at equilibrium. The temperature of the isotherms reaches a minimum at point Z that corresponds to the composition of the azeotrope formed between A and E.

The feed composition is represented by point F in Figure 10.3b. An amount of entrainer is added to the feed, and the mixture is then separated into B and the azeotrope. Line EF represents compositions of mixtures formed by combining the feed with varying amounts of the entrainer. In order to separate the mixture into pure B and the azeotrope, the combined mixture must have a composition that lies on a straight line joining B and Z. Thus, the combined feed composition should correspond to point F_C, the intersection of BZ and EF.

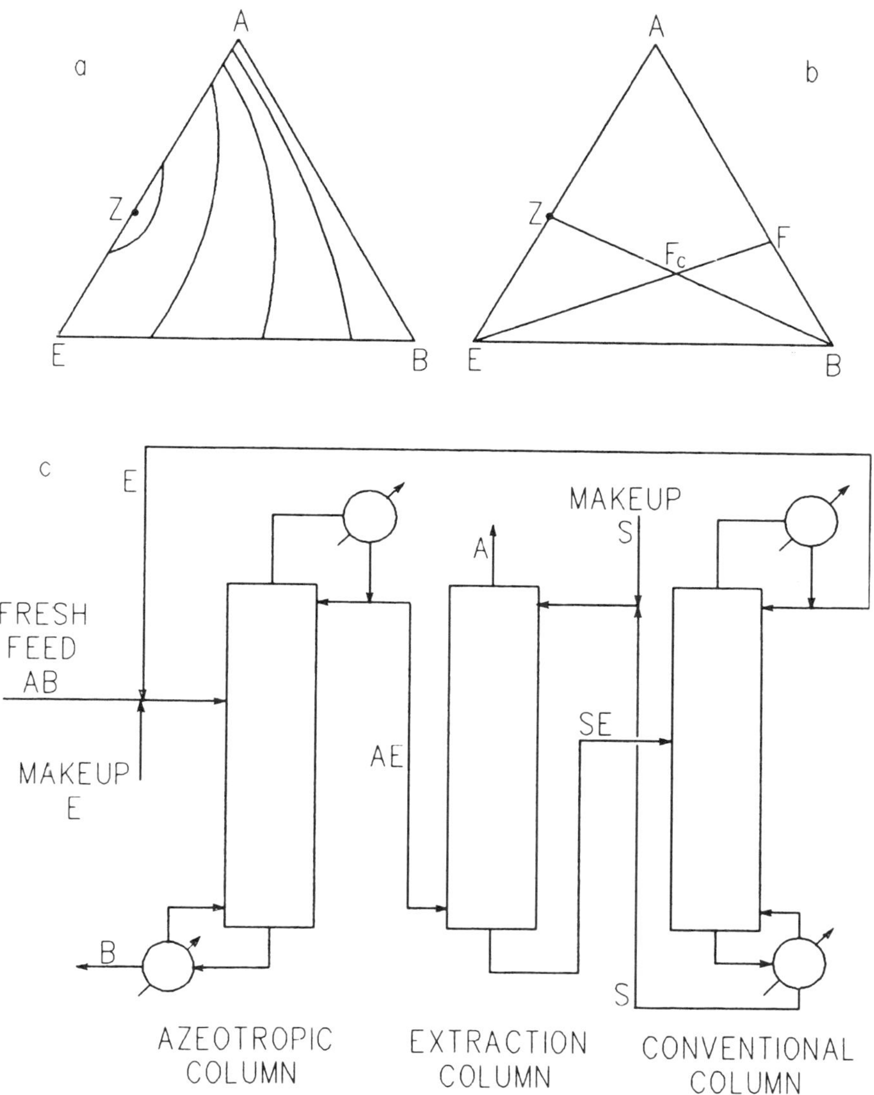

Figure 10.3 Separation by forming an azeotrope with one component.

The amount of entrainer required may be calculated by a material balance on component B. The total feeds (fresh feed + entrainer) equal the total products (azeotrope in the overhead + component B in the bottoms),

$$F_C = F + E = Z + B$$

where F_C is the combined feed rate, F is the fresh feed rate, E is the entrainer rate, Z is the azeotrope product rate, and B is the bottoms rate, assumed to be pure component B. A material balance on component A is written as

$$FX_{AF} = ZX_{AZ}$$

where X_{AF} is the fraction of component A in the fresh feed, and X_{AZ} is the fraction of component A in the azeotrope. It is assumed that all of component A leaves the column with the azeotrope and all of component B leaves in the bottoms. From a material balance on the entrainer

$$ZX_{EZ} = E$$

where X_{EZ} is the fraction of entrainer in the azeotrope. Here again it is assumed that all the entrainer leaves the column with the azeotrope. The above equations are combined to calculate the required ratio of entrainer to fresh feed:

$$\frac{E}{F} = \frac{X_{EZ}X_{AF}}{X_{AZ}}$$

Figure 10.3c is a schematic of a possible process for the separation of such a system. The fresh feed is mixed with the azeotrope-forming entrainer and is then fed to the azeotropic column, where pure component B is taken as column bottoms and the azeotrope AE is taken as overhead. The next step is to separate AE, an azeotrope that cannot be separated by simple distillation. A workable process involves liquid–liquid extraction where a solvent S is used to extract the entrainer, leaving a product that is virtually pure A. The solvent and entrainer are then separated by conventional distillation, the entrainer is recycled and mixed with fresh feed and makeup entrainer, and the solvent is recycled to the liquid–liquid extractor where some makeup solvent may be added.

This process can be used, for instance, to separate benzene and cyclohexane, which are very close boilers. Acetone, which forms an azeotrope with cyclohexane, is added as an entrainer. The cyclohexane is separated from the acetone by extraction with water, which dissolves the acetone. The cyclohexane is practically immiscible in water. The water–acetone solution is separated by simple distillation.

Similar processes are used for separating other close boiling aromatics and paraffins such as the separation of toluene and n-heptane using as entrainer methyl ethyl ketone, which forms a minimum-boiling azeotrope with the heptane. The azeotrope is taken in the overhead and the toluene in the bottoms.

10.2.4 Separation by Forming Two Binary Azeotropes

For certain close boilers or difficult-to-separate binaries, an entrainer may be selected that forms a binary azeotrope with each one of the components to be separated. Figure 10.4 shows a ternary diagram of such a system. Entrainer E forms azeotropes Z_A with A, and Z_B with B. If the difference between the boiling points of the two azeotropes is large enough, they may be separated by fractionation. There should also be a monotonic equilibrium temperature variation with composition from one azeotrope to the other to allow the separation to take place. Temperature ridges in the ternary diagram could result in a maximum in the column temperature profile, which would cause column instability or make the separation impossible. Also, the feasibility of separating the azeotropes leaving the column should be ascertained.

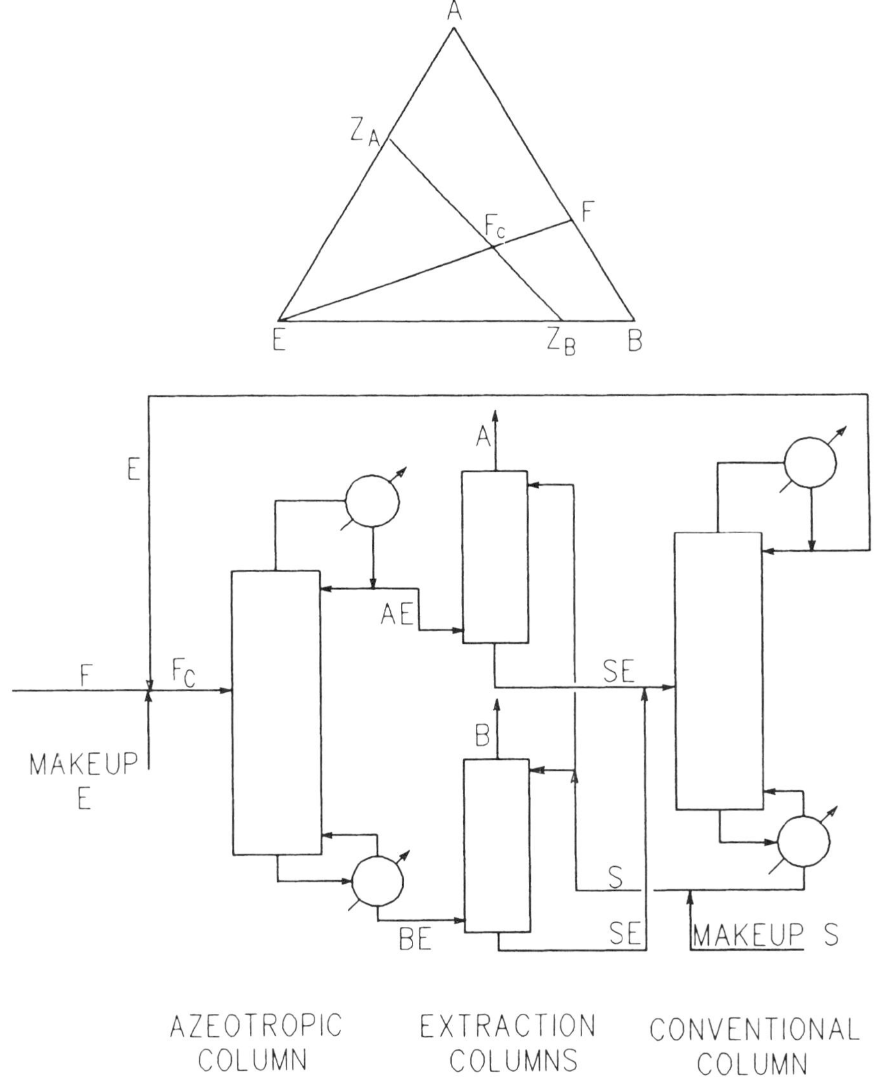

Figure 10.4 Separation by forming two binary azeotropes.

Figure 10.4 also represents graphically the material balance that determines the amount of entrainer that must be added to the feed to generate products Z_A and Z_B. The fresh feed composition is represented by point F. The combined composition of feed and entrainer must lie on EF. The intersection of EF and $Z_A Z_B$ at F_C represents the combined composition of the column feed.

The required amount of entrainer is calculated by material balance. From overall material balance,

$$F_C = E + F = Z_A + Z_B$$

where E, F, and F_C are the fresh feed rate, entrainer rate, and combined rate, and Z_A and Z_B are the rates of azeotropes Z_A and Z_B, respectively. The following equations represent material balances on components A, B, and E:

$$FX_{AF} = Z_A X_{AZA}$$

$$FX_{BF} = Z_B X_{BZB}$$

$$E = Z_A X_{EZA} + Z_B X_{EZB}$$

The mole fraction subscripts represent components A, B, and E in the feed and in the two azeotropes. It is assumed in these approximate calculations that components A and B and the entrainer are totally involved in forming the azeotropes and that the separation between the two azeotropes is perfect. Combining the above equations gives the required ratio of entrainer to feed:

$$\frac{E}{F} = \frac{X_{AF} X_{EZA}}{X_{AZA}} + \frac{X_{BF} X_{EZB}}{X_{BZB}}$$

A typical flowsheet for the separation is also shown in Figure 10.4. The fresh feed is combined with the entrainer then fed to the azeotropic column, where the two azeotropes are separated. Each azeotrope product is sent to a liquid–liquid extraction column to dissolve the entrainer in some suitable solvent. The combined entrainer–solvent solution from both extractors is sent to a distillation column to separate the entrainer and solvent. These are recycled to the azeotropic column and extractors with makeup added to each, as needed.

Typical mixtures that can be separated by a process such as this include mixtures of close-boiling aromatics and paraffins using ethanol or methanol as entrainers. The aromatics such as toluene form the higher-boiling azeotrope while the alcohol and the paraffins form the lower-boiling azeotrope. Water is used to remove the alcohol from the azeotropes by extraction. Finally, the alcohol must be separated from the water and recycled. From this standpoint, methanol is preferred over ethanol because the latter forms an azeotrope with water, thus complicating the ethanol recovery.

10.2.5 Separation by Forming a Ternary Azeotrope

In this process, the entrainer E, added to the close-boiling or azeotrope-forming binary AB, forms a low-boiling ternary azeotrope, ABE. The ternary azeotrope boils at a temperature lower than A or B or their binary azeotrope, if it exists. The ABE azeotrope is removed in the column overhead, while pure component A or B is recovered in the bottoms.

The composition of the ternary azeotrope as well as the composition of the feed determine the relative rates of entrainer and feed required to produce a pure product in the bottoms and the azeotropic composition in the overhead. Figure 10.5 shows a typical ternary diagram for ABE at expected process temperature and pressure. Two liquid phases coexist under the curve, and one liquid phase exists above the curve. The tie lines connect points in the two phases at equilibrium with each other. Point F represents the fresh feed, which is a binary mixture of A and B. Point Z represents the ternary azeotropic composition. A straight line drawn through E and F represents compositions obtained by mixing fresh feed with variable amounts of entrainer. The combined feed composition should be such that it would separate into pure component A and the ternary azeotrope. Therefore, the combined feed composition should fall on a straight line joining A and Z. The amount of entrainer added to the fresh feed should thus yield the composition represented by the intersection point F_C.

The amount of entrainer is calculated by material balance. Using similar nomenclature for rates and compositions as in the previous section, the following equations are written for total component material balances:

$$F_C = E + F = A + Z$$

$$FX_{AF} = A + ZX_{AZ}$$

$$FX_{BF} = ZX_{BZ}$$

$$E = ZX_{EZ}$$

The entrainer-to-feed ratio is calculated by combining the last two equations:

$$\frac{E}{F} = \frac{X_{EZ}X_{BF}}{X_{BZ}}$$

These derivations are based on the assumption of total azeotrope formation and perfect separation between component A and the ternary azeotrope.

A schematic of a flowsheet utilizing the ternary azeotrope phenomenon to separate components A and B is shown in Figure 10.5. The required amount of entrainer may be added either to the fresh feed or directly to the azeotropic column as reflux. In this process the entrainer is added as reflux to the azeotropic column. The ternary azeotrope is taken as column overhead and condensed, whereupon it separates in

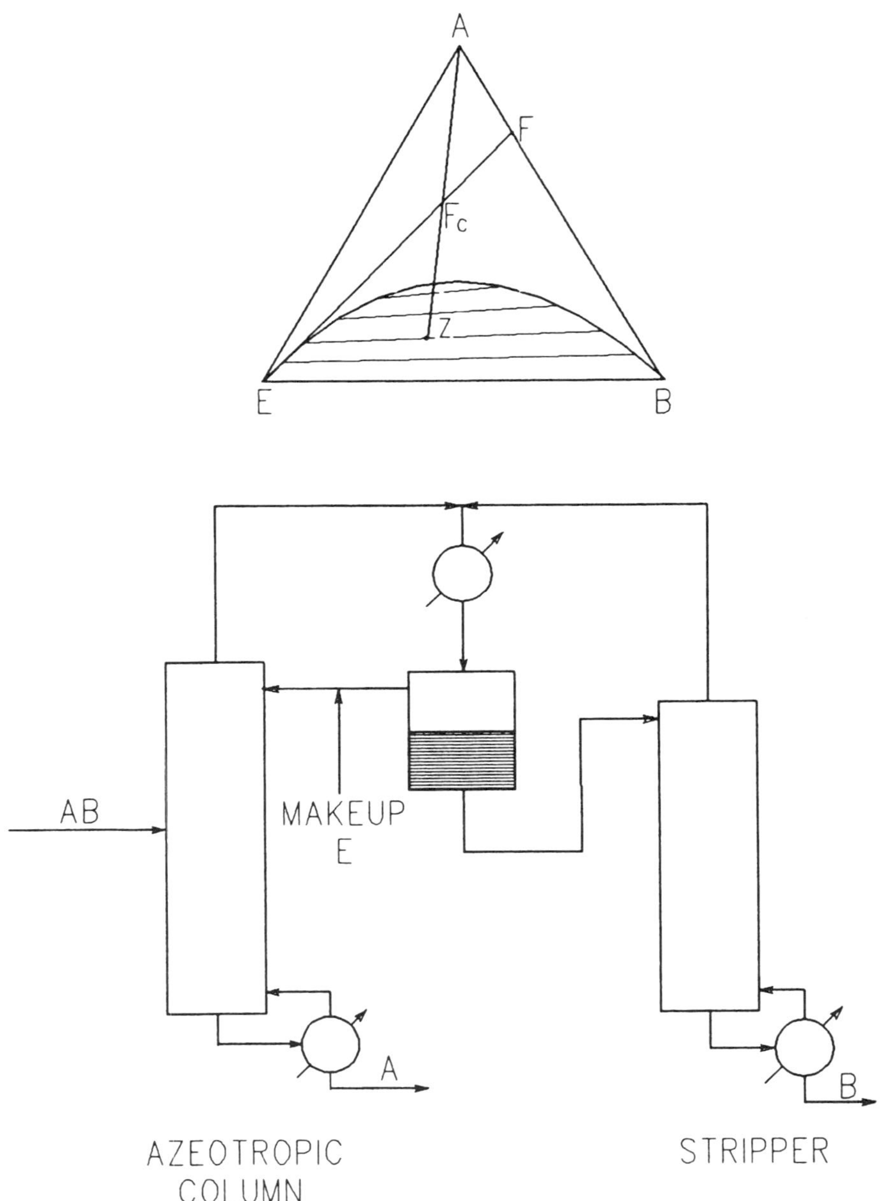

Figure 10.5 Separation by forming a ternary azeotrope.

the receiver into two liquid phases, one rich in the entrainer and the other rich in component B.

Almost pure component A is taken as azeotropic column bottoms. The purity of A is determined by the relative amount of entrainer added. Too little entrainer will not remove sufficient amounts of B in the overhead, thus allowing part of it to flow down with the bottoms. Too much entrainer may result in part of it contaminating the bottoms.

The entrainer-rich liquid from the receiver is returned to the azeotropic column as reflux, and the liquid phase rich in component B is sent to a stripper column, where any amounts of A and E and some of B are removed in the overhead and combined with the azeotropic column overhead. The stripper bottoms is primarily component B.

At steady state, and aside from inevitable small losses, almost all of component A in the fresh feed is recovered as azeotropic column bottoms and B as stripper column bottoms. The entrainer, once introduced into the system, circulates within it indefinitely, and any losses are replaced by fresh makeup. The reflux rate and composition are controlled to maintain the ternary azeotrope composition in the overhead. The reflux composition is controlled by the inventory of entrainer in the receiver and to some degree by the receiver temperature.

Typical of this process is the production of anhydrous ethanol from a concentrated ethanol solution in water, using benzene or cyclohexane or other components as the entrainer.

10.2.6 Separation by Extractive Distillation

In extractive distillation a solvent is utilized that preferentially alters the relative volatilities of the components to be separated. In contrast to the entrainer in azeotropic distillation, the solvent in extractive distillation has a considerably higher boiling point than either component in the feed and is therefore always withdrawn in the column bottoms.

The solvent could depress the volatility of either the lighter or the heavier of the components to be separated. Thus, it is not unusual for the relative volatilities of the components in the feed to be reversed by the addition of the solvent. Because the solvent in extractive distillation is considerably less volatile than the feed components, no azeotropes are formed, and the separation of the solvent from the bottoms product is relatively easy.

The low volatility of the solvent also dictates that it be introduced in the column near the top, above the main feed, in order to maintain the required solvent concentration in as many trays as possible. The column thus contains three sections: the top section between the overhead product and the solvent feed, the middle section between the solvent feed and the main feed, and the bottom section between the main feed and the bottoms product. The top section serves to reflux any solvent that might have risen with the vapor above the solvent feed. The object is to minimize the solvent concentration in the overhead product. The middle section serves to absorb the component whose volatility has been lowered by the solvent and to minimize its concentration in the overhead. The lower section serves to strip out the component with the higher volatility and minimize its concentration in the bottoms.

Figure 10.6 illustrates a typical extractive distillation process consisting of the extractive distillation column and the solvent recovery column. Fresh feed containing the binary AB is introduced around the middle of the extractive distillation column, and the solvent S is introduced near the top. Components A and B are close boilers and/or potentially azeotrope formers that are difficult or impossible to separate by ordinary distillation. Whether as individual components A is more volatile than B or vice versa, in the presence of the solvent, B becomes less volatile due to its higher affinity to the solvent. As a result, essentially pure A is distilled as the overhead of the extractive distillation column. Component B is entrained with the solvent in the bottoms stream, which is sent to the solvent recovery column. The solvent is substantially less volatile than component B, allowing easy separation by ordinary distillation. Practically pure B is recovered in the overhead, and pure solvent in the bottoms. The solvent is recycled to the extractive distillation column with makeup that might be required to compensate for losses.

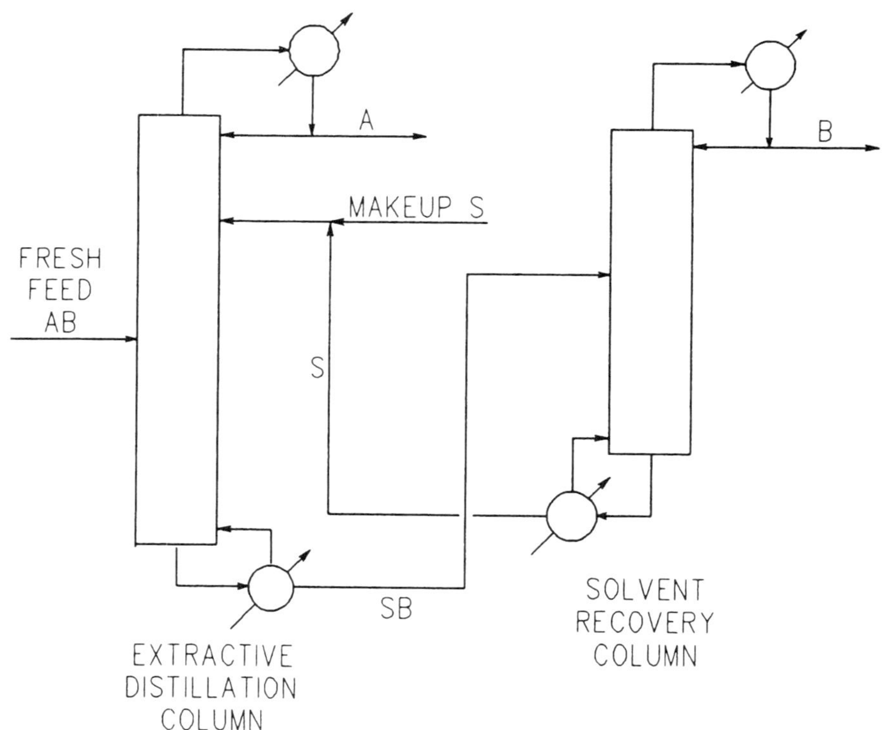

Figure 10.6 Separation by extractive distillation.

Typical mixtures that can be separated by extractive distillation in processes similar to the one described above include cyclohexane and benzene, and toluene and methylcyclohexane, both using phenol as the solvent. In another process, isobutane and 1-butene are separated using furfural as the solvent.

The relative rate of solvent to feed determines the sharpness of separation between the key components, much as the absorbent rate or reflux ratio influences separation in absorbers or conventional columns. An estimate of the required solvent rate may be obtained based on vapor–liquid equilibrium data at different solvent concentrations, as described later in this section.

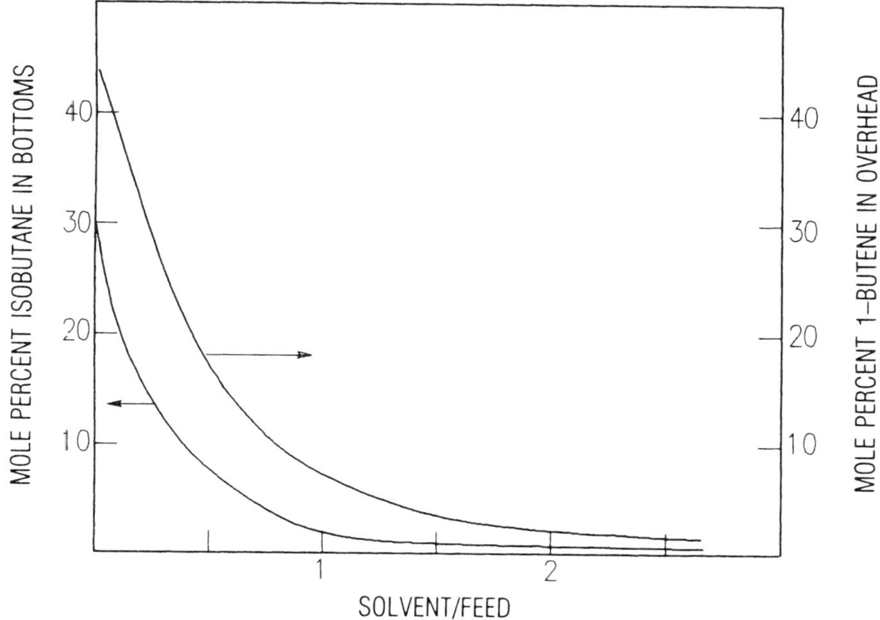

Figure 10.7 Effect of solvent-to-feed ratio on separation in extractive distillation.

The effect of furfural rate on the separation of isobutane and 1-butene was determined by simulation and is shown in Figure 10.7. The column has 23 theoretical stages, a total condenser, and a reboiler. The main feed, containing 40 mole percent isobutane and 60 mole percent 1-butene, is sent to the eleventh tray from the top, and the solvent is introduced at the third tray. The column operates at 75 psia pressure and a temperature ranging from 105°F at the condenser to 108°F in the reboiler. The overhead rate is maintained at 40 mole percent of the isobutane–1-butene feed rate, and the column reflux ratio is maintained at 3. The solvent-to-feed ratio is increased from 0 to 3, and the resulting separation is observed. Figure 10.7 shows a plot of the concentration of 1-butene in the overhead and isobutane in the bottoms as a function of the solvent-to-feed ratio. These concentrations decrease sharply as the solvent ratio is increased from low values but tend to level off asymptotically at a ratio of about 2 to 3.

Graphical Representation

The net effect of adding a solvent in extractive distillation is that new values emerge for the relative volatilities of the binary to be separated. If such data were available for a given system, the vapor and liquid compositions could be plotted on a solvent-free basis (Perry, 1973) to generate Y–X plots similar to ordinary binary Y–X diagrams. In the absence of direct relative volatility or Y–X data, liquid activity coefficient equations (Chapter 1) may be used to predict and plot the Y–X diagrams using computer simulation programs. The relative volatilities depend on the relative amount of solvent added to the binary. Therefore, different Y–X diagrams are obtained for different solvent-to-feed ratios.

The equilibrium curve, together with operating lines for the absorption and stripping sections, could then be used to estimate the number of stages required in each section at a given reflux ratio, as described in Chapters 5 and 6 for binary distillation. Since the Y–X diagram is specific to a given solvent-to-feed ratio, the graphical evaluation of the column is valid for that particular ratio. Through trying diagrams corresponding to a number of ratios, the amount of solvent for a given feed rate may be estimated. As in binary distillation, the graphical representation could be invaluable for qualitative evaluation of the problem prior to rigorous column simulation. A thorough understanding of the effect of different parameters on the column performance can help in better defining the simulation model.

10.3 THREE-PHASE DISTILLATION

The phase behavior of a system determines the nature of a distillation problem. Depending on this behavior, column profiles can range from those characteristic of narrow-boiling to wide-boiling mixtures and from those typical of ideal to highly nonideal systems.

Nonideal behavior is manifested, for instance, in azeotropic and extractive distillation, discussed in this chapter. Another case of nonideality is the condition of vapor–liquid–liquid equilibrium (Chapter 1). If the composition, temperature, and pressure on any column tray or number of trays put the mixture in the vapor–liquid–liquid region, three-phase distillation ensues. Sections 10.2.2 and 10.2.5 describe cases where the formation of two liquid phases makes up part of the separation process.

One of the more common cases of three-phase distillation is where the two liquid phases occur primarily in the condenser. Such occurrence is very common in crude oil and related distillation columns, where water separates in the condenser. When two liquid phases form in the column, one of them is usually taken out as a side draw at the tray where the liquid phase split occurs. This practice helps stabilize the column operation.

The computations in three-phase distillation involve two sets of vapor–liquid equilibrium coefficients, or K-values, derived from the activity coefficients of each component in each liquid phase. The calculations may be simplified, however, in hydrocarbon systems if the second liquid phase is formed by water. In these situations it is possible to assume the aqueous phase to be pure water and account only for water dissolved in the organic phase. The description of rigorous solution methods for three-phase distillation is deferred to Chapter 13. The objective at this point is to consider the effect a liquid phase split can have on distillation.

It has already been shown in Sections 10.2.2 and 10.2.5 how the formation of two liquid phases contributes to the separation of azeotropes or close boilers. The two phases, having distinct compositions, are separated in a single stage by simple decantation, followed by further processing of each phase. These may be considered as special cases of three-phase distillation since the formation of two liquid phases is confined to a single stage: the condenser or the liquid–liquid separator.

Another example of potential three-phase distillation is provided by the methyl ethyl ketone (MEK)–water binary. This system forms two miscible regions and a two-liquid-phase region. Figure 10.8 is a Y–X plot of this binary at 1.0 atm. Below 0.051 mole fraction MEK, a single, water-rich phase exists, and above 0.652 mole fraction MEK, a single, MEK-rich phase exists. Between these two concentrations, two liquid phases coexist.

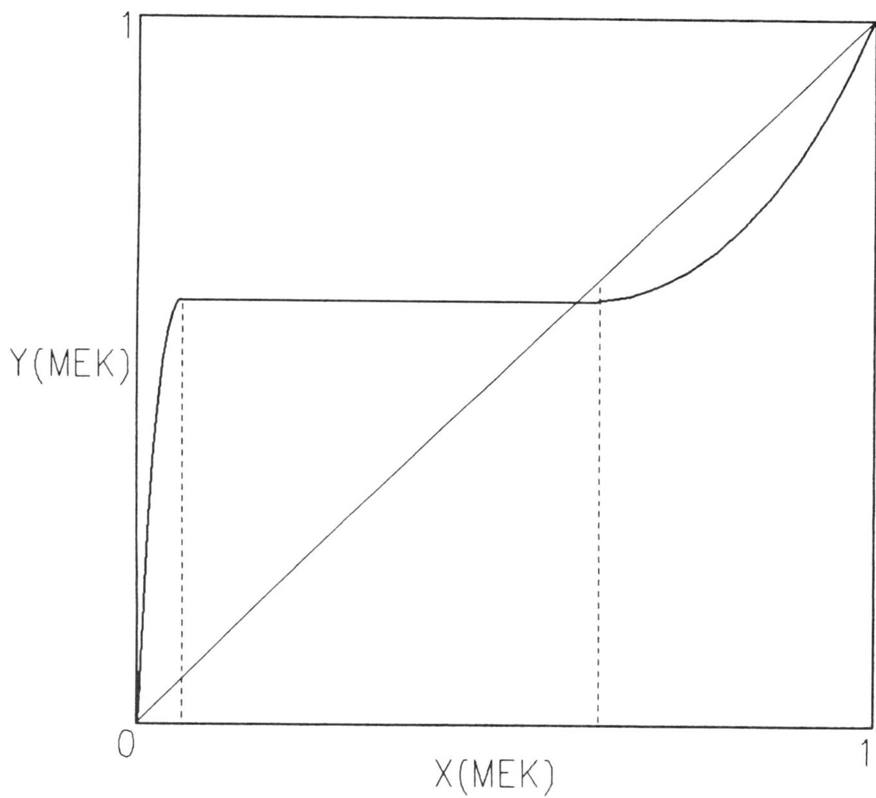

Figure 10.8 Methyl ethyl ketone – water binary.

Consider a column operating at 1.0 atm to be used to produce essentially pure MEK from a mixture with water containing 75 mole percent MEK. The purified MEK is taken as the column bottoms, and the overhead removes almost all of the water in the feed and some of the MEK. If the MEK concentration in the overhead is reduced below 65.2%, two liquid phases will form in the condenser. The water-rich phase is decanted, and the MEK-rich phase is refluxed. This makes for a much more efficient separation since the decanted water contains only about 5.1% MEK. If the overhead MEK concentration is not lowered to the two-phase region, the distillate would contain over 65.2% MEK.

This example involves two liquid phases only in the condenser and is therefore also a special case of three-phase distillation. It is possible to reflux a mixed-liquid phase to the column, creating three-phase conditions on a number of trays. At some tray down the columns, creating three-phase conditions on a number of trays. At some tray down the column, the MEK concentration might become high enough to restore total miscibility. It is obviously advantageous, however, to decant the water phase and reflux only the MEK phase.

REFERENCES

Othmer, D. F., Azeotropic separation, *Chem. Eng. Prog.*, 59, June 1963, 67.
Perry, J. H. and C. H. Chilton, *Chemical Engineers' Handbook, 5th ed.*, New York, McGraw-Hill, 1973.

PROBLEMS

10.1 The feed stream in the azeotropic distillation process described in Figure 10.1 is of variable composition. The column pressures are held constant. What ranges of feed composition can produce a product containing mostly component A and another containing mostly component B using the existing process configuration? With a variable feed composition, describe a control scheme to maintain constant product purities. What strategy may be followed if the feed composition moves outside the feasible range?

10.2 Assuming the process in Figure 10.2 has an external feed of fixed flow rate, composition, and thermal conditions, determine its degrees of freedom. How would you specify the performance of this process?

10.3 For the process shown in Figure 10.3, what control strategy alternatives may be used to maintain a specified purity for product B if the feed composition changes?

10.4 In the process shown in Figure 10.4 for separating a close boiling binary or azeotrope by forming two azeotropes through the addition of an entrainer, describe the sequence of events that would happen if the solvent makeup is cut off. What corrective measures should be taken to maintain product purities until the solvent supply is restored?

10.5 How many degrees of freedom exist for the process shown in Figure 10.5, assuming fixed feed AB? What variables control the purity of products A and B?

10.6 In the extractive distillation process in Figure 10.6, what effect would lowering the solvent rate have on the purity of product A? What changes in operating conditions could be made to maintain the specified purity of A? What potential problems should be considered?

CHAPTER 11

Liquid–Liquid Extraction

This process applies the phenomenon of liquid–liquid equilibria to carry out component separations. When the nonideality of a system causes two immiscible liquid phases to coexist at equilibrium, certain components may be more soluble in one phase than in the other.

Extraction involves the transfer of components between two liquid phases, much as absorption or stripping involves the transfer of components from liquid to vapor or vice versa. As in vapor–liquid multistage separation processes, the device employed to carry out liquid–liquid extraction is usually a counterflow column that performs the function of a number of equilibrium stages interconnected in counterflow configuration. In each equilibrium stage, two inlet liquid streams mix, reach equilibrium, and separate into two outlet liquid streams. As in vapor–liquid columns, the lack of complete equilibrium in liquid–liquid extractors is accounted for by some form of tray efficiency.

The column operation is gravity induced, with the heavier liquid flowing downward and the lighter liquid flowing upward by buoyancy. For the process to be workable, it is therefore required that one liquid phase be of a distinctly higher density than the other. The internal construction of an extractor column could be of several types, including trayed columns, packed columns, or spray columns with or without agitation.

Since liquid–liquid extraction merely accomplishes the transfer of components from one phase to another and not the separation of components, a complete separation process must include additional equipment such as distillation columns to recover the various components. The main parameters that should be considered in the design or performance evaluation of an extraction process include the choice of a solvent, the solvent-to-feed ratio, the number of stages, and any ancillary equipment that may be required to complete the separation.

11.1 EXTRACTION FUNDAMENTALS AND TERMINOLOGY

Liquid–liquid extraction may be represented by a three-component system forming two liquid phases. The solvent and the feed are two essentially immiscible liquids with possible partial mutual solubility. The *extract* component, to be extracted from the feed by the solvent, is soluble in both phases.

11.1.1 Simple Extractors

The basic extraction process is represented in Figure 11.1. The extract is the solvent-rich phase that gets enriched in the extract component, and the *raffinate* is the feed phase that gets depleted in the extract component. If the solvent is lighter than the feed, it must be introduced at the column bottom.

Unlike absorption or stripping, extraction does not involve vaporization or condensation or the heat that accompanies these processes. For this reason extraction is nearly an isothermal process, although some temperature variation could occur as a result of heat of solution.

An extraction column is considered to consist of N equilibrium stages, each having two liquid phases at equilibrium. The conditions of temperature, pressure, and composition are such that no vapor phase exists. The number of independent variables required to define the column constitute the degrees of freedom available at the design phase. They include the variables needed to define the feed and solvent, the number of stages, and the pressure and heat duty on each stage. They may be listed as follows for a feed and solvent with C components and a column with N stages:

Component rates in the feed	C
Component rates in the solvent	C
Temperature and pressure of the feed	2
Temperature and pressure of the solvent	2
Number of stages	1
Stage pressures	N
Stage heat duties	N
Total	2C + 2N + 5

This number of variables is matched by process constraints and performance specifications. The constraints include the existing feed and solvent compositions and thermal conditions. Also, the stage pressures are not truly independent because, once the pressure is set at one stage (by pumping capacity, etc.), the pressure profile is determined by the column hydraulics. If no controlled heat addition or removal exists at the stages, these parameters become constraints, determined by heat losses. A listing of all these constraints follows:

Feed composition	C – 1
Solvent composition	C – 1
Temperature and pressure of the feed	2
Temperature and pressure of the solvent	2
Stage pressures	N
Stage heat duties	N
Total	2C + 2N + 2

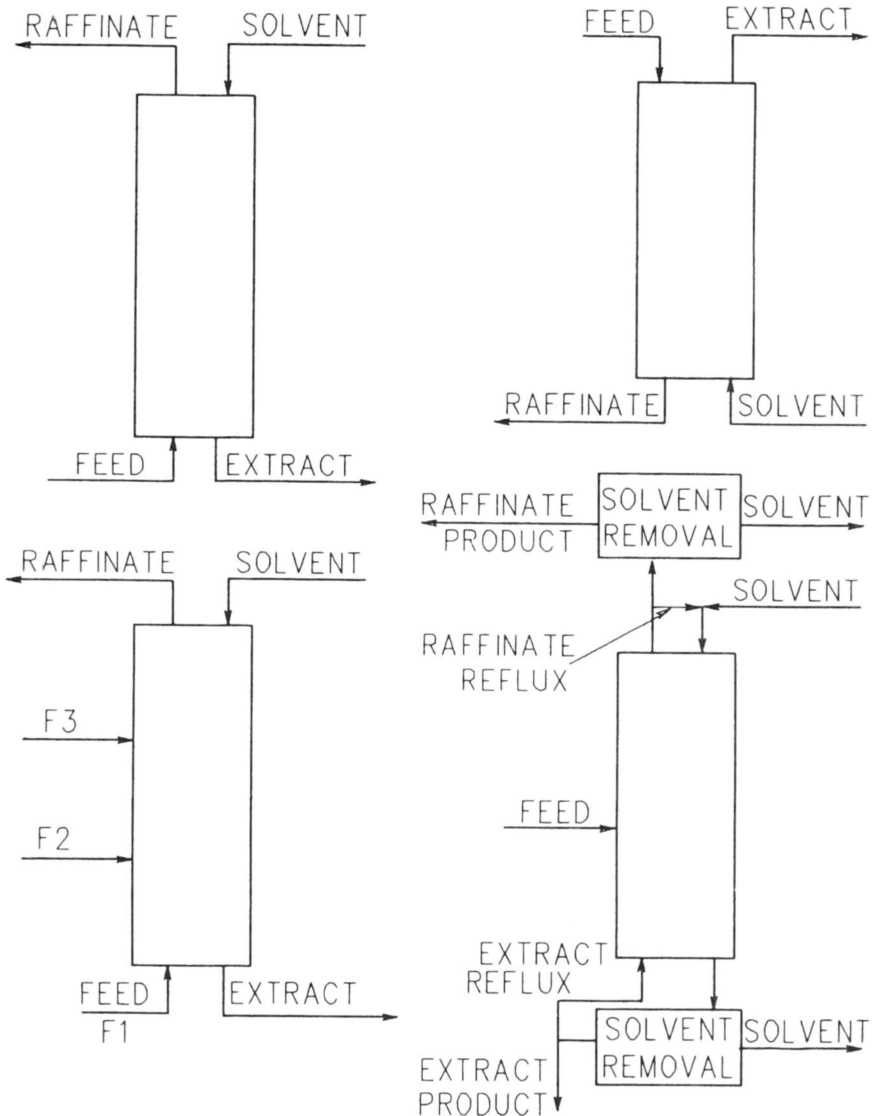

Figure 11.1 Schematics of liquid–liquid extractors.

Thus, at the design phase, the number of variables available to satisfy performance specifications is calculated as

$$(2C + 2N + 5) - (2C + 2N + 2) = 3$$

The three variables are the feed rate, the solvent rate, and the number of stages. For a given feed rate, the engineer must determine the solvent rate and the number of stages required to obtain a given rate of transfer of the extract component from the raffinate to the extract and a given concentration of the extract component in the extract or raffinate.

For an existing column the number of stages is fixed, and the only remaining variables are the feed rate and the solvent rate. For a given feed rate, the column operator can manipulate the solvent rate to achieve a specified extract component recovery or concentration in the extract or raffinate.

The description rule may be applied to confirm the above conclusions concerning the number of independent variables required to describe a liquid–liquid extraction process completely. The number of stages can be set by construction during the design phase. The feed and solvent rates can be controlled independently by external means during column operation. Thus, three variables are required to define the column during design and two during operation.

11.1.2 Multiple Feeds

For a given solvent stream, the concentration of the extract component in a feed stream determines the number of stages between the solvent and the feed that are required to achieve a specified extract component concentration in either product. The higher the extract component concentration in the feed, the more stages are required. If the extract component is to be removed from more than one feed, each with a different extract component concentration, the column could be designed to handle the feed stream with the highest extract component concentration. Alternatively, the feeds could be premixed and the column designed for the mixed feed composition. If a column is designed to handle the high extract concentration feed, the leaner feeds need not be introduced with the rich feed but could be introduced at intermediate stages. Such a setup is shown in Figure 11.1, with feed steams F1, F2, and F3.

Each additional feed adds C + 3 degrees of freedom to the column: 1 component rate for each of C components in the feed and the feed temperature, pressure, and location. In designing a new column, the object is to calculate the optimum number of stages, i.e., the smallest number of stages required to achieve the specified separation for a given feed and solvent. For a multiple-feed column, the object is to calculate the optimum total number of stages, as well as the optimum feed locations; that is, the optimum number of stages between adjacent feeds (including the top feed and solvent), so that the total number of stages is minimized.

The optimum number of stages and feed locations can be determined graphically as described in Section 11.2. For rigorous confirmation of graphical results, especially in most practical situations where systems are not truly ternary but multicomponent, computer simulation becomes necessary. Although generally simulation algorithms are designed for handling fixed configuration columns (i.e., fixed number of stages and feed locations), it is often possible to determine the optimum through multiple solutions or by means of appropriate optimization routines.

11.1.3 *Refluxed Extractors*

An extractor column with both extract and raffinate reflux is shown in Figure 11.1. In this configuration the feed is sent to an intermediate stage in the column.

The purpose of the extract reflux is to contact the extract phase leaving the column with a stream that is richer in the extract component than the feed, thereby generating a higher purity extract product.

Refluxing part of the raffinate, on the other hand, does not increase the removal of the extract component from the raffinate since maximum removal is achieved with pure solvent. The raffinate reflux, which amounts to recirculation of an equilibrium phase back to the same equilibrium stage, does not, in principle, change the product composition. In practice, the raffinate reflux may help in achieving a better approach to phase equilibrium due to premixing with the solvent. The raffinate reflux could, that is, improve the stage efficiency.

The manner in which reflux influences an extractor performance and the extent of its effect depend on the equilibrium characteristics of the particular system. Also, the ranges of reflux and other operating conditions must be restricted to the two-phase region where separation is feasible, both thermodynamically and physically.

11.2 GRAPHICAL REPRESENTATION

Graphical representation of liquid–liquid equilibria provides a convenient pictorial tool for studying extraction problems and doing preliminary extractor calculations. The results can be used as the basis for defining a computer simulation model for rigorous solution and optimization of the process.

Assuming the extraction system is a ternary, the liquid–liquid equilibrium relationships are best represented graphically on a triangular diagram (Section 1.3.5). The general shape of the equilibrium curve or curves depends on the mutual solubilities of the components in the ternary. In many extraction processes, there is total miscibility between the extract component and each of the other components and partial miscibility between the solvent and the raffinate. A ternary having these characteristics is represented on an equilateral triangle in Figure 11.2, where E is the extract component, R the raffinate, and S the solvent. Each of binaries, ER and ES, forms a single liquid phase of all compositions, while binary RS forms two liquid phases between compositions represented by points A and B and a single phase outside this composition range. The ternary mixture forms two liquid phases under curve APB and a single liquid phase outside the curve.

The tie lines shown in Figure 11.2 represent liquid compositions at equilibrium with each other. Thus, point M represents the combined composition of a mixture that separates into two liquid phases at equilibrium with compositions represented by points L and Q. The compositions of the two phases approach each other as the tie lines become shorter. At point P, the *plait* point, the tie line vanishes, and the liquid compositions become identical, forming a single phase.

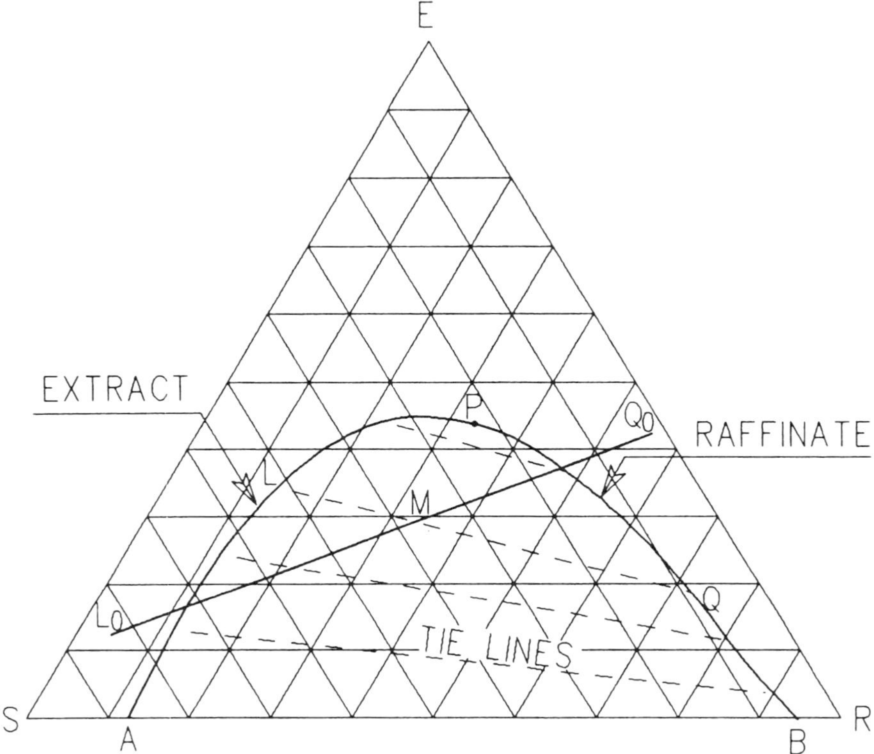

Figure 11.2 Liquid–liquid equilibrium for a ternary.

If more than one binary in the three-component mixture exhibits partial solubility (i.e., regions of immiscibility), additional two-liquid phase regions would appear on the triangle, typically as domes on other sides of the triangle, that may merge in some systems, forming continuous regions of immiscibility. Extractors in practice operate in one particular region of immiscibility and, therefore, for the purpose of graphical calculation of extractors, only that region of immiscibility is of direct interest.

11.2.1 Generating Equilibrium Diagrams

Before any graphical calculation of liquid–liquid extractors can be attempted, reliable liquid–liquid equilibrium data are necessary. These data should include the compositions defining the two-liquid phase envelope and tie lines at temperatures and pressures of interest. Note that a diagram such as Figure 11.2 is specific to one temperature and pressure.

In the absence of experimental data, liquid–liquid equilibria may be predicted from liquid activity coefficient equations (Section 1.3). Some of these equations, such as NRTL and UNIQUAC, are well adapted for predicting liquid–liquid equilibrium compositions. Triangular diagrams may be generated based on computer-predicted activity coefficients much as they would be based on experimental data.

The computer-aided procedure, unless automated by the program, requires running a series of liquid–liquid equilibrium calculations (the equivalent of vapor–liquid "flash" calculations) at constant temperature and pressure. The composition is varied around the equilibrium curve, and the transition points from one phase to two, or vice versa, are noted. As many points as needed are obtained this way to generate the entire equilibrium curve. Also, each time an equilibrium calculation is done in the two-phase region, the compositions of the two phases are recorded. Each pair of data points thus obtained defines a tie line. Data obtained at one temperature and pressure generate one triangular diagram. If so desired, the procedure is repeated at other temperatures and pressures to determine the effect of these variables.

11.2.2 Single-Stage Calculations

The phase rule indicates 3 degrees of freedom for a three-component, two-phase system. Thus, by specifying the temperature, pressure, and concentration of one component in one phase, the state of the system is defined. The component concentration in one phase defines one point on the equilibrium curve and one end of a tie line. The other end is determined thermodynamically either from experimental data or on the basis of liquid activity coefficient predictions methods.

Since all points on a tie line have the same temperature, pressure, component concentration in one phase, and component concentration in the other phase, one more variable must be specified to define the relative amounts of the phases at equilibrium. The specified variable could itself be either the fraction of one phase out of the total or the mixture composition. Note that only two component concentrations are independent in a ternary. In the single-phase region, there are 4 degrees of freedom according to the phase rule. Therefore, if the fraction of one phase or the existence of one or two phases is unknown, four independent variables are required to define the system completely.

The temperature and pressure of the mixture represented by point M in Figure 11.2 are those for which this particular triangular diagram is constructed. The other two variables are the concentrations of two of the components E, R, and S, which fix the location of M. The lengths of the segments of the tie line passing through M determine the ratio of these phases according to the lever arm rule, which is a graphical statement of material balance. With the extract phase designated by L and the raffinate phase by Q, the following component molar balances hold:

$$MX_{MS} = (L + Q)X_{MS} = LX_{LS} + QX_{QS}$$

$$MX_{MR} = (L + Q)X_{MR} = LX_{LR} + QX_{QR}$$

$$MX_{ME} = (L + Q)X_{ME} = LX_{LE} + QX_{QE}$$

where M, L, and Q are molar quantities, and X_{ij} are mole fractions. The first subscript refers to the mixture M, phase L, or phase Q, and the second subscript refers to the solvent S, the raffinate component R, or the extract component E. The equations are rearranged to give the ratio of the phases:

$$\frac{L}{Q} = \frac{X_{MS} - X_{QS}}{X_{LS} - X_{MS}}$$

$$= \frac{X_{MR} - X_{QR}}{X_{LR} - X_{MR}} \tag{11.1}$$

$$= \frac{X_{ME} - X_{QE}}{X_{LE} - X_{ME}} = \frac{QM}{LM}$$

where QM and LM are the segment lengths.

In order to use the equilateral triangular phase diagram to calculate a single-stage extractor, the feed and solvent are first plotted on the diagram as points Q_0 and L_0, respectively (Figure 11.2). Note that, in general, L_0 may not be pure solvent and that the feed, Q_0, may contain some solvent component. The mixture is represented by M on a straight line joining L_0 and Q_0. The lengths of segments L_0M and Q_0M are determined by the lever arm rule:

$$\frac{L_0M}{Q_0M} = \frac{Q_0}{L_0} \tag{11.2}$$

Here L_0 and Q_0 are the molar rates of the solvent and feed.

The tie line through M determines the rates and compositions of the extract, represented by point L, and the raffinate, represented by point Q. The outcome of this process is that the extract component concentration is increased in the extract phase ($X_{LE} > X_{L0E}$) and decreased in the raffinate ($X_{QE} < X_{Q0E}$).

11.2.3 Countercurrent Multistage Calculations

A graphical solution of an extractor can be obtained by representing material balances with operating lines and equilibrium relationships with tie lines on a triangular diagram. The solution is accurate to the extent of the accuracy of the equilibrium data used to generate the diagram. Since a given diagram is constructed for a fixed temperature and pressure, the graphical solution must assume isothermal and isobaric operation of the extractor.

The method assumes equilibrium stages; stage efficiency data would be required to represent an actual column in terms of equilibrium stages. The stage efficiency depends on the physical properties of the liquids and the type of equipment used. It is usually estimated on the basis of experimental data or previous experience with similar columns.

An inherent limitation in using triangular diagrams for solving liquid–liquid extractors is the fact that these diagrams represent strictly ternary systems. Although three primary components are involved in a basic extraction process (the extract component, the raffinate, and the solvent), actual systems are rarely truly ternary and are generally multicomponent. Nevertheless, for many systems the graphical method provides at least a qualitative representation of the process and a visual means for studying the effect of design and operating parameters such as number of stages and flow rates. The final solution of an extractor column usually relies on computer simulation, which can be applied more effectively once a qualitative understanding of the problem is established.

As shown in Figure 11.3, an overall material balance on the column is written as

$$L_0 + Q_{N+1} = L_N + Q_1 = M \qquad (11.3)$$

Thus, if the feed and solvent compositions and flow rates are known, the mixture point M can be determined (Figure 11.4). The solvent and feed points L_0 and Q_{N+1} are first plotted at their known compositions, and a straight line is drawn through them. The mixture point, M, is located on the basis of the lever arm rule:

$$\frac{L_0 M}{Q_{N+1} M} = \frac{Q_{N+1}}{L_0}$$

where $L_0 M$ and $Q_{N+1} M$ are segment lengths, and L_0 and Q_{M+1} are molar flow rates. The second part of Equation 11.3 is now used to determine points L_N and Q_1, corresponding to the raffinate and extract. Both L_N and Q_1 must lie on the equilibrium curve since they are saturated liquids. The concentration of one component in one of these phases must be known, such as in a design situation. If, for instance, the concentration of the extract component in the extract phase is specified, point L_N could be plotted on the diagram. Next, Q_1 is located by drawing a straight line through L_N and M and extending it until it intersects the raffinate side of the equilibrium curve.

Rearranging Equation 11.3 defines the difference point Δ for streams Q_{N+1} and L_N and streams Q_1 and L_0:

$$Q_{N+1} - L_N = Q_1 - L_0 = \Delta \qquad (11.4)$$

The difference point may be located on the basis of either side of this equation by passing a line through either pair and extending it to Δ to satisfy the lever arm rule:

$$\frac{Q_{N+1}\Delta}{L_N \Delta} = \frac{L_N}{Q_{N+1}}$$

or

$$\frac{Q_1 \Delta}{L_0 \Delta} = \frac{L_0}{Q_1} \qquad (11.5)$$

The quantities $Q_{N+1}\Delta$, $L_N\Delta$, $Q_1\Delta$, and $L_0\Delta$ designate segment lengths and Q_{N+1}, L_N, Q_1, and L_0 designate molar flow rates. Since Δ is the same for both pairs of points, it is more easily located as the intersection between the straight lines passing through Q_{N+1} and L_N, and Q_1 and L_0.

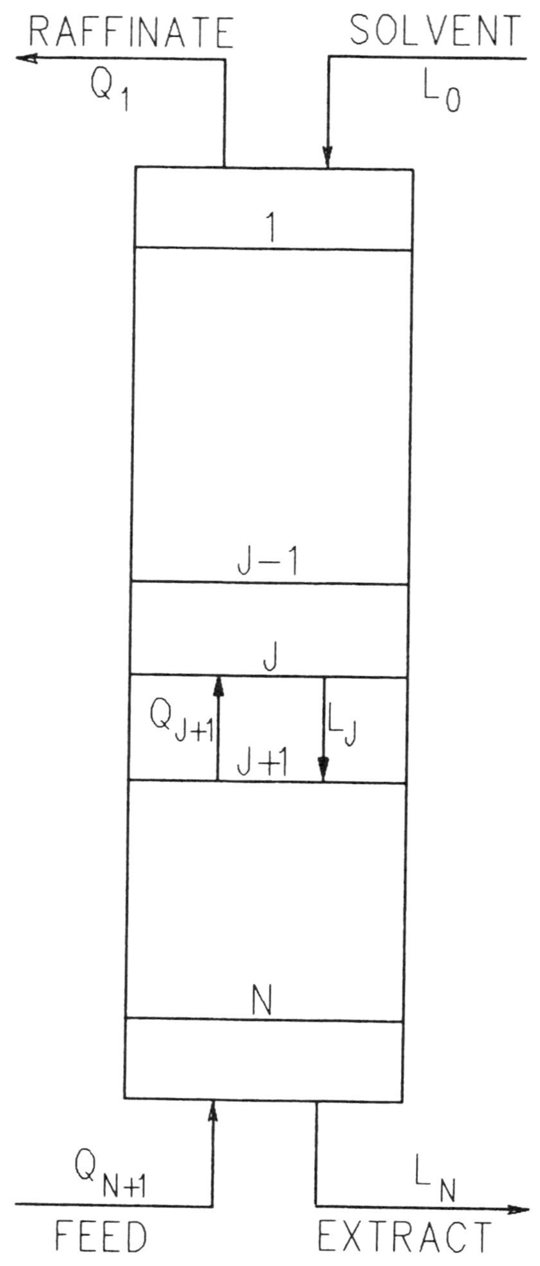

Figure 11.3 Equilibrium stages in a liquid–liquid extractor.

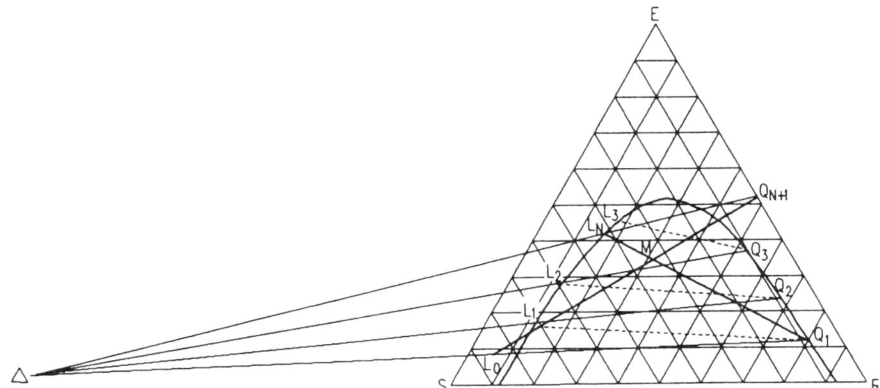

Figure 11.4 Graphical construction of extractor equilibrium stages.

The above procedure establishes the product rates and compositions based on overall material balance and the premise that streams leaving the extractor are saturated streams, falling on the equilibrium curve. To determine the number of equilibrium stages required to achieve the performance specification (such as the concentration of the extract component in the extract product) and to determine the compositions on each stage, a material balance is calculated around the top section of the column through stage j:

$$Q_{j+1} - L_j = Q_1 - L_0 = \Delta \tag{11.6}$$

Note that Δ in this equation is the same as Δ defined in Equation 11.4. It is constant for all passing streams L_j and Q_{j+1} between any two stages j and j + 1.

The graphical construction of equilibrium stages on the triangular ternary phase diagram can start at either the top or bottom of the column. Starting at the top, the first operating line is the line drawn through Δ, L_0, and Q_1 (Figure 11.4). It relates the solvent feed, L_0, to the raffinate product, Q_1. Extract liquid, L_1, leaving stage 1 is at equilibrium with Q_1 and is, therefore, represented by point L_1 at the other end of the tie line through Q_1. The operating line relating passing streams L_1 and Q_2 is obtained by drawing a straight line through Δ and L_1 and extending it to Q_2 on the raffinate side of the equilibrium curve. A tie line drawn through Q_2 locates L_2 on the extract side of the equilibrium curve. Another operating line is drawn through Δ and L_2 to locate Q_3, and so on. The procedure is continued until a tie line is reached that either meets L_N or crosses the operating line between L_N and Q_{N+1}. Thus, according to Figure 11.4, three theoretical stages would surpass the desired performance specifications although two theoretical stages would not be sufficient. The required number of stages could, therefore, include a fraction of a theoretical stage. The actual number of stages is obtained by dividing the number of theoretical stages by the efficiency. Thus, whether or not the theoretical stages include a fractional stage, the resulting calculated number of actual stages could include a fractional stage, which must be rounded up.

The procedure for graphical construction of extractor stages is modified if the set of specifications that define the column is different from the above. In most cases L_0 and Q_{N+1} (the external feeds) are known, but the third specification may not relate to the products. For an existing column, for instance, the number of stages is known, and it is required to calculate the products composition. A trial and error procedure must be used by assuming the Q_1 or L_N location on the equilibrium curve. Once one of these two points is assumed, the other one is located by passing a straight line through M. The stepping off of stages then proceeds in the usual manner to determine the number of stages. If this number does not match the number of theoretical stages corresponding to the existing column, the procedure is repeated with a new location for Q_1 until a match is obtained.

A ternary diagram constructed for a given system can be used as a tool to study the effect of such parameters as the number of stages and solvent-to-feed ratio on the extractor performance.

From Figure 11.4, it can be seen that, as Δ is moved farther away from the triangle, the operating lines become more closely spaced so that more stages are required for the same separation. Conversely, as Δ is moved closer to the triangle, fewer stages are required. According to Equation 11.5, as Δ approaches L_0, L_0/Q_1, the solvent-to-raffinate ratio, increases. Generally, this also implies an increase in the solvent-to-feed ratio. Thus, fewer stages are required with a higher solvent-to-feed ratio.

11.2.4 Multiple Feed and Refluxed Extractors

The formulation of the operating line equations and equilibrium relationships and their graphical implementation is not restricted to the basic extractor configuration but can be applied to columns with multiple feeds and reflux streams. The resulting complexity is handled by breaking up the column into sections, each with its own difference point. This is demonstrated for a column with an intermediate feed, resulting in two sections.

Figure 11.5 depicts a two-section extractor with an intermediate feed going on stage f. The raffinate section consists of stages 1 through $f - 1$ and the extract section consists of stages f through N. An overall material balance on the column is written as

$$Q_{N+1} + F + L_0 = Q_1 + L_N$$

which can be rearranged in two ways:

$$Q_1 - L_0 = Q_{N+1} + F - L_N = \Delta' \tag{11.7}$$

$$Q_{N+1} - L_N = Q_1 - F - L_0 = \Delta'' \tag{11.8}$$

A typical design and optimization problem would be to determine the required number of stages and optimum feed location when the feeds and product compositions are specified. The difference points Δ' and Δ'' are first located in accordance

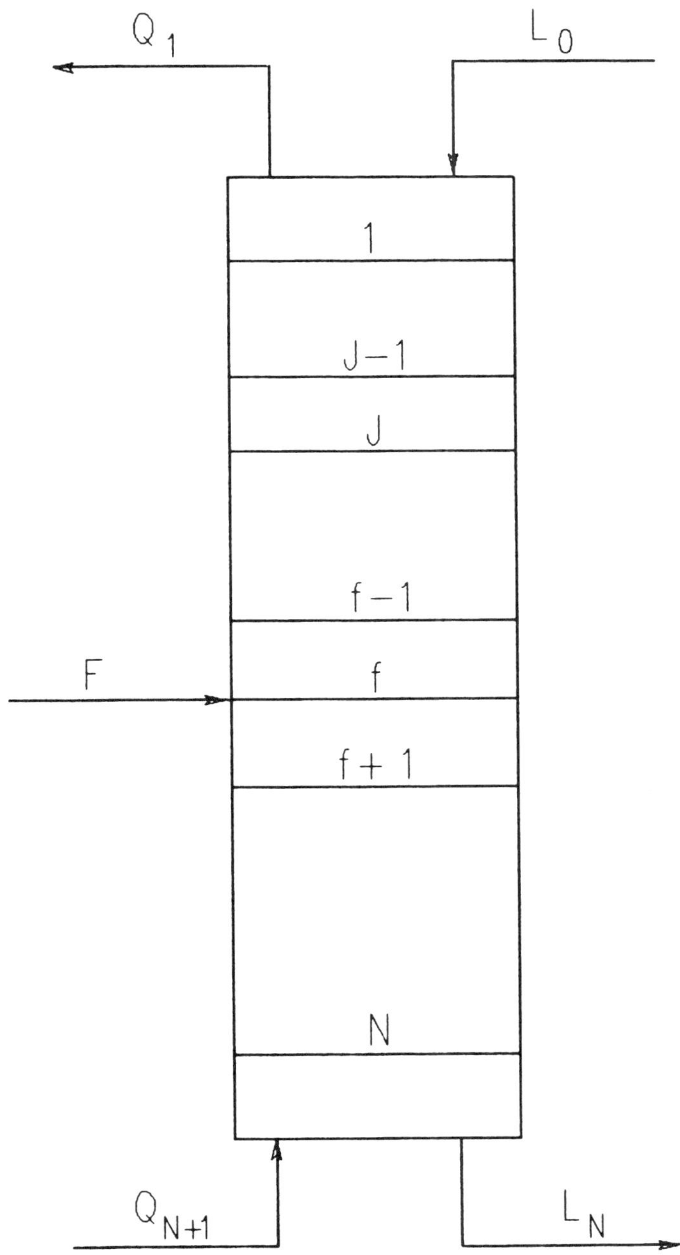

Figure 11.5 Extractor with two feeds.

with Equations 11.7 and 11.8. In a design situation all the feed and product compositions are known and their corresponding points, Q_1, Q_{N+1}, L_0, L_N, and F, are plotted on the ternary diagram. Each difference point is determined by the intersection of lines corresponding to the first and second parts of Equations 11.7 and 11.8. Thus, to determine Δ', a straight line is first drawn through Q_1 and L_0. Next, the mixture point for Q_{N+1} and F is located, and a straight line is drawn through this point and L_N. The intersection of these two straight lines determines Δ'. Δ'' is determined in a similar manner.

A material balance on the top section of the column through any tray j above the feed gives

$$Q_{j+1} - L_j = Q_1 - L_0 = \Delta' \tag{11.9}$$

A material balance on the lower section of the column through any tray j below the feed gives

$$Q_j - L_{j-1} = Q_{N+1} - L_N = \Delta'' \tag{11.10}$$

The graphical construction of operating lines for the equilibrium stages proceeds in the usual manner, using difference point Δ' for the upper column section and Δ'' for the lower column section. The procedure is continued from both ends of the column until the external feed F point is reached on the diagram. This determines the optimum feed location. The required number of stages is the total of the two sections.

11.3 EXTRACTION EQUIPMENT

Extractor equipment considerations are discussed here in the context of their effect on the process performance and not for the purpose of describing detailed design. The main parameter of interest at this point is the number of equilibrium stages that represent the process. Liquid–liquid extraction requires thorough mixing of two liquid phases to achieve thermodynamic equilibrium, followed by complete separation of the phases. The particular equipment selected for a given process is determined, in part, by the mixing and separation characteristics of the phases.

In one arrangement, each stage of the extraction process consists of a mixing vessel followed by a settler where the phases are separated. The mixers could be mechanically agitated. A multistage process consists of a number of such vessels and settlers interconnected in countercurrent flow. In this process each stage could, in principle, be made to approach an equilibrium stage as desired by controlling the agitation and vessel volume (which determines the residence time).

The other type of extractor equipment falls under the general category of columns, with vertical, gravity-induced counterflow. Columns come with a variety of internal designs, including spray columns, packed columns, sieve or bubble cap trayed columns, or baffled columns. Some columns may contain internal mechanical agitators. The heavier phase is introduced at the top of the column and the lighter phase at the bottom. Usually one phase is continuous and the other discontinuous.

In the case of packed columns, one of the phases preferentially wets the packing, and the wetting phase could be either the continuous or the discontinuous phase. Extractor columns usually include a distributor to disperse the discontinuous phase and an interface region to separate the phases.

The efficiency of extractors and the equivalent number of equilibrium stages depends on internal design and other factors. Detailed description of extractor equipment is provided by Najim (1989) and is outside the scope of this book.

NOMENCLATURE

C   Number of components
E   Extract
F   Feed stream
L   Liquid phase of the raffinate
M   Mixture point
N   Number of stages

Q   Liquid phase of the solvent and extract
R   Raffinate
S   Solvent
X   Mole fraction
D   Difference point

SUBSCRIPTS

i   Component designation
j   Stage designation

REFERENCE

Najim, Kaddour, *Process Modeling and Control in Chemical Engineering*, New York, Marcel Dekker, 1989.

PROBLEMS

11.1 A solvent is used to recover component E from a liquid feed containing components R and E by extraction in a single-stage extractor. The compositions and flow rates of the feed and solvent are given below:

Component	Mole Fraction	
	Feed (Q_0)	Solvent (L_0)
S	0.02	0.84
R	0.56	0.04
E	0.42	0.12
Flow Rate, lbmole/hr	100.00	75.00

Using liquid–liquid equilibrium data from Figure 11.2, calculate the flow rates and compositions of the raffinate and extract.

11.2 A pure solvent S is used to remove component E from a binary liquid solution containing 30% mole E and 70% mole R. In a single-stage extractor how many lbmoles solvent per 100 lbmole feed are required to bring the concentration of E in the raffinate to 3% mole? What are the resulting rates and compositions of the raffinate and extract? Use liquid–liquid equilibrium data from Figure 11.2.

11.3 It is required to recover 93% of component E from a binary mixture of components E and R by treating with a solvent in a single-stage extractor. What is the required solvent rate per 100 lbmoles of feed, and what are the product rates and compositions? Assume the liquid–liquid equilibrium data in Figure 11.2 are applicable. The feed and solvent compositions are as follows:

	Mole Fraction	
Component	Feed (Q_0)	Solvent (L_0)
S	0.02	0.95
R	0.59	0.00
E	0.39	0.05

11.4 In the extraction service of Problem 11.2 the solvent rate is limited to 85 lbmole per 100 lbmole of feed. How many extractor stages would be required to achieve the same objective of 3% mole concentration of component E in the raffinate?

11.5 Component E is to be removed by extraction with a pure solvent out of two feed streams containing components E and R. The compositions and flow rates of the feeds and solvent are given below. The concentration of E in the raffinate is specified at 5% mole, and of R in the extract at 15% mole. The main feed is sent to the bottom of the column and the side feed to an intermediate stage. Determine the number of stages required and the optimum location of the side feed. Liquid–liquid equilibrium data may be obtained by modifying Figure 11.2 such that the concentration of E in both phases is the same (the tie lines are parallel to the base of the triangle).

	Mole Fraction		
Component	Main Feed	Side Feed	Solvent
S	0.00	0.00	1.00
R	0.60	0.80	0.00
E	0.40	0.20	0.00
Flow Rate, lbmole/hr	70.00	30.00	90.00

CHAPTER 12

Shortcut Methods

In past chapters the performance of multistage separation processes was analyzed qualitatively on the basis of fundamental principles developed earlier. The object was to gain an understanding of the different types of separation processes and columns and the factors that affect their performance. Chapters 5 and 6 employed graphical and semiquantitative methods to represent a limited set of separation processes, namely binary distillation. Chapters 10 and 11 also used combinations of graphical and analytical methods applied to binary or ternary systems to represent specific classes of nonideal separations.

This chapter treats multicomponent separations using approximate or "shortcut" methods that rely on certain simplifying assumptions to solve the column equations. Some of these techniques consider limiting conditions of total reflux and minimum reflux with an infinite number of trays. The results can only approximate real processes where the reflux rate and the number of trays are finite. Other shortcut simplifying assumptions may or may not be valid for a given problem. Nevertheless, analyzing a distillation problem on the basis of these methods, where applicable, is useful for preliminary estimations and for determining practical column operating limits. Some shortcut methods are capable of calculating the required number of stages for a given separation problem, whereas rigorous methods usually assume a fixed number of stages. Further, the information derived from shortcut calculations may be used for evaluating alternative trade-offs of number of trays vs. reflux rate for achieving desired separation criteria. In spite of the simplifications, however, these methods are still quite computation-intensive and in most cases must be solved using computer programs. However, due to the much shorter computing time, these methods are useful in situations where computing time is crucial, such as in on-line, real-time applications. The shortcut solution can often be used as the foundation for progressing toward a rigorous solution of the problem. In many applications the two methods are used in conjunction with each other to attain efficient and reliable simulation strategies.

12.1 COLUMNS AT TOTAL REFLUX

The concept of total reflux was discussed in Chapters 3 and 6. The conditions of total reflux may be approached either by operating the column at finite feed and product rates with a very large reflux rate or by cutting off the feed and products and maintaining internal boilup and reflux by adding and removing heat at the reboiler and condenser.

By using a simplified model of binary distillation with two stages, it was shown in Chapter 3 that maximum separation is achieved at total reflux. Also, the Y–X diagram was applied in Chapter 6 to verify that, for a given number of stages, maximum separation of a binary mixture is obtained at total reflux. A corollary to this statement is that, for a specified separation, the required number of stages is least at total reflux. The performance of multicomponent columns at total reflux is discussed next.

12.1.1 Model Description

The model considered here is a single-feed, multicomponent, multistage column with a partial condenser and reboiler, a vapor overhead product, and a liquid bottoms product. The feed is assumed to be of fixed rate, composition, and thermal conditions. The column pressure is also assumed fixed. The column has N stages, including the condenser and the reboiler, and the feed is introduced on stage f. A schematic of the model is shown in Figure 12.1. The model assumes a column diameter, condenser, and reboiler large enough to handle the internal reflux and boilup required to approach the conditions of total reflux. The column reflux or stripping fluid may not be supplied by external streams such as in absorbers or strippers, where the concept of total reflux is meaningless.

It will become evident in the mathematical derivation that the feed location in a total reflux column is immaterial. Hence, although the model is developed for a single feed, multiple feeds can be handled just as well, as long as they are not intended to serve as external reflux or stripping streams. The model may be extended to simulate multiple product columns as described in Section 12.1.5.

The other assumption in the model relates to the vapor–liquid equilibrium coefficients, or K-values. The K-values at a given pressure are assumed to be a function of temperature only, and not of composition. It is further assumed that the temperature dependence of the K-values for the different components is similar, i.e., the ratio of the K-values of any pair of components is independent of temperature. Thus, the relative volatilities, defined as the ratios of K-values of any two components, are assumed constant throughout the column.

The feed, distillate, and bottoms flow rates are F, D, and B, respectively. The stages are numbered from the top down, the condenser being stage 1 and the reboiler stage N. The total vapor molar flow leaving stage j is designated as V_j and the total liquid molar flow leaving stage j as L_j. The vapor flow of component i leaving stage

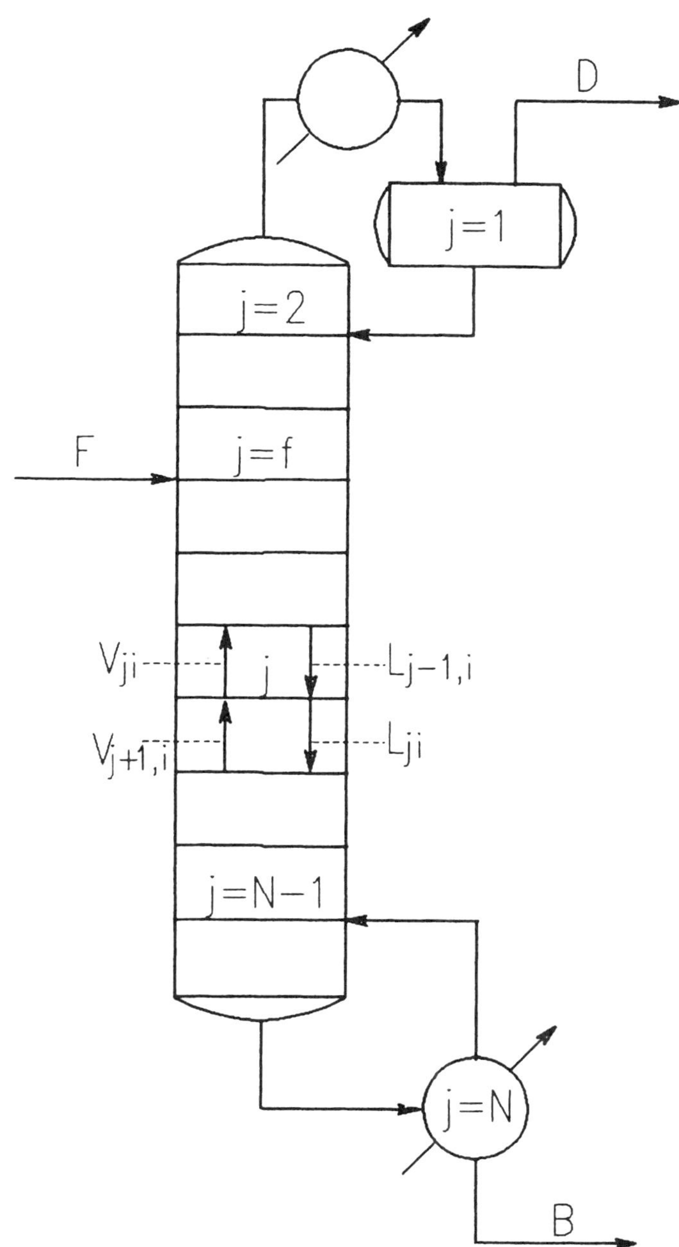

Figure 12.1 Column schematic.

j is V_{ji}, and the liquid flow of component i leaving stage j is L_{ji}. Finally, the mole fractions of component i in the vapor and liquid leaving stage j are designated as Y_{ji} and X_{ji}, respectively. In general, the first subscript refers to the stage number and the second subscript refers to the component number.

12.1.2 Mathematical Representation

From the above definitions, it follows that

$$V_{ji} = V_j Y_{ji}$$
$$L_{ji} = L_j X_{ji}$$

(12.1)

And, from the definition of the vapor–liquid distribution coefficient (Chapter 1),

$$K_{ji} = \frac{Y_{ji}}{X_{ji}}$$

(12.2)

As indicated in Figure 12.1, an overall material balance around trays 1 through j for j between 1 and f (above the feed tray) is written as

$$V_{j+1} = L_j + D$$

(12.3)

For j between f and $N - 1$, the material balance must include the feed:

$$F + V_{j+1} = L_j + D$$

(12.4)

Dividing each of Equations 12.3 and 12.4 by V_{j+1} gives

$$\frac{L_j}{V_{j+1}} + \frac{D}{V_{j+1}} = 1$$

and

$$\frac{F}{V_{j+1}} + 1 = \frac{L_j}{V_{j+1}} + \frac{D}{V_{j+1}}$$

At total reflux, L_j and V_{j+1} become infinitely large compared to D and F and approach each other in magnitude. Hence, at total reflux,

$$\frac{D}{V_{j+1}} = 0$$

and

$$\frac{L_j}{V_{j+1}} = 1$$

These relations apply throughout the column, i.e., for j between 1 and $N - 1$.

A material balance on component i around trays 1 through j takes the form

$$V_{j+1}Y_{j+1,i} = L_j X_{ji} + DY_{li}$$

$$(1 \le j < f)$$

(12.5)

and

$$FZ_i + V_{j+1}Y_{j+1,i} = L_j X_{ji} + DY_{li}$$

$$(f \le j \le N - 1)$$

(12.6)

where Z_i is the mole fraction of component i in the feed. Division of either of Equations 12.5 or 12.6 by V_{j+1} and substitution of the total reflux expressions derived above results in the following:

$$Y_{j+1,i} = X_{ji}$$

$$(1 \le j \le N - 1)$$

(12.7)

This equation applies throughout the column at total reflux. Equation 12.2 is written for tray $j + 1$,

$$K_{j+1,i} = \frac{Y_{j+1,i}}{X_{j+1,i}}$$

and, combined with Equation 12.7 to eliminate $Y_{j+1,i}$,

$$X_{j+1,i} = \frac{X_{ji}}{K_{j+1,i}}$$

(12.8)

Equation 12.8 is next applied to trays $j = 1, 2, 3, \ldots, N - 1$ as follows:

$$X_{2i} = \frac{X_{li}}{K_{2i}} \qquad (j = 1)$$

$$X_{3i} = \frac{X_{2i}}{K_{3i}} = \frac{X_{li}}{K_{2i}K_{3i}} \qquad (j = 2)$$

$$X_{Ni} = \frac{X_{li}}{K_{2i}K_{3i}\ldots K_{Ni}} \qquad (j = N - 1)$$

Substituting $X_{1i} = Y_{1i}/K_{1i}$ (Equation 12.2) in the last equation results in the following:

$$X_{Ni} = \frac{Y_{1i}}{K_{1i} K_{2i} \dots K_{Ni}} \tag{12.9}$$

Equation 12.9 relates the bottoms composition, X_{Ni}, to the distillate composition, Y_{1i}, at total reflux. The derivation of this equation on the basis of material balances and equilibrium relationships is equivalent to the graphical solution for binary mixtures described in Chapter 5. However, in the binary graphical solution the material balance is represented by the operating curves, not necessarily at total reflux, while in the present derivation the material balance equation is obtained on the basis of total reflux.

If the molar flow rates of component i in the distillate and bottoms are designated as d_i and b_i respectively, the following relationships hold:

$$d_i = DY_{1i}$$

$$b_i = BX_{Ni} \tag{12.10}$$

Substituting the values of Y_{1i} and X_{Ni} from Equation set 12.10 into Equation 12.9, the following is obtained:

$$\frac{b_i}{d_i} = \frac{B/D}{\displaystyle\prod_{j=1}^{N} K_{ji}} \tag{12.11}$$

where the multiplication is carried out for j = 1, 2, …, N. The relative volatility, α_{ji}, for component i is defined relative to some "reference component," r:

$$\alpha_{ji} = \frac{K_{ji}}{K_{jr}} \tag{12.12}$$

K_{ji} is now substituted in Equation 12.11 to give

$$\frac{b_i}{d_i} = \frac{B/D}{\displaystyle\prod_{j=1}^{N} \alpha_{ji} K_{jr}} \tag{12.13}$$

The factor $\prod_{j=1}^{N} K_{jr}$ may be replaced by $(K_r)^N$, where K_r is the K-value of reference component r at average column conditions. Also, since constant relative volatility is assumed, the product of the relative volatilities may be replaced by $(\alpha_i)^N$. Equation 12.13 may thus be rewritten as

$$\frac{b_i}{d_i} = \frac{B/D}{\alpha_i^N K_r^N}$$

(12.14)

A column overall material balance is written as

$$F = B + D$$

(12.15)

and a material balance on component i around the column is given by

$$f_i = d_i + b_i$$

(12.16)

where f_i is the flow rate of component i in the feed. The values of D and d_i from Equations 12.15 and 12.16 are now substituted in Equation 12.14 to give

$$b_i = \frac{f_i}{1 + \dfrac{F - B}{B}\left(\alpha_i^N K_r^N\right)}$$

(12.17)

This is a set of equations applicable to $i = 1, \ldots, C$, where C is the total number of components. Equation 12.17 is one form of the Fenske equation (M.R. Fenske, 1932) for a total reflux distillation column. The number of stages N, is the minimum required to achieve the separation corresponding to component flow rates f_i and b_i and bottoms rate B. In this equation, N, the number of stages, is also known as the *fractionation index*.

12.1.3 Degrees of Freedom

In Equation set 12.17, F and f_i are known quantities. The relative volatilities, α_i, are assumed constant for each component. At a fixed pressure, K_r is a function of temperature only. The bottoms total flow rate, B, may be replaced by the sum of b_i, the component flow rates in the bottoms. There are therefore $C + 2$ variables in Equation 12.17: b_i ($i = 1, 2, \ldots, C$), N, and K_r (or an average column temperature). Equation 12.17 is a set of C independent equations for $i = 1, 2, \ldots, C$. With $C + 2$ variables and C equations, the column has 2 degrees of freedom, i.e., two variables must be specified in order to define the column performance. The two specifications could be any two of the above variables or functions thereof. For instance, the specifications could be N and B, or b_1 and b_h, the light and heavy key component flow rates in the bottoms, or Y_{Dh} and X_{Bl}, the mole fractions of the heavy key component in the distillate and the light key component in the bottoms, etc. In each case, with two variables specified, all the other variables can be calculated from Equation set 12.17.

12.1.4 Solution Methods

The set of nonlinear Equations 12.17 may be solved by different techniques, depending on what pair of variables is specified. An iterative algorithm is outlined here for the case where the number of stages, N, and the bottoms rate, B, are specified.

Step 1. Assume an average column temperature, T_r, and calculate the reference component vapor–liquid equilibrium coefficient at average conditions, K_r, using some K-value correlation of the form $K_r = K_r(T_r)$. Also, at the same temperature, use a similar correlation to calculate all the component K-values. Calculate the component relative volatilities.

Step 2. Calculate b_i from Equation 12.17 for $i = 1, \ldots, C$.

Step 3. Calculate the sum of b_i.

Step 4. If the sum of b_i equals the specified B within a designated tolerance, a solution has been obtained. Otherwise, go to the next step.

Step 5. Update K_r using some convergence technique. A simple method might be to recalculate K_r as follows:

$$K_r \left(new \right) = K_r \left(old \right) \frac{\sum b_i}{B}$$

Calculate a new average temperature, T_r, corresponding to the new K_r. At the new temperature, calculate updated component K-values and relative volatilities.

The calculations are repeated at step 2 until a solution is reached. Also, based on the calculated distillate and bottoms compositions, the distillate and bottoms temperatures may be calculated as the dew point and bubble point temperatures, respectively (Chapter 2).

Example 12.1 Total Reflux Calculations

It is required to calculate the maximum separation possible for a hydrocarbon mixture using a ten-theoretical-stage column. The column pressure is 75 psia. The feed is saturated liquid at 75 psia. The feed component flow rates are as follows:

	Component	f_i, mole/hr
1.	Ethane	12
2.	Propane	48
3.	n-Butane	25
4.	n-Pentane	15

The bottoms flow rate is specified at 40 mole/hr. The separation takes place between propane and n-butane, the light and heavy key components. Using n-pentane as the reference component, the following relative volatilities are given:

$$\alpha_1 = 13.304$$

$$\alpha_2 = 5.553$$

$$\alpha_3 = 2.315$$

$$\alpha_4 = 1.000$$

Equation 12.17 is written for $N = 10$ and $B = 40$:

$$b_i \frac{f_i}{1 + 1.5(\alpha_i K_r)^{10}}$$

An initial estimate for K_r is required, which could either be an arbitrary value or an estimate based on a K-value temperature-dependent prediction method. Since the column temperature is not known, an arbitrary initial estimate is used for K_r. The reference component is pentane, the heaviest component in the mixture. Therefore, K_r should have a fairly small value, anywhere between 0 and 1. Using a starting value $K_r = 0.1$, b_i is calculated from the above equation for $i = 1, 2, 3, 4$. The calculated bottoms rate is the sum of b_i, which equals 88.244. Since the specified rate is 40, a second iteration is used, with an updated value of K_r:

$$K_r = (0.1)(88.244)/40 = 0.221$$

The iterations are continued until convergence is reached at $K_r = 0.275$, where the calculated bottoms rate is 40.06. The converged component flow rates in the bottoms are given in Table 12.1, along with the corresponding flows in the distillate.

Table 12.1 Stream Component Flow Rates for Example 12.1

Component	Feed, f_i lbmole/hr	Bottoms, b_i lbmole/hr	Distillate, d_i lbmole/hr
Ethane	12.00	0.00	12.00
Propane	48.00	0.46	47.54
n-Butane	25.00	24.60	0.40
n-Pentane	15.00	15.00	0.00
Totals	100.00	40.06	59.94

The solution method described above is workable only if the specifications are N and B. In general, the two specifications, designated as G_1 and G_2, could be any function of the variables in Equation set 12.17. Thus, G_1 and G_2 may be expressed as

$$G_1(\overline{b}_i, N, K_r) = 0$$

$$G_2(\overline{b}_i, N, K_r) = 0$$

(12.18)

Equations 12.17 and 12.18 comprise $C + 2$ equations which must be solved for the $C + 2$ variables. The entire set of equations could be solved simultaneously using the Newton-Raphson technique. In an alternative method which simplifies the calculations, Equation set 12.18 is expressed as functions of N and B:

$$G_1(N, B) = 0$$

$$G_2(N, B) = 0$$

(12.19)

Equation set 12.19 is equivalent to Equation set 12.18 since N and B are implicit functions of b_i, N, and K_r. N and B are now the independent variables, and b_i and K_r the dependent variables. The calculations alternate between a Newton–Raphson solution of Equation set 12.19 and an iterative solution of Equation set 12.17 as described previously. The algorithm is outlined as follows:

Step 1. Assume initial estimates for N and B.

Step 2. Solve Equation set 12.17 using the algorithm described previously for the case where N and B are specified.

Step 3. Check if the specifications, Equation set 12.18, are satisfied within specified tolerances. If they are, the current values of b_i, N, and K_r constitute a converged solution. If Equation set 12.18 is not satisfied, proceed to the next step.

Step 4. The values N and B are updated by adding corrections ΔN and ΔB calculated from a Taylor series expansion of Equation 12.19:

$$\frac{\partial G_1}{\partial N} \Delta N + \frac{\partial G_1}{\partial B} \Delta B + G_1 = 0$$

$$\frac{\partial G_2}{\partial N} \Delta N + \frac{\partial G_2}{\partial B} \Delta B + G_2 = 0$$

(12.20)

The partial derivatives, as well as G_1 and G_2, are calculated numerically using Equations 12.17 and 12.18 at current values of N and B. Equation set 12.20 is then solved for ΔN and ΔB, which are used to update N and B. The calculations are repeated at step 2 until a solution is obtained.

12.1.5 Multiple Products

As in two-product columns, total reflux in multiple product columns is a limiting condition where the column internal liquid and vapor flows are very large compared to each of the products and feed(s). Multiproduct columns are considered to consist of column sections defined by the product locations. Each section is bounded by two products, one at its top and the other at its bottom. Thus, a column with s sections has s + 1 products. As in two-product columns, multiproduct columns operating at total reflux achieve the maximum separation possible with a given number of stages in each section and for a given set of product rates. Conversely, if the separation between the different products is specified, the minimum trays required in each section are evaluated by total reflux calculations.

Since at total reflux the ratio of any feed or product rate to internal liquid or vapor flow is infinitesimal, Equation 12.7, derived for two-product columns, applies to multiproduct columns as well. This conclusion may be ascertained by carrying out component material balances over different column sections, similar to the material balances represented by Equations 12.5 and 12.6. The equilibrium relation, Equation 12.2, holds regardless of the number of products. The derivations that lead to Equation 12.9 can therefore be generalized to multiproduct columns at total reflux, and that equation may be rewritten for component i in any column section as follows:

$$X_{li} = \frac{Y_{hi}}{K_{hi} K_{h+1,i} \cdots K_{li}} \tag{12.21}$$

This equation relates the liquid composition at the lowest stage in a column section, X_{li}, to the vapor composition at the highest stage in the section, Y_{hi}, in terms of the product of the K-values at each stage in the section.

Equation 12.21 is converted to a more usable form for a column section, as in the case of a two-product column. Constant relative volatilities are assumed for each section or throughout the column, and a reference component equilibrium coefficient, K_r^s, is used for each section at average section conditions. Substituting

$$K_{ji} = \alpha_i K_r^s$$

in Equation 12.21 and rearranging, the following equation is obtained:

$$\frac{Y_{hi}}{X_{li}} = \left(\alpha_i K_r^s\right)^{l-h+1} \tag{12.22}$$

Also, since

$$Y_{hi} = X_{hi} K_{hi} = X_{hi} \alpha_i K_r^s$$

Equation 12.22 may be expressed in terms of the component mole fractions in the liquid leaving the top and bottom trays in the section:

$$\frac{X_{hi}}{X_{li}} = \left(\alpha_i K_r^s\right)^{l-h} \tag{12.23}$$

If the liquid side draw from tray j has a molar flow rate of L_j^s and the component flow in L_j^s is L_{ji}^s, then

$$L_{ji}^{s} = L_{j}^{s} X_{ji}$$

Equation 12.23 is now rewritten in terms of product and component rates:

$$\frac{L_{hi}^{s}}{L_{li}^{s}} = \frac{L_{h}^{s}}{L_{l}^{s}} \left(\alpha_{i} K_{r}^{s} \right)^{l-h}$$

The top product from section s is also the bottom product from section s − 1: $L_{h}^{s} = L_{l}^{s-1}$, $L_{hi}^{s} = L_{li}^{s-1}$, and the above equation becomes

$$\frac{L_{li}^{s-1}}{L_{li}^{s}} = \frac{L_{l}^{s-1}}{L_{l}^{s}} \left(\alpha_{i} K_{r}^{s} \right)^{l-h} \tag{12.24}$$

Equation 12.24 is applicable to all column sections that are bounded by side products on both sides. The uppermost and lowest sections are bounded by the overhead or bottoms on one side and a side draw on the other. If a partial condenser is used, Equation 12.22 should be applied to the uppermost section. Expressed in terms of stream and component flow rates, the equation for the uppermost section takes the form

$$\frac{d_{i}}{L_{li}^{s}} = \frac{D}{L_{l}^{s}} \left(\alpha_{i} K_{r}^{s} \right)^{l} \tag{12.25}$$

Integer 1 designates the tray number where the highest side product is drawn. For the lowest column section, Equation 12.23 is rewritten as

$$\frac{L_{hi}^{s}}{b_{i}} = \frac{L_{h}^{s}}{B} \left(\alpha_{i} K_{r}^{s} \right)^{N-h} \tag{12.26}$$

where N is the total number of column stages, and h is the tray number where the lowest side product is drawn.

In the general case where the number of sections, s, is greater than two, the column has s − 2 of Equations 12.24 plus Equations 12.25 and 12.26 for each component i. If there are two sections (one side product), only Equations 12.25 and 12.26 apply. A column component material balance results in one more set of equations:

$$f_{i} = d_{i} + \sum_{s} L_{i}^{s} + b_{i} \tag{12.27}$$

The summation is over all the side products. The stage subscript of L_{i}^{s} is dropped because side products are drawn only at the end stage of each section.

For a column with s sections, the total number of independent equations includes s sets of Equations 12.24, 12.25, and 12.26 and one set of Equation 12.27. If C is the number of components, the total number of independent equations is $C(s + 1)$. The variables are listed below:

d_i	$i = 1, ..., C$	C
L_i^s	$i = 1, ..., C, s = 1 ..., s - 1$	$C(s - 1)$
b_i	$i = 1, ..., C$	C
N^s	Number of trays in each section	s
K_r	Reference component K-value	s

The quantities D, B and L^s are not considered independent variables since they may be replaced by the sums of d_i, b_i, and L_1^s, respectively. There is a total of $C(s + 1)$ + 2s variables. Therefore, the column has 2s degrees of freedom and, in order to define its performance, 2s variables must be specified.

If N^s in each section and any s of the product rates D, B, and L^s are specified, the solution algorithm could proceed along the steps hereby outlined:

Step 1. Assume an average temperature for each section and calculate K_r^s and relative volatilities for all the sections. Use these values in Equations 12.24, 12.25, and 12.26.

Step 2. Equations 12.24, 12.25, 12.26, and 12.27 are now a set of linear equations. Solve them for d_i, b_i, and L_1^s using a matrix inversion method.

Step 3. Check the equalities

$$D = \sum_i d_i$$

$$B = \sum_i b_i$$

$$L^s = \sum_i L_i^s$$

If they are satisfied within a specified tolerance, a solution has been reached. If not, the K_r^s vector is updated as described in Section 12.1.4, and the calculations are repeated beginning at step 1. An updated temperature profile is used, back calculated from the new K_r^s values.

If the specifications are not N^s, D, B, and L^s, they may be expressed as functions thereof:

$$G_n\left(N^s, B, L^s\right) = 0$$

$$n = 1, 2, ..., 2s$$

(12.28)

Equation set 12.28 is solved along with Equations 12.24, 12.25, 12.26, and 12.27. As in the method described for two-product columns, the calculations alternate between a Newton–Raphson solution of Equation set 12.28 and an iterative solution of the rest of the equations as described above.

12.2 MINIMUM REFLUX RATIO

The operation of a distillation column at minimum reflux was discussed in Chapters 5 and 6 in the context of binary systems. Using the Y–X diagram, it was shown that the reflux ratio is directly related to the slope of the rectifying section operating line. As the operating line approaches the equilibrium curve, the reflux ratio becomes smaller. The operating lines may not cross the equilibrium curve; therefore, the minimum reflux corresponds to the smallest slope of the rectifying section operating line that will not result in either of the operating lines crossing the equilibrium curve. At minimum reflux, one of the operating lines touches the equilibrium curve or both of them touch it at their intersection (Figure 12.2A and B). The location of the point where contact occurs depends on the shape of the equilibrium curve and also on the feed and product compositions and the feed thermal conditions. The number of stages that can be stepped off goes up indefinitely as the operating line approaches the equilibrium curve, and all these stages have constant composition. Figure 12.2A shows an infinite number of trays in the rectifying section, and Figure 12.2B shows an infinite number of trays around the feed tray.

The concept of minimum reflux also applies to multicomponent mixtures. It is defined as the reflux ratio below which a specified separation is infeasible, irrespective of the number of trays. At minimum reflux ratio, an infinite number of trays would be required to achieve the specified separation. With infinite trays there must exist in the column at least one section where the vapor and liquid compositions do not change from tray to tray.

The definition of minimum reflux is linked to a specification of separation between two components. In multicomponent mixtures the specified components are the light key and the heavy key components. Minimum reflux is meaningless if the column is specified in a manner that does not define a particular separation between two components, such as a specification of the number of trays and a product rate. Also, minimum reflux in this context is associated with single-feed, two-product columns. For a given feed composition, the minimum reflux depends on the key components, their separation specification, and the feed thermal conditions.

In rigorous column solution methods, the number of trays is usually assumed fixed and, therefore, the concept of minimum reflux does not apply. For binary mixtures, minimum reflux is probably best determined by graphical methods such as the Y–X diagram method. For multicomponent systems the Underwood equation (Underwood, 1948), derived on semirigorous grounds, provides an estimate of the minimum reflux corresponding to a specified separation of two key components in a given stream. The assumptions used in the derivation are similar to the Fenske assumptions for calculating minimum trays, namely constant relative volatilities and constant liquid and vapor flows in the column. The Underwood equation should, therefore, be used only for columns and mixtures where these assumptions are valid.

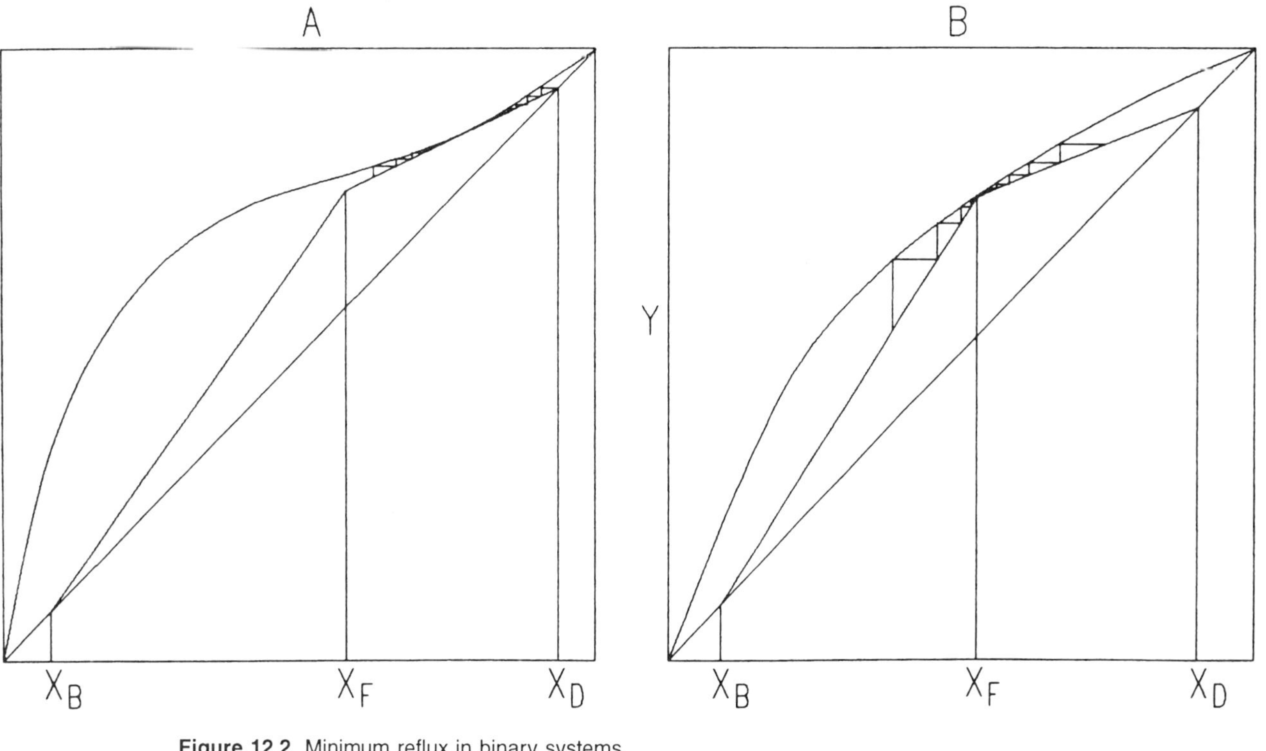

Figure 12.2 Minimum reflux in binary systems.

The equation in its final form is presented here. The reader is referred to the original article (Underwood, 1948) for detailed derivation. The minimum reflux ratio, R_m, is given by the equation

$$R_m = \sum_{i=1}^{c} \frac{\alpha_i X_{Di}}{\alpha_i - \theta} - 1 \qquad (12.29)$$

where α_i is the relative volatility, and X_{Di} are the mole fractions in the distillate at minimum reflux. The composition must be estimated independently, usually using the Fenske total reflux equation. The parameter θ has a value between the relative volatilities of the light and heavy key components. It is determined from the equation

$$\sum_{i=1}^{c} \frac{\alpha_i Z_i}{\alpha_i - \theta} = 1 - q \qquad (12.30)$$

where Z_i is the mole fraction of component i in the feed, and q is the feed thermal condition, as defined in Section 5.2.1.

The number of trays in an actual column is, of course, never infinite, and columns will never achieve the specified separation if operated at minimum reflux ratio. What the estimation of minimum reflux does provide is a limiting condition or a reference point. The operating reflux ratio is commonly expressed as a factor multiplied by the minimum reflux ratio. This factor must be greater than 1 in order to make the specified separation possible. The minimum reflux is also used for correlating the required number of trays and reflux ratio as described below.

12.3 COLUMN DESIGN AND PERFORMANCE ANALYSIS

The minimum trays, N_m, and minimum reflux ratio, R_m, estimated by the above shortcut methods, represent limiting conditions that require either infinite (total) reflux or infinite trays to achieve a specified separation. They also serve as correlating parameters for predicting the reflux ratio required to achieve the specified separation with a given number of theoretical trays. The Gilliland correlation (Gilliland, 1940) is presented as a plot of $(N - N_m)/(N + 1)$ vs. $(R - R_m)/(R + 1)$ in Figure 12.3. R is the reflux ratio required to achieve the separation with N theoretical trays. Thus, if the separation is given or specified, the minimum trays may be calculated by the Fenske method, the minimum reflux may be calculated by the Underwood method, and the reflux required with a given number of trays may be obtained from the Gilliland correlation.

In a design situation the engineer can use the reflux vs. trays relationship to evaluate the economics of designing a column with more trays and lower reflux vs. fewer trays and higher reflux. The higher cost associated with a column with a large number of trays is offset by a reduced column diameter that would be required with a lower reflux ratio. The utility consumption for the reboiler and condenser is also

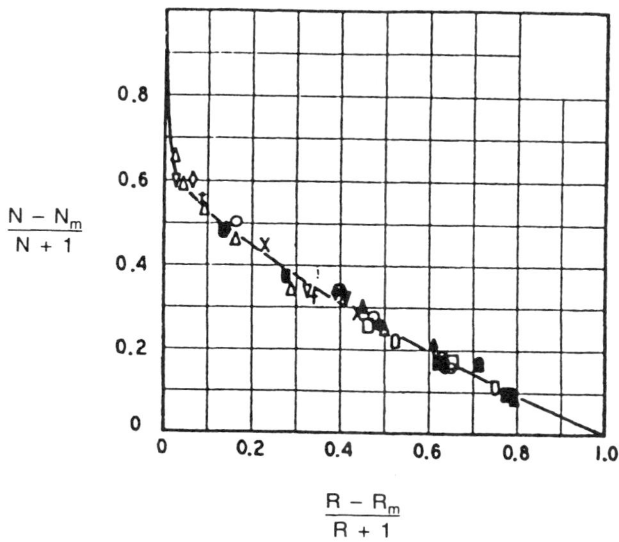

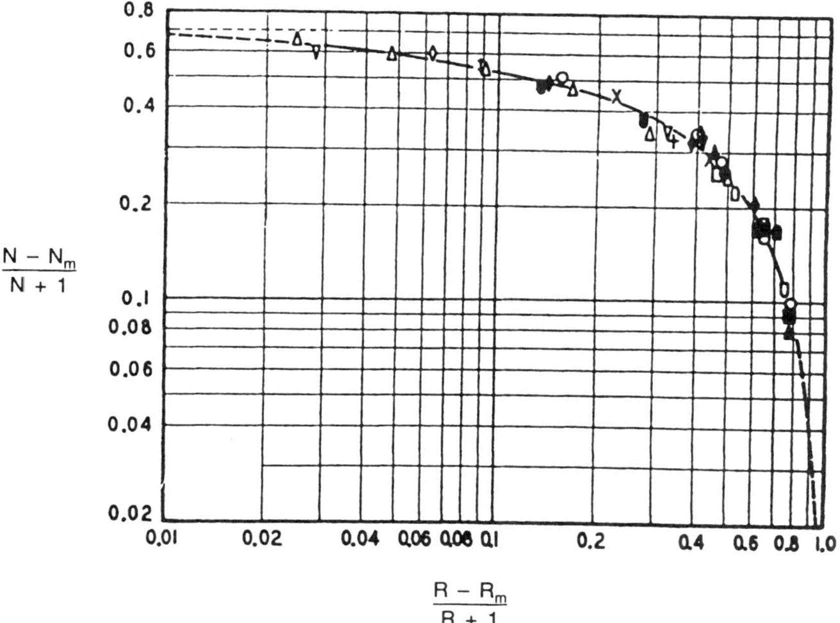

Figure 12.3 Gilliland correlation. (Reprinted with permission from E. R. Gilliland, *Ind. Eng. Chem.*, 32, 1220, 1940. American Chemical Society.)

reduced at the lower reflux. With fewer trays the column diameter and utility consumption are higher since a higher reflux ratio is required.

The calculations, aided by computer simulation programs, are started by making rough estimates of the overhead and bottoms compositions, based on the desired separation. The condenser pressure is determined so that it ensures condensation of the overhead. That is, the condenser pressure should be such that the estimated dew point temperature of the overhead stream should be higher than the temperature of the cooling medium at hand. The column pressure should be higher than the condenser pressure to allow for vapor-flow pressure drops across the column and condenser. Once the operating pressure is set, it is held constant in the first calculation pass.

Next, the relative volatilities, α, are determined as averages of estimated α's obtained from K-value calculations at overhead and bottoms conditions. These conditions include the operating pressure, the estimated overhead and bottoms compositions, and the estimated dew point and bubble point temperatures. With a starting set of relative volatilities, the minimum trays, N_m, and the overhead and bottoms compositions at total reflux are calculated by the Fenske method (Equation 12.17) for the specified separation. The new compositions may be used to recalculate more accurately the temperatures, pressures, and relative volatilities. The process is repeated until the α's stabilize.

The final compositions and relative volatilities are then used to calculate the minimum reflux, R_m, using the Underwood method (Equations 12.29 and 12.30). The calculated values of N_m and R_m are next applied in the Gilliland correlation to determine a suitable combination of trays and reflux, N and R, consistent with economic and design considerations.

The shortcut methods may also be used for the approximate analysis of the performance of an existing column. Here, the number of trays, N, is fixed, and the objective is to determine the reflux ratio required to meet a specified separation. The Fenske and Underwood methods (Equations 12.17, 12.29, and 12.30) are used to calculate the minimum trays and minimum reflux, N_m and R_m. The operating reflux ratio corresponding to the given number of trays is then read from the Gilliland chart (Figure 12.3). The internal vapor and liquid rates are calculated from the reflux ratio and product rates. A check must be made to determine if the existing column can handle the calculated vapor and liquid traffic.

Example 12.2 Minimum Reflux Calculations

Using the Underwood equations, determine the minimum reflux required to achieve the separation obtained in Example 12.1 at total reflux. The separation is specified as propane (light key) mole fraction of 0.0115 in the bottoms and n-butane (heavy key) mole fraction of 0.0067 in the overhead, as calculated from Table 12.1. Use the relative volatilities given in Example 12.1 and the compositions in Table 12.1. Assume the feed is saturated liquid, i.e., q = 1.

First calculate θ from Equation 12.30. Substituting the values of α_i and q, this equation becomes

$$\frac{(13.304)(0.12)}{13.304 - \theta} + \frac{(5.553)(0.48)}{5.553 - \theta} + \frac{(2.315)(0.25)}{2.315 - \theta} + \frac{(1)(0.15)}{1 - \theta} = 0$$

The value of θ must be somewhere between the key components' relative volatilities. A trial and error solution gives $\theta = 2.86$.

The minimum reflux is computed directly from Equation 12.29, using the overhead composition at total reflux:

$$R_m = \frac{(13.304)(0.200)}{13.304 - 2.86} + \frac{(5.553)(0.793)}{5.553 - 2.86} + \frac{(2.315)(0.007)}{2.315 - 2.86} + \frac{(1)(0)}{1 - 2.86} - 1 = 0.86$$

The specified separation could thus be achieved with an infinite number of trays at a reflux ratio of 0.86. As shown in Example 12.1, the same separation could be achieved with ten trays at total reflux.

The reflux ratio required for a given number of trays may now be estimated from the Gilliland chart (Figure 12.3). If, for instance, $N = 15$, then

$$\frac{N - N_m}{N + 1} = \frac{15 - 10}{15 + 1} = 0.3125$$

From the chart, this corresponds to

$$\frac{R - R_m}{R + 1} = 0.39$$

or $R = 2$.

The column vapor and liquid traffic at this reflux ratio are estimated from the relationships $V_r = L_r + D$ and $R = L_r/D$, where V_r and L_r are the vapor and liquid molar flows in the column rectifying section. The distillate rate, D, is 60 lbmole/hr (Example 12.1). Therefore,

$$L_r = RD = (2)(60) = 120 \text{ lbmole/hr}$$

and

$$V_r = L_r + D = 120 + 60 = 180 \text{ lbmole/hr}$$

Since the feed is saturated liquid, the liquid flow in the stripping section is

$$L_s = L_r + F = 120 + 100 = 220 \text{ lbmole/hr}$$

By material balance, the stripping section vapor flow is the same as the rectifying section, $V_s = V_r = 180$ lbmole/hr. The column diameter must be checked for its ability to handle the vapor and liquid traffic (Chapter 14).

12.4 MODULAR SHORTCUT METHODS

The assumptions of total reflux or minimum reflux in the shortcut methods described above restrict their application to a limited class of columns. Another shortcut method is described in this section which is flexible enough to be generalized to a broad range of multistage, multicomponent separation processes. The method is based on the column section, a module that could be used as a building block to model various column configurations. If such a module is to remain in the "shortcut" domain where computing time would not exceed about one tenth of the rigorous computing time, it is obvious that simplifying assumptions would still have to be made.

The concepts of this generalized shortcut approach are outlined in this section along with methods for calculating the model parameters.

12.4.1 Column Sections

A column section is a group of contiguous adiabatic vapor–liquid (or liquid–liquid) equilibrium stages whose only feeds and products are the following: a liquid feed to the top stage, a vapor feed to the bottom stage, a liquid product from the bottom stage, and a vapor product from the top stage (Figure 12.4). A simple absorber, stripper, or extractor can itself be considered a column section. Although the calculations developed here are focused on vapor–liquid column sections, they could be modified to apply to liquid–liquid extractors.

Based on Figure 12.4, a mass balance for component i around tray j is written as

$$L_{j-1}X_{j-1,i} + V_{j+1}Y_{j+1,i} - L_jX_{ji} - V_jY_{ji} = 0 \tag{12.31}$$

As in other shortcut methods, the liquid and vapor molar flow rates are assumed constant in the column section:

$$L_{j-1} = L_j = L$$

$$V_j = V_{j+1} = V$$

In addition, the present method assumes a constant equilibrium coefficient for each component in the column section:

$$K_{ji} = K_i$$

This assumption is more restrictive than the assumption of constant relative volatilities, or relative K-values, that is used in the Fenske and Underwood methods. The payback for this assumption is the ability to generalize the model to different degrees of column complexity. The success of the method is dependent on proper evaluation of effective K-values or other model parameters that would represent actual behavior of the column section. The equilibrium coefficient is commonly lumped with the vapor and liquid molar flows in the column to define the stripping factor,

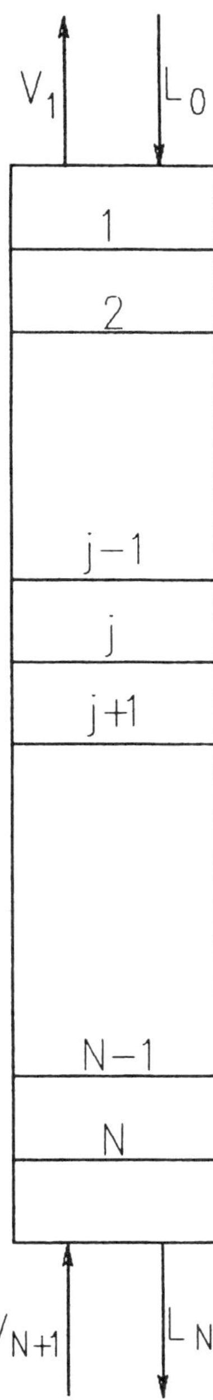

Figure 12.4 Schematic of a column section.

$$S_i = K_i \frac{V}{L}$$

Since K_i, V, and L are all assumed constant in the column section, the stripping factors are also constant.

The assumption of constant liquid and vapor molar flow may, in many situations, be grossly in error, especially in processes where mass transfer between the phases takes place mostly in one direction, such as in absorption and stripping. The method may, nevertheless, be used satisfactorily in such situations if appropriate stripping factors can be determined. This topic is addressed in more detail further along in this section.

By substituting $Y_{ji} = K_i X_{ji}$ in Equation 12.31 and replacing $K_i V/L$ with S_i, the following equation is obtained:

$$S_i X_{j+1,i} - (1 + S_i)X_{ji} + X_{j-1,i} = 0 \tag{12.32}$$

This equation is now written for $j = 1, 2, ..., N - 1$:

$$S_i X_{2i} - (1 + S_i)X_{1i} + X_{0i} = 0$$

$$S_i X_{3i} - (1 + S_i)X_{2i} + X_{1i} = 0$$

$$...$$

$$S_i X_{Ni} - (1 + S_i)X_{N-1,i} + X_{N-2,i} = 0$$

By successive substitution, and applying Equation set 12.1, and rearranging, the following equation can be derived (Smith, 1963; Kremser, 1930):

$$\left(V_{N+1,i} + L_{0i}\right)\left(1 - S_i^N\right) + L_{0i}\left(S_i^N - S_i\right) - L_{Ni}\left(1 - S_i^{N+1}\right) = 0 \tag{12.33}$$

This equation relates the rate of component i leaving the column section in the liquid product to the rates of component i entering the column section in the liquid and vapor feeds. The rate of component i leaving the column section in the vapor product is calculated by component balance around the column section:

$$(V_{N+1,i} + L_{0i}) - (V_{1i} + L_{Ni}) = 0 \tag{12.34}$$

An overall enthalpy balance is now written as follows:

$$(H_{N+1}V_{N+1} + h_0 L_0) - (H_1 V_1 + h_N L_N) = 0 \tag{12.35}$$

where H and h are the molar enthalpies of the vapor streams and liquid streams, respectively. They are either rigorously calculated compositional enthalpies or simplified functions of temperature only (Chapter 1).

Two more relationships must be satisfied in the column section: the vapor product must be at its dew point and the liquid product at its bubble point (Chapter 2):

$$\sum_{i=1}^{C} \left(\frac{V_{1i}}{V_1 K_{1i}} - 1 \right) = 0 \tag{12.36}$$

$$\sum_{i=1}^{C} \left(\frac{L_{Ni} K_{Ni}}{L_N} - 1 \right) = 0 \tag{12.37}$$

Here, also, the K-values could be either rigorous, composition-dependent, or simple functions of temperature.

Equations 12.33, 12.34, 12.35, 12.36, and 12.37 comprise 2C + 3 equations:

Equation 12.33	C equations
Equation 12.34	C equations
Equation 12.35	1 equation
Equation 12.36	1 equation
Equation 12.37	1 equation
	2C + 3 equations

With a fixed number of stages in the column section, fixed stage pressures, and zero heat duties or heat losses and with the feeds completely defined, the remaining variables associated with the column section as a whole are the following:

V_{1i}	Component rates in the vapor product:	C
L_{Ni}	Component rates in the liquid product:	C
T_1	Top tray temperature:	1
T_N	Bottom tray temperature:	1
V/L	Vapor-to-liquid ratio in the column:	1
		2C+3

With equal number of variables and equations, the problem is completely defined. The K-values and enthalpies are calculated physical properties and are not considered as independent variables. The vapor-to-liquid ratio, V/L, is a model variable that must be calculated to satisfy the set of equations.

Equations 12.33 through 12.37 can be solved either simultaneously or iteratively using one or more iterative loops (Smith and Brinkley, 1960).

In the iterative procedure described here, Equations 12.33 and 12.34 are first solved with an assumed set of stripping factors, S_i. Given the component rates in the feeds, the component rates in the products are calculated directly in this step.

Next, Equations 12.36 and 12.37 are solved using the calculated vapor and liquid products' compositions to determine the top and bottom tray temperatures. These are essentially dew point and bubble point temperature calculations that could themselves be iterative (Chapter 2). Temperatures must be found at which the K-values are such that Equations 12.36 and 12.37 are satisfied. If ideal solutions are assumed where the K-values are functions of temperature only, Equations 12.36 and 12.37 could be solved directly.

Using the calculated top and bottom tray temperatures, Equation 12.35 is now solved for V_1 and L_N. The first and second terms in this equation are known quantities. Moreover, V_1 or L_N could be eliminated using overall material balance:

$$V_1 + L_N = V_{N+1} + L_0$$

where V_{N+1} and L_0 are known. The streams' molar enthalpies are calculated either by ideal solution or composition-dependent methods (Chapter 1), and Equation 12.35 is solved directly for V_1 and L_N.

The final step is to update V/L and the stripping factors using some convergence technique. A simple procedure is to calculate arithmetic averages of V, L, and K_i:

$$V = (V_1 + V_{N+1})/2$$

$$L = (L_0 + L_N)/2$$

$$K_i = (K_{1i} + K_{Ni})/2$$

Values for K_{1i} and K_{Ni} are taken from the dew point and bubble point calculations. The stripping factors are then calculated as

$$S_i = K_i(V/L)$$

The calculations are restarted by solving Equations 12.33 and 12.34 with the new values of the stripping factors. The procedure is repeated until convergence is achieved, that is, until the variation of S_i or any other parameter from one iteration to the next becomes smaller than some acceptable tolerance.

12.4.2 Reduced Model

The assumption of constant stripping factors is necessary for deriving the analytical solution of multistage separation in a column section, Equation 12.33. The simplicity of this approach and the resulting great savings in computing time are highly desirable.

The accuracy of the method hinges on proper determination of effective stripping factors. The method used in the previous section for calculating the stripping factors is based on solving the group of equations that describe the column section. The method can be enhanced by tuning the stripping factors and other model parameters to improve the model accuracy. The objective is to find model parameters that would make the shortcut results match actual performance data or rigorous solutions.

One approach involves periodic updating of the shortcut model parameters using the results of a rigorous model (Jirapongphan et al., 1980; Trevino-Lozano et al., 1984). The method is especially appropriate in situations where repetitive model solutions are required as the feed streams and other process conditions change, such as in process control or optimization. The rigorous solution provides a base case which serves to calculate the shortcut model parameters. The shortcut model is then used to solve the column section as conditions move away from the base case. A good shortcut model should be capable of predicting the column section performance at conditions reasonably distanced from the base case, using the same base case parameters. Additional periodic rigorous updates of the shortcut model parameters may be required as operating conditions move farther away from the base case.

The shortcut model is developed in terms of reduced parameters that are not strongly dependent on stream compositions, temperature, and pressure. The shortcut model, represented by Equations 12.33 through 12.37, is solved in conjunction with reduced equations for calculating enthalpies, vapor–liquid equilibrium coefficients, and effective stripping factors based on the rigorous base case.

The enthalpies, which in the rigorous model are generally calculated by compositional methods appropriate to the mixture under consideration, are expressed in the shortcut model as a linear function of temperature only:

$$H = a + b(T - T_r) \tag{12.38}$$

where T_r is a reference temperature. The compositional effect on the enthalpy is small compared to the temperature effect, especially since moderate deviations from the base case are not expected to cause large changes in the composition. Parameters a and b are evaluated for each stream from rigorous model enthalpies calculated at T_r and one other temperature.

The vapor–liquid equilibrium coefficients are expressed in terms of the relative volatility, α, and a reference K-value at the top and bottom stages:

$$K_{1i} = \alpha_{1i} K_{1r}$$
$$K_{Ni} = \alpha_{Ni} K_{Nr} \tag{12.39}$$

The K_i values may be taken directly from the rigorous model and the reference K-values calculated as weighted logarithmic averages:

$$\ln K_{1r} = \sum_{i=1}^{c} Y_{1i} \ln K_{1i}$$
$$\ln K_{Nr} = \sum_{i=1}^{c} X_{Ni} \ln K_{Ni} \tag{12.40}$$

The relative volatilities are calculated from Equation set 12.39, and are held constant for the same base case. Departures of the reference K-values from the base case are calculated from the relationships

$$\ln\left(K_{1r}P\right) = A_1 + B_1\left(\frac{1}{T_1} - \frac{1}{T_1^*}\right)$$

$$\ln\left(K_{Nr}P\right) = A_N + B_N\left(\frac{1}{T_N} - \frac{1}{T_N^*}\right)$$

(12.41)

where T_1^* and T_N^* are the base case top and bottom temperatures, and P is the average column section pressure. Parameters A_1 and B_1 and A_N and B_N are calculated for the top and bottom trays from rigorous K_r values at the base case temperature and one other temperature, with the composition held constant. If the relative volatilities are constant, Equation set 12.39 is used once again to calculate the K-values away from the base case.

Effective component stripping factors, S_i, are also calculated from base case rigorous data using Equation 12.33. A reference stripping factor, S_r, is calculated as some average of S_i, using an equation similar to Equation 12.40. Reduced parameters β_i are then calculated from the defining equation

$$S_i = \beta_i S_r$$

(12.42)

Once parameters β_i are calculated for a given base case, they are assumed constant until a new rigorous calculation is performed. Justification for this assumption is that, as conditions deviate from the base case, the component effective stripping factors will change, although their variation relative to the reference stripping factor will be much smaller.

The reference stripping factor, S_r, is calculated from effective component stripping factors only the first time in each base case calculation in order to determine parameters β_i. In subsequent shortcut calculations of deviations from the base case, S_r is calculated from the reduced model equations.

The reduced model is completely defined by the column section mass balances, energy balances, and separation and equilibrium relations (Equations 12.33 through 12.37) and the physical property models (Equations 12.38 through 12.42). The solution strategy starts by solving the rigorous model to calculate V_{1i}, V_1, L_{Ni}, L_N, T_1, and T_N for given feeds L_0 and V_{N+1}, number of stages N, and pressure profile (or constant average section pressure, P). Equation set 12.33 is then solved for S_i, an average stripping factor S_r is calculated, and Equation set 12.42 is solved for β_i. The rigorous model results are also used to calculate K_{1i} and K_{Ni}, and Equations 12.40 and 12.39 are solved for K_{1r}, K_{Nr}, α_{1i}, and α_{Ni}. Next, the rigorous model temperature is perturbed to calculate parameters a and b in Equation 12.38 for streams V_1, V_{N+1}, L_0, and L_N and parameters A_1, B_1, A_N, and B_N in Equation set 12.41.

The physical property parameters as derived here can be used with the other equations of the reduced model to calculate the column section. The reduced model should be able to reproduce the base case results exactly. The calculations can be carried out iteratively using a scheme similar to the one described in Section 12.4.1. The calculations start with the base case stripping factor S_r, to solve for the product stream component rates. Equations 12.36 and 12.37 are then solved with the reduced model K-value equations (Equations 12.39, 12.40, and 12.41) to calculate the top and bottom tray temperatures. The enthalpy balance (Equation 12.35), coupled with the enthalpy model (Equation 12.38), is solved for the product vapor and liquid flow rates. The new product component rates are finally used in Equation set 12.33 to calculate S_i; then S_r is calculated from Equation 12.42. The computations are repeated until S_r converges within an acceptable tolerance.

With this method the column section products can be calculated if the feed streams are given. The method cannot directly calculate the column section to satisfy performance specifications such as product composition or quality. Performance specifications could be handled by superimposing an external iterative loop that would vary a fixed model parameter such as the feed rate or temperature to meet the specification.

12.4.3 Complex Configurations

By its definition, the column section is limited in practical application to modeling simple absorbers, strippers, and liquid–liquid extractors — processes with one feed and one product at each end and with no heat addition or removal. Given a reliable method for solving the column section, however, it could be used as a module to build more complex configurations.

A distillation column, for instance, would be modeled with a column section for the stripper section, a column section for the rectifying section, and single equilibrium stages for the feed tray, the condenser, and the reboiler. In order to solve the distillation column separation equations analytically, two sets of Equation 12.33, each with the appropriate stripping factors for the corresponding section, would have to be solved simultaneously along with a component balance around the feed tray, the condenser, and reboiler equations. Such a solution does exist for conventional distillation and for certain extraction problems (Smith and Brinkley, 1960).

With the analytical approach it is possible to solve only a limited number of column configurations, with little flexibility regarding model parameter tuning. The alternative is to model the process using column sections, equilibrium stages (or general flash blocks), mixers, and splitters and to solve these modules sequentially or by simultaneous numerical techniques. Depending on the sequence chosen, certain streams may have to be initialized to start the computations. Each time a module is to be solved, the streams feeding it must be either known or initialized. In principle, the modular approach could be applied to any complexity configuration. Coupled with the parameter updating technique described in Section 12.4.2, reliable modeling is possible for broad ranges of applications.

In building complex multistage separation models from column sections and other blocks, the question arises as to the number of variables that must be specified in a given situation to define the problem completely. That is, the number of degrees of freedom must be determined. Rather than tally the number of equations, variables, etc., to determine the degrees of freedom, these can be inferred from the process configuration. Under conditions of fixed pressure, fixed number of stages, and fixed feeds, a column section has 0 degrees of freedom. A single equilibrium stage at a fixed pressure and with fixed feeds and heat duty also has no degrees of freedom. A mixer falls in the same category. The degrees of freedom of a splitter (fixed feeds and pressure) equal the number of products minus one. Each time one of the fixed parameters is allowed to vary, 1 degree of freedom is created. The following are examples of different types of multistage separation processes modeled with column sections and other ancillary modules.

Reboiled Stripper

This process may be modeled with a column section and an equilibrium stage as shown in Figure 12.5. If the pressure profile; feed F; reboiler duty, Q, and number of stages, N, are all fixed, this process has 0 degrees of freedom. An operator of such a column would have no control over the column products. If one parameter, such as the reboiler duty, is allowed to vary, one other parameter such as the bottoms rate, B, could be controlled and must be specified in order to define the problem completely.

The computations could proceed by the initial solving of the column section with an assumed vapor feed, V_{N+1}, to calculate V_1 and L_N. The reboiler is then solved with feed stream, L_N, and either the reboiler duty, Q, or the bottoms rate, B, fixed. The column section is solved again with the updated stream, V_{N+1}. The calculations are repeated until the product streams stabilize within an acceptable tolerance.

Distillation Column with a Partial Condenser

As shown in Figure 12.6, this process is modeled with six modules. The feed tray is calculated in module 1, which actually is a combination of a mixer and an equilibrium stage. The mixer serves to mix the external feed, F, liquid from the rectifying section, and vapor from the stripping section. The equilibrium stage calculates the vapor going to section 2 and the liquid going to section 3. The condenser and reboiler are modeled with equilibrium stages 4 and 5. Since part of the condensed liquid is refluxed and the other part is taken as a distillate product, a splitter, unit 6, is used to model this part of the process.

With fixed feeds, heat duties, pressure profiles, and number of stages, all the modules except unit 6 have 0 degrees of freedom. Unit 6 is a splitter with two products and, therefore, has 1 degree of freedom. Hence, the entire column as defined has 1 degree of freedom and requires one more specification to be completely defined. If parameters such as the condenser and reboiler duties are allowed to vary, the number of specifications must increase to match the number of variables.

In modular solution the computational sequence is usually arbitrary but may depend on the way the column is specified. In the present example, if the only degree of freedom is the one coming from the splitter and the reflux rate, R, is specified, the modules could be solved according to the sequence 1, 2, 4, 6, 3, 5. With this

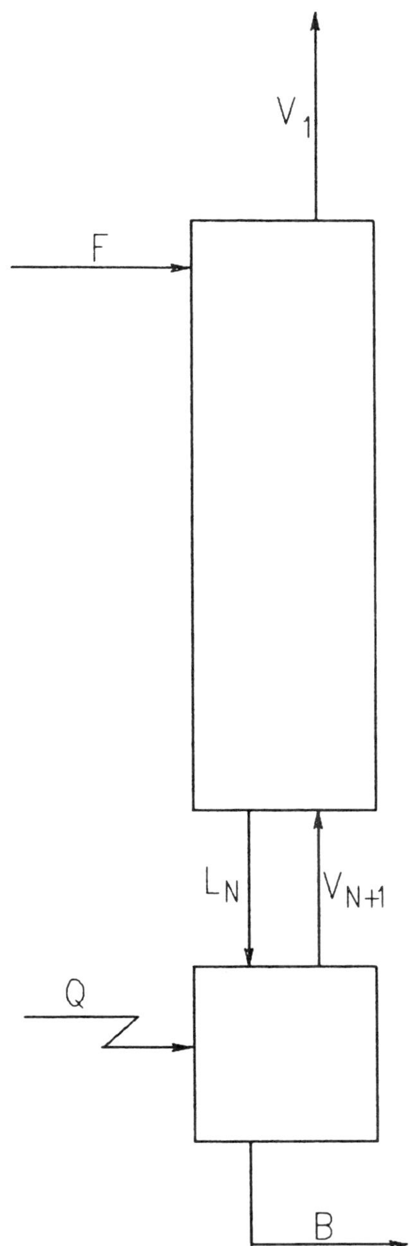

Figure 12.5 Reboiled stripper.

sequence, streams L2, V3, R, and V5 would have to be initialized. Different streams would have to be initialized with a different computational sequence. Convergence is reached when all the streams have stabilized within tolerance.

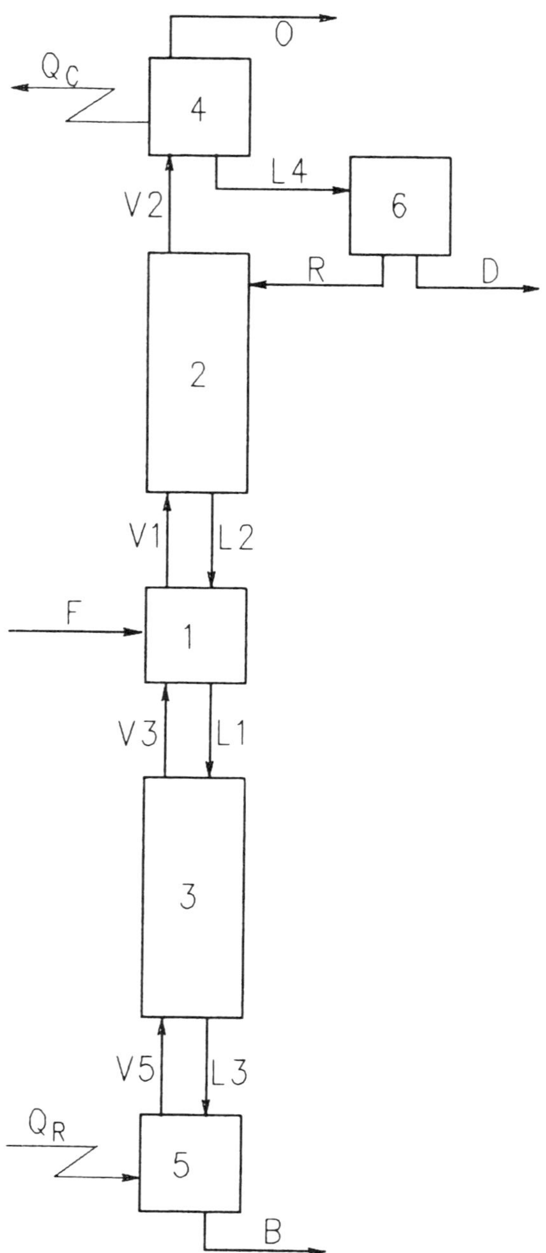

Figure 12.6 Distillation column with a partial condenser.

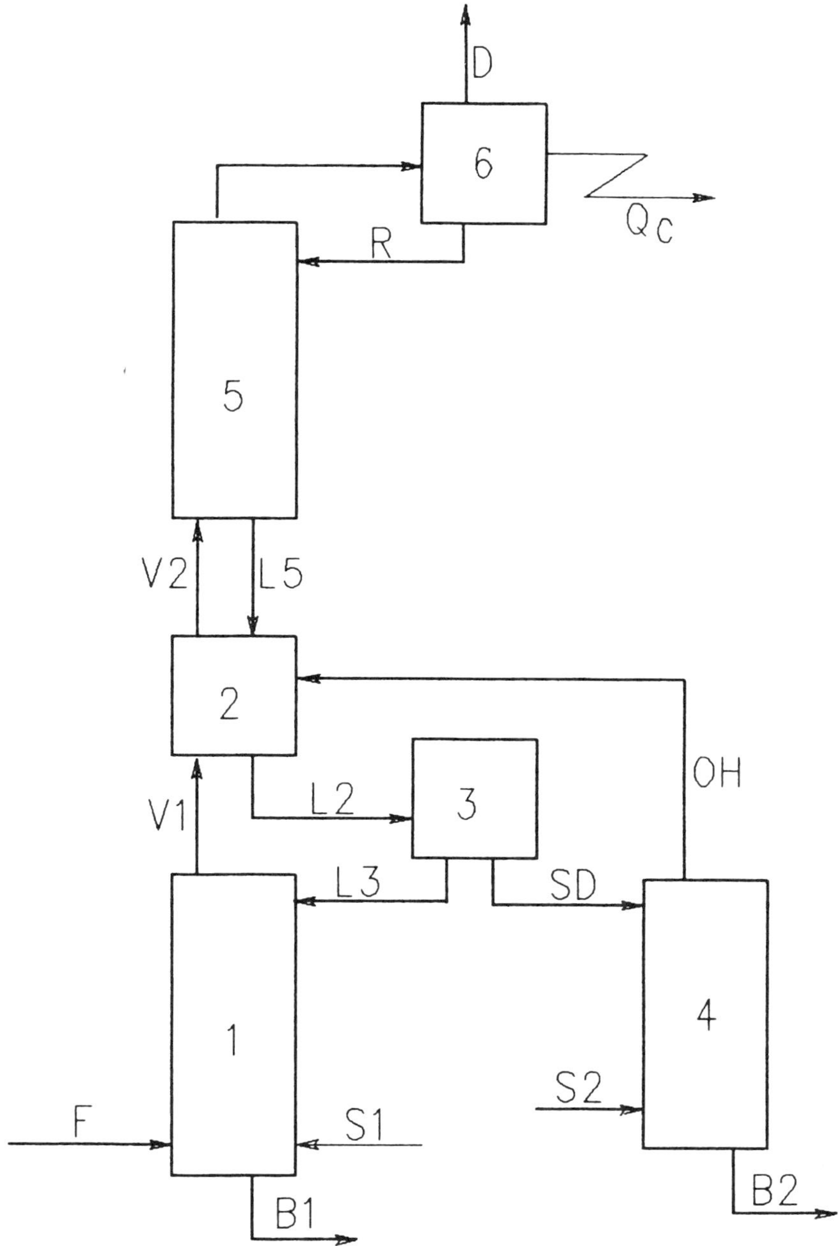

Figure 12.7 Multicolumn system.

Multicolumn System

Figure 12.7 shows a model of a main column and a sidestripper. Vapor feed F goes to the bottom of the main column, which is steam stripped with stream S1, using no reboiler. The column has a partial condenser with vapor distillate D. Side draw SD is taken from around the middle of the main column and steam stripped in the sidestripper, with the overhead vapor, OH, returned to the draw tray in the main column.

Module 1, a column section, represents the section in the main column below the draw tray. No separate module is needed at the bottom since no reboiler is used. Module 2 models the draw tray and consists of a mixer and an equilibrium stage. Module 3 is a splitter that takes the liquid from the draw tray and splits it into side draw SD and the remaining liquid flowing down to the bottom column section. The sidestripper and the upper column section are modeled with column sections, modules 4 and 5, and the condenser is modeled with an equilibrium stage, module 6. Using computational sequence 1, 2, 3, 4, 5, 6 requires initialization of streams L3, L5, OH, and R.

NOMENCLATURE

B	Bottoms stream designation or molar flow rate	N	Number of theoretical stages
b_i	Molar flow rate of component i in the bottoms	N_m	Minimum number of stages
		P	Pressure
C	Number of components in column streams	Q	Heat duty
		q	Feed thermal condition
		R	Reflux ratio
		R_m	Minimum reflux ratio
D	Distillate or overhead stream designation or molar flow rate	s	Column sections
		S	Stripping factor
d_i	Molar flow rate of component i in the distillate	T	Temperature
		V_j	Vapor molar flow rate leaving tray j
F	Feed stream designation or molar flow rate	V_{ji}	Component i vapor molar rate leaving tray j
f_i	Molar flow rate of component i in the feed	X_{ji}	Mole fraction of component i in liquid leaving tray j
G	General column specification	Y_{ji}	Mole fraction of component i in vapor leaving tray j
H	Vapor molar enthalpy	Z_i	Mole fraction of component i in feed
h	Liquid molar enthalpy	α	Relative volatility
K	Vapor–liquid equilibrium coefficient	β	Stripping factor parameter
		θ	Parameter intermediate between relative volatilities of light key and heavy key
L_j	Liquid molar flow rate leaving tray j		
L_{ji}	Component i liquid molar rate leaving tray j		

SUBSCRIPTS

B	Bottoms designation	h	Heavy key component designation or highest stage in a section
D	Distillate designation		
f	Feed tray designation	i	Component designation
j	Stage designation	r	Reference component or rectifying section
l	Light key component designation or lowest stage in a section	s	Side product or stripping section

SUPERSCRIPT

s   Column section designation

REFERENCES

Fenske, M. R., Fractionation of straight-run Pennsylvania gasoline, *Ind. Eng. Chem.*, 24, 482, 1932.

Gilliland, E. R., *Ind. Eng. Chem.*, 32, 1220, 1940.

Jirapongphan, S., J. F. Boston, H. I. Britt, and L. B. Evans, "A Nonlinear Simultaneous Modular Algorithm for Process Flowsheet Optimization," Paper presented at AIChE 73rd Annual Meeting, Chicago, November 1980.

Kremser, A., *Nat. Petrol. News*, 22, May 21, 1930.

Smith, B. D., *Design of Equilibrium Stage Processes*, New York, McGraw-Hill, 1963.

Smith, B. D. and W. K. Brinkley, *AIChE Journal*, 6(3), 446, 1960.

Trevino-Lozano, R. A., T. P. Kisala, and J. F. Boston, *Computers and Chemical Engineering*, 8(2), 105, 1984.

Underwood, A. J. V., *Chem. Eng. Progr.*, 44, 603, 1948.

PROBLEMS

12.1 For the separation problem described in Examples 12.1 and 12.2, estimate the number of trays required if the condenser can handle a maximum vapor rate of 150 lbmole/hr and the overall tray efficiency is 65%.

12.2 A packed column is to be used for the separation of propane and butane as described in Examples 12.1 and 12.2. The column uses a partial condenser and reboiler, and is 36 ft high. Assuming 3 ft of packing is equivalent to an equilibrium stage, estimate the reflux rate required to achieve the specified separation.

12.3 Determine the optimum column size for the separation problem described in Examples 12.1 and 12.2. The following information is given:

Capital cost of the column (expected life of ten years): $C_C(\$) = 500\,N_a\sqrt{d}$
N_a = Actual number of trays
d = Column diameter, ft
Overall tray efficiency = 65%
Energy cost = $\$2/10^6$ Btu
Cooling water cost = $\$2/10^7$ Btu
Average heat of condensation/vaporization = 7500 Btu/lbmole
Average vapor molar volume in the column = 125 ft³/lbmole
Target vapor velocity in the column = 3 ft/sec

Assume the feed to the column is saturated liquid.

12.4 (a) Based on the Fenske equation, derive an expression for the minimum number of stages for a binary mixture in terms of the separation specifications. Assume constant relative volatility.

(b) A saturated liquid binary mixture of components 1 and 2 is to be separated in a distillation column into a distillate product (mostly component 1) and a bottoms product (mostly component 2). The feed composition and separation specifications are as follows:

| | Mole Fraction | | |
Component	Feed	Distillate	Bottoms
1	0.40	0.90	0.05
2	0.60	0.10	0.95

Assuming a constant relative volatility of 1.8, calculate the minimum number of theoretical stages and the minimum reflux ratio.

12.5 A distillation column is used to separate a three-component mixture into three products: the distillate, a side draw, and the bottoms. It is proposed to monitor the product compositions by measuring the product rates, using total reflux calculations. The minimum number of stages equivalent to the actual column are four stages above the sidedraw and five below it. Calculate the product compositions corresponding to the following feed composition and product rates. The relative volatilities, assumed constant, are also given.

| | | Relative Volatility | |
Component	Feed Mole Fraction	Top Section	Bottom Section
1	0.25	1.60	1.50
2	0.40	1.30	1.25
3	0.35	1.00	1.00

Feed flow rate	= 100.0 lbmole/hr
Distillate flow rate	= 25.0 lbmole/hr
Side draw rate	= 40.0 lbmole/hr
Bottoms flow rate	= 35.0 lbmole/hr

CHAPTER **13**

Rigorous Methods

The rigorous column solution methods described in this chapter are applicable to multistage vapor–liquid separation processes in general, including distillation, absorption, stripping, etc. Consideration is mostly confined to steady-state processes where none of the parameters is time-dependent at solution. (The *relaxation* method takes a dynamic route during iteration.) The methods involve simultaneous solution of a large number of nonlinear equations that represent mass and energy balances and phase equilibrium relations.

In a rigorous solution, no assumptions are made to simplify the equations as in shortcut methods. It is noted, however, that certain physical assumptions must be made in order to complete the mathematical model. Thus, in the mathematical model, a tray is considered an equilibrium stage. This leads to the concept of the tray efficiency, a parameter used for converting actual trays to theoretical stages, and vice versa, in design and performance evaluations. Further, the model is only as good as the physical properties prediction methods used for estimating phase equilibrium coefficients and enthalpies (Chapter 1).

Because of the complexity of the equations, the only practical way for solving them rigorously is by means of numerical solution algorithms implemented on computer programs. These programs have become necessary for predicting the performance of multistage separation processes. An understanding of the solution methods is important for efficient and intelligent use of the programs.

The model for which the solution methods are developed represents a fixed configuration column. Thus, parameters such as number of trays, feed and draw tray locations, and heater and cooler locations are all fixed. The column is completely specified by defining a number of additional performance specifications equal to the degrees of freedom of the column. The model is then solved to determine all the other parameters.

13.1 MODEL DESCRIPTION

The model is shown in Figure 13.1 for a generalized multistage vapor–liquid counterflow column. Each stage j may have an external feed, F_j; a liquid product, L_j^s; a vapor product, V_j^s; and a heat duty, Q_j. Also, the net liquid transferred from stage j to stage j + 1 is designated as L_j, and the net vapor transferred from stage j to stage j – 1 is designated as V_j. The stage temperature is T_j, and its pressure is P_j. Stages are numbered from the top down — from j = 1 at the condenser or top stage to j = N at the reboiler or bottom stage. Since in the generalized model each stage can have liquid and/or vapor products and a heat duty, there is no need for any special designation of the condenser or reboiler.

The component molar rates and fractions and total rates are interrelated as follows:

$$L_{ji} = L_j X_{ji}$$

$$V_{ji} = V_j Y_{ji}$$

$$F_{ji} = F_j Z_{ji} \qquad (13.1)$$

$$L_{ji}^s = L_j^s X_{ji}$$

$$L_{ji}^s = V_j^s Y_{ji}$$

The initial thermal conditions of external feed streams are designated by superscript f; thus, the initial temperature, pressure, and enthalpy of a stream going to stage j are designated as T_j^f, P_j^f, and H_j^f, respectively. Regardless of its phase, single or mixed, a stream going to stage j is assumed to mix with the contents of that stage, reach phase equilibrium, then separate into vapor and liquid. A column pumparound is simulated by a pair of streams consisting of a product and a feed with the appropriate constraining equations that tie them together.

Since all stages in the model have a general designation, the equations are written for any stage (j = 1, 2, ..., N). The equations required to completely define a stage include material balance, heat balance, and phase equilibrium relationships. The material balance around stage j is written in terms of total flows and mole fractions:

$$L_{j-1}X_{j-1,i} - \left(L_j + L_j^s\right)X_{ji} + V_{j+1}Y_{j+1,i} - \left(V_j + V_j^s\right)Y_{ji} + F_j Z_{ji} = 0 \qquad (13.2)$$

The mole fractions must add up to unity:

$$\sum_i X_{ji} - 1 = 0$$

$$\sum_i Y_{ji} - 1 = 0 \qquad (13.3)$$

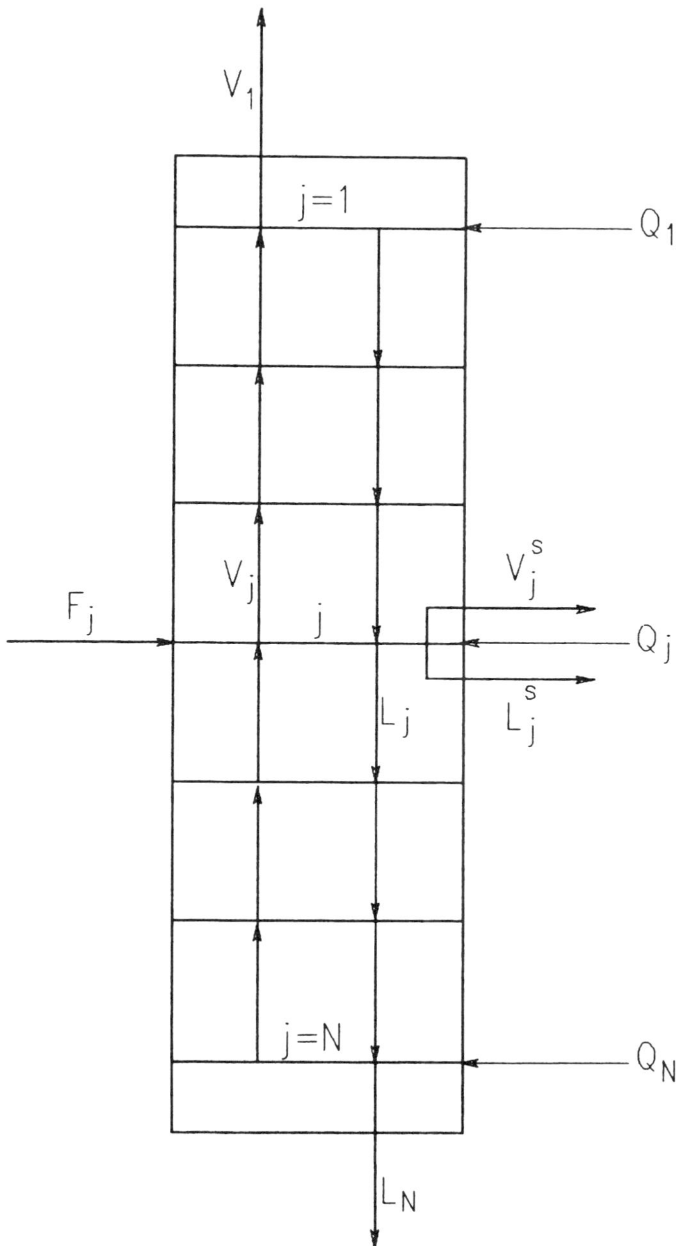

Figure 13.1 Generalized column model.

Note that Equation 13.2 can be summed up over all the components, then combined with Equation set 13.3 to yield overall material balances on each tray. The condition that the sum of the feed mole fractions, Z_{ji}, must equal 1 follows from these balances.

Further, the overall tray material balance, summed over trays 1 through N, results in a column overall material balance.

The heat balance around stage j is given by

$$L_{j-1}h_{j-1} - \left(L_j + L_j^s\right)h_j + V_{j+1}H_{j+1} - \left(V_j + V_j^s\right)H_j + F_jH_j^f + Q_j = 0 \qquad (13.4)$$

Liquid and vapor enthalpies are designated by h and H, respectively, and a feed enthalpy is expressed as H^f.

The phase equilibrium relations are written as

$$Y_{ji} - K_{ji}X_{ji} = 0 \qquad (13.5)$$

The fluid properties, K_{ji}, H_j, and h_j, are either supplied data or predicted by some thermodynamic methods that correlate them as functions of temperature, pressure, and composition (Chapter 1). Expressed mathematically, the data or correlations may be written as follows:

$$K_{ji} = K_{ji}(T_j, P_j, S_{ji}, Y_{ji}) \qquad (13.6)$$

$$H_j = H_j(T_j, P_j, Y_{ji}) \qquad (13.7)$$

$$h_j = h_j(T_j, P_j, X_{ji}) \qquad (13.8)$$

In all the above equations, j = 1, 2, …, N, the number of stages, and i = 1, 2, …, C, the number of components. The total number of independent equations describing the model is arrived at as follows:

Equation set (13.2)	CN equations
Equation set (13.3)	2N equations
Equation set (13.4)	N equations
Equation set (13.5)	CN equations
Equation set (13.6)	CN equations
Equation set (13.7)	N equations
Equation set (13.8)	N equations
Total	(3C + 5)N equations

The model variables include all the variables in Equations 13.2 through 13.8. A quick inspection indicates that there are more variables than the total number of equations. The number of variables minus the number of equations equals the degrees of freedom of the model. In order to define the column performance, a number of parameters equal to the degrees of freedom must be specified. In general, these specifications are expressed as functions of the column variables:

$$G(\overline{w}) = 0 \qquad (13.9)$$

where G represents the specifications expressed in terms of the variables vector, $\overline{w}$. The model is mathematically defined if the total number of equations, Equations 13.2 through 13.9, is equal to the total number of variables, w. A column solution is reached when a set of values for w is found that satisfies Equations 13.2 through 13.9.

The specification functions, Equation set 13.9, are frequently simply a constant. For instance, the mole fraction of component i in the vapor from the top stage (the distillate) may be specified as $Y_{1i} = 0.95$. Certain other parameters such as tray pressures and feed compositions are considered in most solution algorithms as fixed quantities, thus reducing the number of variables.

Examine the number of equations and variables in the generalized model as applied to a specific column. Consider a column with a single feed, a condenser and a reboiler, a vapor overhead, and a bottoms product. There are no side products and the only heaters or coolers are the reboiler and condenser. The feed is assumed to be of fixed rate, composition, and thermal conditions. Under these conditions the variables in Equations 13.2 through 13.8 are the following:

L_j	N variables
V_j	N variables
X_{ji}	CN variables
Y_{ji}	CN variables
h_j	N variables
H_j	N variables
Q_j $j = 1, j = N$	2 variables
K_{ji}	CN variables
T_j	N variables
P_j	N variables
Total	$2 + (3C + 6)N$ variables

The number of variables exceeds the number of equations by $N + 2$. If the N tray pressures are fixed, the column has 2 degrees of freedom and requires two specifications to define its performance, as concluded in earlier chapters for this type of column.

13.2 SOLUTION METHODS

The equations that make up the generalized column model, Equations 13.2 through 13.9 are highly complex and nonlinear, and their number can run in the tens of thousands. (In a 10-component, 30-tray column with two specifications, the number of equations would be $3 \times 10 + 5) \times 30 + 2 = 1052$.) It is clear that analytical solution of these equations is impossible, the alternative being numerical, iterative techniques using computer programs. Many algorithms have been proposed for solving the equations, a good review of which is presented by Wang et al. (1980). Due to the complexity of the equations, most algorithms are prone to convergence difficulties in certain column situations. Contributing to these difficulties is the large variation in the relative magnitudes of the variables, round-off errors, and the sparse matrices that result from the equations.

Certain algorithms may be better suited for certain types of columns. Convergence characteristics depend on whether the column is of a predominantly distillative type or absorption/stripping type, or both (Chapters 7, 8, and 9). In distillation columns, where mixtures are relatively narrow boiling, the separation is effected primarily as a result of the variation of equilibrium coefficients with temperature. In absorbers and strippers, the mixtures are wide boiling, and the separation is more influenced by the heat of absorption associated with the transfer of components from one phase to the other. It may thus be seen that different equations in the column model have varying weights for each particular type of column. Algorithms that are designed for one type may not function very effectively for another. Complex columns could contain sections where the equilibrium equations are predominant and others where the heat balances are predominant. Thus, algorithms designed for solving the generalized column model should be capable of handling this variety of situations.

13.2.1 VLE Columns

The differences among the solution methods result from several considerations, including the choice of independent variables, the grouping and arrangement of the equations, and the convergence path. Some methods are based on equation decoupling, while others solve the entire system of equations simultaneously.

In the equation decoupling methods, the model equations (13.2 through 13.9) are grouped into smaller sets of equations, which are solved one set at a time. The solution of one set of equations is substituted in the other sets to form iterative loops. Each algorithm has its unique way of sequencing the calculations. Within this group of methods, some algorithms are better suited for specific types of columns. In general, these methods are adapted for handling only limited types of column performance specifications.

Following is an overview of some of the methods, considered representative of the main algorithmic categories and presented mostly in historical order.

Method of Thiele and Geddes

Although not widely used anymore, this method is presented because it was the forerunner for many of the more recent and more efficient methods. It is one of the earlier methods for rigorous numerical solution of distillation columns. It was first described by Thiele and Geddes (1933) and later detailed by Lyster et al. (1959).

The method is basically designed for solving conventional distillation columns with one feed, F; a distillate, V_1; a bottoms product, L_N; a condenser with duty, Q_1; and a reboiler with duty, Q_N. In the generalized model of Figure 13.1, all the feeds except one, all the products except V_1 and L_N, and all the duties except Q_1 and Q_N are set to 0. The method becomes numerically unstable for columns other than conventional distillation. Another drawback of the method is that it can handle limited types of performance specifications. In the outline that follows, the specifications are the distillate rate, V_1, and the reflux rate, L_1. It is assumed that the number of stages and the column pressure profile are known.

The stage temperatures are considered the independent variables. The calculations are started by assuming a set of stage temperatures, T_j; vapor flows, V_j; and liquid flows, L_j. The equilibrium coefficients, K_{ji}, are estimated based on the assumed tray temperatures. Equations 13.1 and 13.5 are combined to give

$$L_{ji} = A_{ji} V_{ji}$$

where

$$A_{ji} = \frac{L_j}{V_j K_{ji}}$$

In the rectifying section, the above equation is divided by V_{1i}:

$$\frac{L_{ji}}{V_{1i}} = A_{ji} \frac{V_{ji}}{V_{1i}} \tag{13.10}$$

Next, a component material balance is written around the top of the column, including any tray j above the feed. The resulting equation is divided by V_{1i}:

$$\frac{V_{j+1,i}}{V_{1i}} = \frac{L_{ji}}{V_{1i}} + 1 \tag{13.11}$$

Replacing j by j + 1 in Equation 13.10 and eliminating $V_{j+1,i}/V_{1i}$ from Equation 13.11, the following is obtained:

$$\frac{L_{j+1,i}}{V_{1i}} = A_{j+1,i}\left(\frac{L_{ji}}{V_{1i}} + 1\right) \tag{13.12}$$

This equation is written for j = 1, and L_{1i}/V_{1i} is replaced by R/K_{1i}, where R is the reflux ratio (L_1/V_1):

$$\frac{L_{2i}}{V_{1i}} = A_{2i}\left(\frac{R}{K_{1i}} + 1\right)$$

The assumed liquid and vapor rates are used to evaluate A_{2i} and, since the reflux ratio is specified, L_{2i}/V_{1i} may be computed. Its value is then substituted in Equation 13.12 written for j = 2, from which L_{3i}/V_{1i} is calculated. The procedure is continued through the stage just above the feed.

The equivalent of Equation 13.12 for the stripping section is derived in a similar manner by combining the equilibrium and component material balance equations. The result is

$$\frac{L_{j-1,i}}{L_{Ni}} = \frac{1}{A_{ji}} \frac{L_{ji}}{L_{Ni}} + 1 \tag{13.13}$$

Written for stage $j = N$, the reboiler, Equation 13.13 becomes

$$\frac{L_{N-1,i}}{L_{Ni}} = \frac{1}{A_{Ni}} + 1$$

After $L_{N-1,i}/L_{Ni}$ has been calculated, Equation 13.13 is applied in succession to calculate the remaining liquid component flow ratios, L_{ji}/L_{Ni}, from the bottom up through the feed stage. As in the rectifying section, the equilibrium coefficients, K_{ji}, are calculated from the assumed stage temperatures; then the K_{ji}s and the assumed stage vapor and liquid flows are used to evaluate the factors A_{ji}. The ratio of component i in the products, L_{Ni}/V_{1i}, is determined by combining Equations 13.12 and 13.13 and matching the downward and upward stage calculations at the feed stage. For a bubble point feed, the result is as follows (Lyster et al., 1959):

$$\frac{L_{Ni}}{V_{1i}} = \frac{L_{f-1,i}/V_{1i} + 1}{L_{f-1,i}/V_{Ni} - 1} \tag{13.14}$$

where subscript f designates the feed tray.

The distillate component flow rates may now be calculated from an overall column component material balance:

$$V_{1i} = \frac{FZ_i}{1 + L_{Ni}/V_{1i}} \tag{13.15}$$

At convergence, the summation of V_{1i} over all components should equal the specified distillate rate, V_1. If this test is not satisfied, then new, corrected values of V_{1i} must be found. A factor θ is defined such that

$$V_1 - \sum_{i=1}^{C} \frac{FZ_i}{1 + \theta\left(L_{Ni}/V_{1i}\right)} = 0 \tag{13.16}$$

This equation is solved iteratively to determine θ. Once θ has been evaluated, the corrected distillate and bottoms component rates are calculated as follow:

$$V'_{1,i} = \frac{FZ_i}{1 + \theta\left(L_{Ni}/V_{1i}\right)} \tag{13.17}$$

and

$$L'_{Ni} = \theta \frac{L_{Ni}}{V_{1i}} V'_{1i} \tag{13.18}$$

where the prime denotes corrected flow rates.

The stage liquid compositions are calculated from the corrected component liquid flow rates. In the rectifying section,

$$X_{ji} = \frac{\left(L_{ji}/V_{1i}\right)V'_{1i}}{\displaystyle\sum_{i=1}^{C}\left(L_{ji}/V_{1i}\right)V'_{1i}} \tag{13.19}$$

And in the stripping section,

$$X_{ji} = \frac{\left(L_{ji}/L_{Ni}\right)L'_{Ni}}{\displaystyle\sum_{i=1}^{C}\left(L_{ji}/L_{Ni}\right)L'_{Ni}} \tag{13.20}$$

The updated stage temperatures are obtained by calculating the bubble point temperature on each stage on the basis of the new liquid compositions given by Equations 13.19 and 13.20. The updated compositions and temperatures are used to calculate updated total liquid and vapor flows on each stage using enthalpy balances as described in the modified Thiele–Geddes method in the next section.

The entire procedure is repeated until convergence is achieved. The use of θ to correct the component flow rates tends to guide the computations toward faster convergence by forcing an overall material balance in each iteration. Convergence is considered reached when the difference between the specified and calculated distillate rates is within a preset tolerance.

Modified Thiele–Geddes Method

This method also considers the stage temperatures as the independent variables. The algorithm is applied to a single-feed, two-product column with a partial condenser and reboiler. As in the original Thiele-Geddes method, the problem definition is such that the feed component flow rates, f_{ji}, are known and fixed. The column pressure profile is also fixed, as well as its configuration, which defines the number of stages and feed location. In addition, one product rate (the distillate) and one internal flow (such as the reflux rate, L_1) are specified. The solution method, outlined below, is described in detail by Holland (1975).

1. Assume a set of stage temperatures and vapor rates.
2. Calculate stage liquid rates by applying total material balance on each stage.

3. Estimate equilibrium coefficients on the basis of a composition independent correlation.
4. Combine component material balance and phase equilibrium relations to calculate vapor rates of each component on each stage.
5. Calculate liquid rates of each component on each stage.
6. Calculate mole fractions of each component in the liquid and vapor on each stage.
7. Calculate a set of corrected mole fractions to satisfy overall material balance.
8. Calculate a set of improved stage temperatures based on an updated set of equilibrium coefficients.
9. Using the latest computed sets of stage temperatures and compositions, apply enthalpy balances to update the vapor flow rates and compute the condenser and reboiler duties.
10. If the new vapor rates and temperatures agree with the previous set within a preset tolerance, the problem is converged. Otherwise, repeat the calculations beginning at step 4.

The first step is to assume a set of temperatures, T_j, and vapor flows, V_j. The temperatures may be obtained by linear interpolation between condenser and reboiler temperatures, determined by dew point and bubble point calculations of products estimated by shortcut methods or on the basis of past experience with similar columns. The vapor rates are estimated from the specified distillate and reflux rates. Constant vapor rates are assumed above and below the feed. The change in the vapor profile at the feed tray is estimated on the basis of the feed vapor fraction.

In step 2, the stage liquid rates, L_j, are calculated by total material balance on each stage, beginning at the condenser. Since L_1, the reflux rate, and V_1, the distillate rate, are both specified, V_2 is calculated from a material balance on stage 1, the condenser:

$$V_2 = V_1 + L_1$$

Next, a material balance on stage 2 yields

$$L_2 = L_1 - V_2 + V_3$$

In this equation V_3 is taken from the estimates in step 1. The calculations are continued through the bottom stage. At the feed tray the feed must be taken into account in the material balance.

The K-values in step 3 may be estimated using an equation based on Raoult's law or Henry's law (Chapter 1) such as

$$K_{ji} = \frac{P_i^0\left(T_j\right)}{P_j}$$

where P_j is the tray pressure, and $P_i^0(T_j)$ is an expression for the vapor pressure or Henry's law constant of component i at tray temperature T_j.

With a given set of V_j, L_j, and K_{ji}, step 4 applies component material balances and phase equilibrium relations to calculate component vapor rates using the *tridiagonal matrix* method. The formulation of equations at this point is general, in that each stage can have an external feed and vapor and liquid side products. The results may be applied to the special case of the modified Thiele–Geddes model by zeroing all the nonexistent feeds and products. Equations 13.1, 13.2, and 13.5 are combined and written for $j = 1, 2, \ldots, N$. The resulting equation for component i on tray j is as follows:

$$A_{j-1,i}V_{j-1,i} - B_{ji}V_{ji} + V_{j+1,i} = -F_{ji} \qquad (13.21)$$

where

$$A_{ji} = \frac{L_j}{K_{ji}V_j} \qquad (13.22)$$

and

$$B_{ji} = A_{ji}\left(1 + \frac{L_j^s}{L_j}\right) + \frac{V_j^s}{V_j} + 1 \qquad (13.23)$$

When Equation 13.21 is written for stages $j = 1, 3, \ldots, N$, the following set of equations results:

$$
\begin{array}{llllll}
-B_{1i}V_{1i} & +V_{21} & & & & = -F_{1i} \\
A_{1i}V_{1i} & -B_{2i}V_{2i} & +V_{3i} & & & = -F_{2i} \\
& A_{2i}V_{2i} & -B_{3i}V_{3i} & +V_{4i} & & = -F_{3i} \\
& & \cdot & \cdot & \cdot & \cdot \\
& & \cdot & \cdot & \cdot & \cdot \\
& & \cdot & \cdot & \cdot & \cdot \\
& A_{N-2,i}V_{N-2,i} & -B_{N-1,i}V_{N-1,i} & +V_{Ni} & = -F_{N-1,i} \\
& & A_{N-1,i}V_{N-1,i} & -B_{Ni}V_{Ni} & = -F_{Ni}
\end{array}
$$

$$(13.24)$$

The equations are solved for the vector $V_{1i}, V_{2i}, \ldots, V_{Ni}$. Inspection of this set of equations indicates that their coefficient matrix is made up of zero elements everywhere except along the three central diagonals, hence the name tridiagonal matrix. The matrix may be solved by recurrence formulas using the Thomas algorithm (Carnahan et al., 1964) or its modified version (Boston et al., 1972).

The solution of Equation set 13.24 gives the vapor flows of component i on all stages. The equation set must be solved repeatedly for each component to determine the vapor flows of all components on all trays.

The component liquid rates (step 5) are calculated from the component vapor rates using Equation 13.1 substituted into Equation set 13.5:

$$L_{ji} = \frac{L}{V_j K_{ji}} V_{ji}$$

or

$$L_{ji} = A_{ji} V_{ji}$$

The mole fractions (step 6) are calculated directly from Equation set 13.1.

The mole fractions are adjusted in step 7 to satisfy both the overall material balance for each component and the distillate rate specification. The overall component material balance requires that

$$FZ_i = V'_{1i} + L'_{Ni} \tag{13.25}$$

The prime indicates adjusted values. The feed stage subscript is dropped because there is only one feed in this model. The distillate rate specification requires that

$$\sum_{i=1}^{C} V'_{1i} = V_1 \tag{13.26}$$

where V_1 is the specified distillate rate. A correction factor θ is defined, relating corrected to calculated flow rates as follows:

$$\frac{L'_{Ni}}{V'_{1i}} = \theta \frac{L_{Ni}}{V_{1i}} \tag{13.27}$$

The multiplier, θ, must be determined so that both Equations 13.25 and 13.26 are satisfied. The following equation is obtained by eliminating L'_{Ni}/V'_{1i} between Equations 13.27 and 13.25 and rearranging:

$$V'_{1i} = \frac{FZ_i}{1 + \theta\left(L_{Ni}/V_{1i}\right)} \tag{13.28}$$

Summation over components i results in

$$\sum_{i=1}^{C} V'_{1i} = V_1 = \sum_{i=1}^{C} \frac{FZ_i}{1 + \theta\left(L_{Ni}/V_{1i}\right)} \tag{13.29}$$

The one unknown in this equation is θ, which is calculated by an iterative technique such as Newton–Raphson's. The liquid and vapor mole fractions are then updated using the following equations:

$$X_{ji} = \frac{\left(L_{ji}/V_{1i}\right)V'_{1i}}{\sum_{i=1}^{C}\left(L_{ji}/V_{1i}\right)V'_{1i}} \tag{13.30}$$

$$Y_{ji} = \frac{\left(V_{ji}/V_{1i}\right)V'_{1i}}{\sum_{i=1}^{C}\left(V_{ji}/V_{1i}\right)V'_{1i}} \tag{13.31}$$

The phase equilibrium coefficients are next calculated rigorously based on the latest mole fractions and stage temperatures (Chapter 1). These K-values are used to update the stage temperatures (step 8). A reference (hypothetical) component, r, is used to define the relative volatility:

$$\alpha_i = \frac{K_i}{K_r} = \frac{Y_i}{X_i K_r}$$

Rearranged and written for stage j, this expression becomes

$$Y_{ji} = K_{jr}\alpha_{ji}X_{ji}$$

Both sides are summed up over all the components on stage j, resulting in the following expression for K_{jr}:

$$K_{jr} = \frac{1}{\sum_{i=1}^{C}\alpha_{ji}X_{ji}} \tag{13.32}$$

Equation 13.32 is used to calculate an updated K_{jr} for each stage based on current α_{ji}'s and X_{ji}'s. The stage temperatures are then calculated from some temperature-dependent correlation of K_r (Holland, 1975).

Enthalpy balances (Equation 13.4) are used in step 9 to update the vapor flows and to calculate the condenser and reboiler duties. First apply the enthalpy balance to stage 1 (the condenser):

$$V_2H_2 = V_1H_1 + L_1h_1 + Q_1 \tag{13.33}$$

where V_1 is the distillate vapor rate, V_2 is the vapor rate from stage 2, and L_1 is the reflux rate. The vapor molar enthalpy on stage j is designated as H_j, and the liquid molar enthalpy as h_j. Q_1 is the condenser duty. The term V_2H_2 is evaluated from component enthalpies:

$$V_2H_2 = \sum_i V_{2i}H_{2i} \tag{13.34}$$

where, in ideal solutions, H_{2i} is the vapor enthalpy of pure component i on stage 2, and in nonideal solutions it is the partial molar enthalpy (Chapter 1). Using a component material balance, V_{2i} is replaced by $V_{1i} + L_{1i}$:

$$\begin{aligned}
V_2H_2 &= \sum_i V_{1i}H_{2i} + \sum_i L_{1i}H_{2i} \\
&= V_1\sum_i Y_{1i}H_{2i} + \sum_i X_{1i}H_{2i}
\end{aligned} \tag{13.35}$$

Elimination of V_2H_2 between Equations 13.33 and 13.35 yields an expression for the condenser duty:

$$Q_1 = V_1\sum_i Y_{1i}H_{2i} + L_1\sum_i X_{1i}H_{2i} - V_1H_1 - L_1h_1 \tag{13.36}$$

The values of V_1 and L_1 are known, and the current stage compositions and temperatures are used to calculate the enthalpies in Equation 13.36.

Once the condenser duty is calculated, L_j is calculated in the rectifying section in a similar manner as above by eliminating $V_{j+1}H_{j+1}$. The vapor rate V_{j+1} is then calculated from a total material balance, $V_{j+1} = V_1 + L_j$. The procedure is extended to the stripping section with appropriate inclusion of the feed. Finally, the reboiler duty is calculated by an overall enthalpy balance.

The final step is to check for convergence, which is verified by one of two criteria: either the variation of the V_j's and T_j's from the previous iteration is within a given tolerance, or Equations 13.2 through 13.5 are satisfied within tolerance.

The modified Thiele–Geddes method was mainly designed for conventional distillation columns although it has been generalized to handle complex columns (Holland, 1963). It is limited in the types of performance specifications it can handle and could be numerically unstable, especially for wide-boiling or nonideal mixtures (Wang et al., 1980).

Method of Wang and Henke

This method (Wang et al., 1966) is similar to the modified Thiele–Geddes method. Its authors introduced the tridiagonal matrix method described in the previous section for solving the component balance and equilibrium equations to determine the stage compositions. No "forcing technique" is employed to correct the compositions comparable to using the factor θ in the modified Thiele–Geddes method.

From the calculated stage compositions, the stage temperatures are determined by bubble point calculations. The calculations are carried out iteratively using Muller's algorithm (Wang et al., 1966). The authors have determined that convergence by this method is more reliable than the Newton–Raphson technique.

In the next step the enthalpies of the stage vapor and liquid streams are calculated by some generalized method (Chapter 1). The enthalpy balance relations (Equation 13.4) are then solved to compute updated vapor flows. The results are used in the next iteration to solve the tridiagonal matrix, and the remaining steps are repeated until convergence is reached. The problem is considered converged when the maximum stage temperature variation from iteration to iteration is within a given tolerance.

The method has been successfully employed for many column types, although convergence problems can occur for nonideal mixtures where the K-values are strongly composition dependent or for highly complex columns. The method is also limited in the types of performance specifications it can handle.

Method of Tomich

This method (Tomich, 1970) differs from the foregoing methods mainly in that the summation statements and the enthalpy balances (Equations 13.3 and 13.4) are solved simultaneously. The benefits of the simultaneous solution are twofold. First, distillation columns and absorbers and columns that are hybrids of both types of processes can all be solved with the same method. Second, different types of column performance specifications can be incorporated in the simultaneous solution of the equations. The method is also computationally stable and efficient because it uses Broyden's modification of the Newton–Raphson technique for solving the equations (Broyden, 1965). A brief description of the method follows:

1. Assume initial temperature and vapor profiles, T_j and V_j (the independent variables).
2. Compute the liquid profile, L_j, from a total material balance on each stage.
3. Solve for the vapor and liquid compositions on each stage by setting up a tridiagonal matrix as described in the modified Thiele–Geddes method (Equation 13.24). The terms A_{ji} and B_{ji} in the tridiagonal solution require knowledge of the equilibrium coefficients, K_{ji}. In the first iteration K_{ji} is estimated from a composition-independent correlation of temperature and pressure. In subsequent iterations the current compositions are used in rigorous calculations of K_{ji}.

4. Solve Equations 13.3 and 13.4 simultaneously for a new set of T_j and V_j. At this point, any performance specifications that the column may be required to meet (Equation 13.9) are expressed as functions of T_j and V_j and solved simultaneously with Equations 13.3 and 13.4.
5. If the convergence criteria are met, a solution has been reached and the calculations are stopped. Otherwise, the calculations are repeated beginning at step 2, using the updated values of T_j and V_j.

Step 4 requires some further elaboration. Equation set 13.3 is combined and written as follows:

$$S_j = \sum_{i=1}^{C} Y_{ji} - \sum_{i=1}^{C} X_{ji} = 0 \tag{13.37}$$

If the left-hand side of Equation 13.4 is designated as E_j, it may be written as

$$E_j = 0 \tag{13.38}$$

For a given set of variables T_j and V_j, it is possible to calculate the corresponding values of L_j, X_{ji}, and Y_{ji} as described in steps 2 and 3. With these quantities known, the enthalpies can be calculated. Thus, S_j and E_j are implicit functions of T_j and V_j (the independent variables). The specification functions, Equation 13.9, can also be expressed in terms of X_{ji}, Y_{ji}, H_j, h_j, and other stage variables as well as some performance variables. The specification functions are therefore also implicit functions of T_j and V_j.

In this method (Tomich, 1970) the summations and enthalpy balances, Equations 13.37 and 13.38, are rewritten as functions of T_j and V_j.

$$S_j(T_j, V_j) = 0 \tag{13.39}$$

$$E_j(T_j, V_j) = 0 \tag{13.40}$$

If the only column specifications are the duties, Q_j, and the product rates, Equations 13.39 and 13.40 completely define the column. At convergence, S_j and E_j are equal to 0 within tolerance. If other specifications are required, they must be expressed as functions of T_j and V_j and solved along with Equations 13.39 and 13.40. The following steps are described for the case where no specifications other than the duties and product rates are given. At a given iteration, k, Equations 13.39 and 13.40 are expanded in a Taylor series truncated after the first derivative to calculate the functions at the next iteration. The new functions are set to 0:

$$S_j^{k+1} = S_j^k + \sum_{j=1}^{N} \left(\frac{\partial S_j^k}{\partial T_j} \right) \Delta T_j + \sum_{j=1}^{N} \left(\frac{\partial S_j^k}{\partial V_j} \right) \Delta V_j = 0 \tag{13.41}$$

$$E_j^{k+1} = E_j^k + \sum_{j=1}^{N}\left(\frac{\partial E_j^k}{\partial T_j}\right)\Delta T_j + \sum_{j=1}^{N}\left(\frac{\partial E_j^k}{\partial V_j}\right)\Delta V_j = 0 \qquad (13.42)$$

The values of S_j^k, E_j^k and their derivatives are calculated from Equations 13.37 and 13.38 (combined with Equation 13.4), the derivatives by finite difference methods. Equations 13.41 and 13.42 are solved for ΔT_j and ΔV_j by inverting the Jacobian matrix, the matrix of partial derivatives of S_j and E_j with respect to T_j and V_j. The independent variables, T_j and V_j, are then updated for the next iteration:

$$T_j^{k+1} = T_j^k + \alpha\Delta T_j \qquad (13.43)$$

$$V_j^{k+1} = V_j^k + \alpha\Delta V_j \qquad (13.44)$$

where α is an adjustment factor. The value of α can range from -1 to $+1$ and is determined such that the sum of error squares at iteration $k + 1$ is less than that at the previous iteration:

$$\sum_{j=1}^{N}\left[\left(S_j^{k+1}\right)^2 + \left(E_j^{k+1}\right)^2\right] < \sum_{j=1}^{N}\left[\left(S_j^k\right)^2 + \left(E_j^k\right)^2\right] \qquad (13.45)$$

Once the new T_j's and V_j's are calculated, the iterative procedure is repeated at step 2.

In the classical Newton–Raphson technique, the Jacobian matrix is inverted every iteration in order to compute the corrections ΔT_j and ΔV_j. The method of Tomich, however, uses the Broyden procedure (Broyden, 1965) in subsequent iterations for updating the inverted Jacobian matrix.

Method of Naphtali and Sandholm

This method was aimed at overcoming some of the weaknesses of other methods. While certain methods may be better suited for wide-boiling or narrow-boiling mixtures, this method (Naphtali et al., 1971), which solves all the model equations simultaneously, is an attempt to handle all types of columns using the same algorithm. The method is also capable of directly solving for various types of performance specifications by including them in the solution algorithm. The original article (Naphtali et al., 1971) also describes how the Murphree tray efficiencies can be incorporated directly into the calculations.

The simultaneous solution uses the Newton–Raphson method, which is based on linearizing the model equations. Two characteristics are inherent in this method. Since the equations are highly nonlinear, the success of linearization usually requires good starting values. On the other hand, as the solution is approached, the linearized equations become progressively more accurate and convergence is accelerated.

The Newton–Raphson simultaneous solution procedure is formulated in a generalized terminology by representing all the model equations (Equations 13.2 through 13.9) by a function vector, g, and all the variables by a variable vector, w. The system of equations is written as

$$\overline{g}(\overline{w}) = 0 \tag{13.46}$$

Naphtali and Sandholm (1971) grouped the equations by stages, writing Equation 13.46 in its expanded form as follows:

$$g_{11}\left(w_{11}, \ldots, w_{1i}, \ldots, w_{1,2C+1}, \ldots, w_{ji}, \ldots, w_{N,2C+1}\right) = 0$$

$$\vdots$$

$$g_{1i}(\ldots) = 0$$

$$\vdots$$

$$g_{1,2C+1}(\ldots) = 0 \tag{13.47}$$

$$\vdots$$

$$g_{ji}(\ldots) = 0$$

$$\vdots$$

$$g_{N,2C+1}(\ldots) = 0$$

It is recalled that the first subscript designates the stage and the second one the component. Variables w_{j1} through w_{jC} represent V_{ji}, the component vapor rates leaving stage j. $w_{j,C+1}$ is the stage temperature T_j and variables $w_{j,C+2}$ through $w_{j,2C+1}$ represent L_{ji}, the component liquid rates leaving stage j. Function g_{j1} is the enthalpy balance on stage j, functions g_{j2} through $g_{j,C+1}$ are the component material balances on stage j, and functions $g_{j,C+2}$ through $g_{j,2C+1}$ are the phase equilibrium relationships on stage j. These are therefore N(2C + 1) equations with N(2C + 1) unknowns.

The equations are specifically ordered in the manner shown in Equation set 13.47, namely the equations for each stage are grouped together, beginning at the top stage and going sequentially stage by stage down the column. It turns out that the Jacobian matrix resulting from this grouping has a block tridiagonal structure, which lends itself to simple solution using a Gaussian elimination technique.

Equation set 13.47 is first linearized using a Taylor series expansion truncated after the first derivative. The resulting equations are used to calculate the corrections Δw to the variables w at each iteration:

$$\frac{\partial g_{11}}{\partial w_{11}} \Delta w_{11} + \ldots + \frac{\partial g_{11}}{\partial w_{N,2C+1}} \Delta w_{N,2C+1} = -g_{11}$$

$$\vdots \qquad \vdots \qquad \vdots \qquad \vdots \tag{13.48}$$

$$\frac{\partial g_{N,2C+1}}{\partial w_{11}} \Delta w_{11} + \ldots + \frac{\partial g_{N,2C+1}}{\partial w_{N,2C+1}} \Delta w_{N,2C+1} = -g_{N,2C+1}$$

The functions on the right-hand side and their partial derivatives are evaluated from the values of the variables at the current iteration. The partial derivatives are evaluated, in general, numerically by finite differences. When convergence is reached, the functions g_{ji} are equal to 0 within tolerance. The matrix of derivatives, or Jacobian, may be written in the form shown in Matrix 13.1.

$$\begin{bmatrix}
\dfrac{\partial g_{11}}{\partial w_{11}} & \cdots & \dfrac{\partial g_{11}}{\partial w_{1,2C+1}} & \dfrac{\partial g_{11}}{\partial w_{21}} & \cdots & \dfrac{\partial g_{11}}{\partial w_{2,2C+1}} & \dfrac{\partial g_{11}}{\partial w_{31}} & \cdots & \dfrac{\partial g_{11}}{\partial w_{3,2C+1}} & \cdots \\
\vdots & & \vdots & \vdots & & \vdots & \vdots & & \vdots & \\
\dfrac{\partial g_{1,2C+1}}{\partial w_{11}} & \cdots & \dfrac{\partial g_{1,2C+1}}{\partial w_{1,2C+1}} & \dfrac{\partial g_{1,2C+1}}{\partial w_{21}} & \cdots & \dfrac{\partial g_{1,2C+1}}{\partial w_{2,2C+1}} & \dfrac{\partial g_{1,2C+1}}{\partial w_{31}} & \cdots & \dfrac{\partial g_{1,2C+1}}{\partial w_{3,2C+1}} & \cdots \\
\dfrac{\partial g_{21}}{\partial w_{11}} & \cdots & \dfrac{\partial g_{21}}{\partial w_{1,2C+1}} & \dfrac{\partial g_{21}}{\partial w_{21}} & \cdots & \dfrac{\partial g_{21}}{\partial w_{2,2C+1}} & \dfrac{\partial g_{21}}{\partial w_{31}} & \cdots & \dfrac{\partial g_{21}}{\partial w_{3,2C+1}} & \cdots \\
\vdots & & \vdots & \vdots & & \vdots & \vdots & & \vdots & \\
\dfrac{\partial g_{2,2C+1}}{\partial w_{11}} & \cdots & \dfrac{\partial g_{2,2C+1}}{\partial w_{1,2C+1}} & \dfrac{\partial g_{2,2C+1}}{\partial w_{21}} & \cdots & \dfrac{\partial g_{2,2C+1}}{\partial w_{2,2C+1}} & \dfrac{\partial g_{2,2C+1}}{\partial w_{31}} & \cdots & \dfrac{\partial g_{2,2C+1}}{\partial w_{3,2C+1}} & \cdots \\
\dfrac{\partial g_{31}}{\partial w_{11}} & \cdots & \dfrac{\partial g_{31}}{\partial w_{1,2C+1}} & \dfrac{\partial g_{31}}{\partial w_{21}} & \cdots & \dfrac{\partial g_{31}}{\partial w_{2,2C+1}} & \dfrac{\partial g_{31}}{\partial w_{31}} & \cdots & \dfrac{\partial g_{31}}{\partial w_{3,2C+1}} & \cdots \\
\vdots & & \vdots & \vdots & & \vdots & \vdots & & \vdots & \\
\dfrac{\partial g_{3,2C+1}}{\partial w_{11}} & \cdots & \dfrac{\partial g_{3,2C+1}}{\partial w_{1,2C+1}} & \dfrac{\partial g_{3,2C+1}}{\partial w_{21}} & \cdots & \dfrac{\partial g_{3,2C+1}}{\partial w_{2,2C+1}} & \dfrac{\partial g_{3,2C+1}}{\partial w_{31}} & \cdots & \dfrac{\partial g_{3,2C+1}}{\partial w_{3,2C+1}} & \cdots \\
\vdots & & \vdots & \vdots & & \vdots & \vdots & & \vdots &
\end{bmatrix}$$

Matrix 13.1

The Jacobian matrix is made up of matrix blocks, shown enclosed in boxes. A matrix block is a submatrix that includes the derivatives of the functions of one stage with respect to the variables of that stage or any other one stage. The matrix blocks enclosed in solid boxes contain derivatives of functions of one stage with respect to variables on stages other than that stage or the adjacent ones. Since the functions of a given stage involve only variables of that stage and the adjacent ones, all the derivatives in the solid boxes are 0. As a consequence, the Jacobian matrix has a block tridiagonal structure. All matrix blocks other than the three central diagonals are 0. With this structure the matrix can be readily solved by a Gaussian elimination scheme (Naphtali et al., 1971).

Once the Jacobian matrix is solved for the corrections Δw, the straight Newton–Raphson method could be applied to update the variables for the next iteration:

$$w^{k+1} = w^k + \Delta w$$

However, if current values of the variables are far from solution, applying the entire corrections may result in divergence. For this reason, the corrections are multiplied by an adjustment factor which can have a value between 0 and 1:

$$w^{k+1} = w^k + \alpha\Delta w$$

Two methods are suggested for determining α: either start with 1 and keep halving it until a smaller sum of error squares is obtained or initiate a search for the value of α that minimizes the sum of error squares. This modification to the Newton–Raphson procedure helps stabilize the calculations, thereby highly enhancing the likelihood of convergence.

Method of Wang and Oleson

The method of Wang and Oleson (Wang et al., 1980) also solves the model equations (Equations 13.2 through 13.9, plus the specification equations) simultaneously by linearization. Since the resulting Jacobian is a large and sparse matrix, the matrix inversion procedure is crucial in these simultaneous methods. Whereas Naphtali and Sandholm grouped the equations by stages, resulting in a block tridiagonal matrix, Wang and Oleson partitioned the matrix by equation types.

The Newton–Raphson technique is also modified in this method in that a damping factor, α, between 0 and 1 is applied to the corrections as above. The way α is calculated, however, is different from the Naphtali–Sandholm method.

This technique, combined with Fenske shortcut calculations for generating initial estimates of temperature profiles and stage liquid or vapor flow rates, is a robust method that can solve a large percentage of different types of separation processes. The algorithm also has provision for handling inequality specifications (Brannock et al., 1977). For each inequality specification, an alternate equality specification is required to ensure a unique solution. In this manner, so-called over-constrained problems may be solved since inequality specifications are not subject to degrees of freedom restrictions.

"Inside-Out" Methods

A more recent approach to the rigorous solution of multistage separation processes has been developed with three primary objectives: to be flexible in handling performance specifications, to require little information for generating initial estimates, and to have high computational speed.

The techniques employed for achieving these goals are embodied in a method described by Russell (1983). The method, based on the original article by Boston and Sullivan (1972), consists of an inner loop, where column tray compositions and flows and temperature profiles are calculated on the basis of simplified K-value and enthalpy calculations, and an outer loop, where these properties are calculated rigorously. The outer loop (and hence, the entire column) is considered converged when the K-values and enthalpies used in the inner loop are the same as those calculated rigorously in the outer loop. The method is fast because the rigorous thermodynamic calculations are relegated to the outer loop, where they are not performed as often. The term "inside-out" is commonly applied to this type of method because of the strategy of confining the time-consuming rigorous thermodynamic calculations to the outer loop. It should be noted that the overall solution is rigorous since the outer, rigorous loop must be converged before a solution is reached.

The method, outlined in the following steps, is described in more detail in the original article (Russell, 1983):

1. Solve the mass balance equations. The basic component mass balances, Equation set 13.21, are solved in terms of a simplified K-value model. Also, stripping factors are used instead of the temperatures and vapor flows as independent variables. Other independent variables include product withdrawal factors and variable feed rate factors, which are scaled values of the variable product and feed rates.

   Equation set 13.21 is first rewritten in terms of liquid component rates, total vapor and liquid rates, and the K-values. In the simplified K-value model, the component K-values are expressed in terms of a base component K-value, K_{jb}, and the relative volatility, α_{ji}:

$$K_{ji} = a_{ji} K_{jb}$$

   The base component K-value is considered a function of temperature only, written as

$$\ln K_{jb} = A - (B/T)$$

   The stripping factors, defined as

$$S_j = K_{jb} \frac{V_j}{L_j}$$

are then incorporated into the component mass balances to replace K_{jb}, V_j, and L_j. Rather than use the stripping factors themselves as independent variables, relative stripping factors, S_{rj}, are used, which are defined as follows:

$$S_j = S_{rj}S_b$$

S_b is a parameter that is adjusted so that the calculated component rates satisfy overall material balance.

The calculations require starting values for the inner loop convergence parameters, which are the relative stripping factors, the product withdrawal factors, and the variable feed rate factors. Using these starting values, the component mass balance equations are solved while adjusting S_b to satisfy overall material balance. Coefficients A and B of the K_b model are determined for each tray by evaluating K-values at two different temperatures.

The solution of the component mass balances gives the component flow rates for both phases on all the trays and for all feeds and products. This information serves to calculate the temperatures of all trays and streams, using the K_b model. The enthalpies are then calculated using either rigorous compositional methods or simplified models. Also, the total stream rates are calculated by summing up the component rates.

2. The next major step is to update the inner loop convergence parameters in order to satisfy the stage heat balances and the specifications on the product rates or qualities. For this purpose a Jacobian matrix is required which consists of the partial derivatives of the enthalpies and specifications with respect to the inner loop convergence parameters. If the Jacobian matrix has not already been calculated in previous inner loop iterations, the partial derivatives are evaluated by perturbing each inner loop convergence parameter and recalculating the enthalpies and specifications as in step 1. The inverse of the Jacobian matrix and the errors in the calculated variables (heat balances and specifications) are used to calculate the changes in the convergence parameters. These changes are used to update the convergence parameters, which are then used to recalculate the component balances and the enthalpies and product specifications as in step 1. If the resulting average of errors in these calculated variables is not lower than the previous errors average, the changes in the convergence parameters are reduced and the calculations repeated.

In subsequent iterations in the inner loop, the Jacobian matrix is not usually recalculated; instead, the matrix inverse is updated using Broyden's formula (Broyden, 1965). The iterations in the inner loop are repeated until the errors in the calculated variables are within an acceptable tolerance.

3. The solution obtained in step 2 is consistent with all the mass, energy, and phase equilibrium relationships and satisfies all the specifications. It is based, however, on a simplified K-value and, possibly, enthalpy calculation methods. To ensure a rigorous solution, an outer loop is used where rigorous relative volatilities and enthalpies are calculated based on the compositions obtained in the inner loop. The rigorous results are used to update the simplified K-value and, if used, the simplified enthalpy model coefficients. The updated models are used to solve anew the inner loop. The outer loop is considered converged when the rigorously calculated K-values and enthalpies match the values used in the latest inner loop iteration.

Convergence by Dynamic Iteration

In this method, the component material balance, enthalpy balance, and phase equilibrium equations are written for component i on stage j at unsteady state:

$$L_{j-1,i} + V_{j+1,i} - L_{ji} - V_{ji} + F_{ji} - L_{ji}^{s} - V_{ji}^{s} = \frac{dm_{ji}}{dt} \tag{13.49}$$

$$L_{j-1}h_{j-1} - \left(L_{j} + L_{j}^{s}\right)h_{j} + V_{j+1}H_{j+1} - \left(V_{j} + V_{j}^{s}\right)H_{j} + F_{j}H_{j}^{f} + Q_{j} = \frac{d\left(m_{j}C_{pj}T_{j}\right)}{dt} \tag{13.50}$$

$$\frac{V_{ji}/V_{j}}{L_{ji}/L_{j}} = K_{ji} \tag{13.51}$$

Equation 13.49 is the unsteady state counterpart of Equations 13.1 and 13.2 combined. The term on the right-hand side of Equation 13.49 is the variation of the moles of component i on stage j per unit time. In this equation, m_{ji} is the holdup of component i, i.e., the amount of component i in both phases on stage j at time t. Equation 13.50 is the unsteady state form of Equation 13.4. In Equation 13.50, m_{j}, the sum of m_{ji}, is the total moles in both phases on stage j. C_{pj} is an average heat capacity for both phases. Thus, the term on the right-hand side of Equation 13.50 represents the variation of the total enthalpy on stage j per unit time. The equilibrium relationship is written in its usual, steady-state from, Equation 13.51, since phase equilibrium is assumed to exist at all times.

Although the fundamental equations are written for unsteady state conditions, the method is not, as presented here, concerned with a quantitative prediction of the transient performance of the column. The unsteady state analysis is merely used to define a convergence path that corresponds to the transition from unsteady- to steady-state conditions. As the column moves toward steady state, the terms on the right-hand side of Equations 13.49 and 13.50 approach zero and the equations reduce to steady-state relationships. Thus, reaching steady state is equivalent to reaching a converged solution. This is commonly referred to as the *relaxation* method.

The method has been applied to absorbers (Khoury, 1980), where all heat duties, side draws, and all feeds except the liquid feed at the top of the column and the vapor feed at the bottom are set to zero. If the variations of m_{ji} and $m_{j}C_{pj}T_{j}$ are broken down into liquid and vapor constituents,

$$-\frac{dm_{ji}}{dt} = \Delta L + \Delta V$$

and

$$-\frac{d\left(m_j C_{pj} T_j\right)}{dt} = \Delta h + \Delta H$$

then Equations 13.49 and 13.50, applied to the absorber model, may be written as

$$L_{ji} + V_{ji} = L_{j-1,i} + \Delta L + V_{j+1,i} + \Delta V \qquad (13.52)$$

$$L_j h_j + V_j H_j = L_{j-1} h_{j-1} + \Delta h + V_{j+1} H_{j+1} + \Delta H \qquad (13.53)$$

The subscripts for ΔL, ΔV, ΔH, and Δh have been dropped for simplicity. If two successive iterations are made to simulate two successful points in time in the approach to steady state and if quantities with superscript k represent iteration k, the following expressions will hold:

$$(L_{j-1,i} + \Delta L)^k = (L_{j-1,i})^{k-1}$$

$$(V_{j+1,i} + \Delta V)^k = (V_{j+1,i})^{k-1}$$

$$(L_{j-1} h_{j-1} + \Delta h)^k = (L_{j-1} h_{j-1})^{k-1}$$

$$(V_{j+1} H_{j+1} + \Delta H)^k = (V_{j+1} H_{j+1})^{k-1}$$

Substitution of these expressions in Equations 13.52 and 13.53 yields

$$(L_{ji})^k + (V_{ji})^k = (L_{j-1,i})^{k-1} + (V_{j+1,i})^{k-1} \qquad (13.54)$$

$$(L_j h_j)^k + (V_j H_j)^k = (L_{j-1} h_{j-1})^{k-1} + (V_{j+1} H_{j+1})^{k-1} \qquad (13.55)$$

As in other methods, the calculations are started with an assumed temperature and vapor profile, T_j and V_j, from which the liquid profile, L_j, and the K_{ji}'s are estimated. The tridiagonal matrix, Equation 13.24, is then solved only at the start to generate initial estimates for the component liquid and vapor flows, L_{ji} and V_{ji}. Following this, liquid and vapor enthalpies, h_j and H_j, are calculated for all the stages.

The calculated values are used as initial estimates corresponding to iteration $k - 1 = 0$. With known quantities on the right-hand side, Equations 13.54 and 13.55 are written for the first iteration $k = 1$ as follows:

$$L_{ji} + V_{ji} = C_j \qquad (13.56)$$

Enthalpy balance:

$$L_j h_j + V_j H_j = E_j \qquad (13.57)$$

where C_j and E_j are evaluated from initial estimates. The equations are grouped by stage, i.e., Equations 13.56, 13.57, and the equilibrium relationship, Equation 13.51, are solved simultaneously for each stage. This grouping of equations is equivalent to single equilibrium stage calculations, which may be solved by methods described in Chapter 2. The solutions generate new values for C_j and E_j in Equations 13.56 and 13.57 for the next iteration. At this point the component flow rates may be adjusted, if necessary, to meet overall column material balance. The procedure is repeated until the maximum variation of C_j and E_j from iteration to iteration is within an acceptable tolerance.

The method may be generalized to other column configurations. Various computational techniques may be incorporated in the algorithm to reduce computing time, such as skipping rigorous compositional K-value calculations on iterations where the change in composition is not large. The method is stable, although convergence tends to slow down asymptotically as the solution is approached. For this reason, once the column is stable and close to solution, the computations may be switched over to one of the other methods for final convergence.

13.2.2 Chemical Reactions in Multistage Separation

Chemical reactions and phase separation can occur simultaneously in multistage separation processes as in reactive distillation and absorption. This phenomenon is found in several operations in the petroleum, chemical, and petrochemical industries.

One example is the purification of gases by removing certain components by reactive absorption as in amine treatment (De Leye et al., 1986). In other applications, chemicals are produced by contacting gases and liquids in multistage processes where reaction products are concentrated in the same operation. Examples of these processes include the production of nitric acid (Koukolic et al., 1968) and chlorinations.

In reactive distillation the reaction products are separated from the reactants in order to enhance the product yields as in esterification processes. Reactive distillation is also used to facilitate the separation of very close boilers, such as isomers, by the addition of a reactant that reacts preferentially with the components to be separated (Saito et al., 1971; Tierney et al., 1982).

Another multistage separation process that is accompanied by chemical reactions is the stripping of hydrogen sulfide and/or ammonia from water in a sour water stripper. This is a common process for lowering the levels of sour gases in effluent water. The problem is usually to determine the amount of boilup or stripping steam required to lower the sour gas concentration in the bottoms to acceptable levels. The electrolytic reactions taking place in the column must be considered in order to predict the column performance correctly (Chen et al., 1982).

One or more reactions could take place in one or more stages in the column, and different reactions could take place in different stages. They could either be equilibrium or kinetically controlled reactions and could occur in either phase. The trays are assumed to act as continuous, well-mixed reactors as well as phase separation devices. For catalytic reactions, a catalyst is placed on the trays where reactions are intended to take place.

To simulate multistage separation processes with reactions, the reaction equations must be included in the model equations. The component balances, Equation

13.2, will include an additional term that represents the rate of generation or disappearance of components by kinetic reactions. Another term is included in Equation 13.2 to represent composition changes due to equilibrium reactions. The kinetic reaction rate is calculated by a power law expression, which involves a holdup term computed from tray geometry and fluid hydraulics. The equilibrium conversion is calculated from the equilibrium constant, which may be determined from experimental data or calculated from Gibbs free energies.

If the elemental reference state is used to calculate stream enthalpies, no heat of reaction calculation is necessary, and the same enthalpy balance, Equation 13.4, applies.

Several algorithms have been proposed to solve the model equations with chemical reactions (Holland, 1981; Saito et al., 1971; Tierney et al., 1982). Venkataraman et al. (1990) applied the "inside-out" method for this purpose.

13.2.3 Three-Phase Distillation

Nonideal situations that can result in three-phase distillation are discussed in Section 10.3. This phenomenon, where the liquid splits into two phases, could occur on a number of trays or in the entire column. From a practical standpoint and for better control of the column operation, one of the liquid phases is usually withdrawn when it forms.

The basic three-phase distillation model is similar to the general model described in Section 13.1 and Figure 13.1. The only difference is the potential existence of two liquid phases flowing from each tray j to tray j + 1, designated as L_j' and L_j'', and two liquid draws from each tray j, designated as $L_j^{s'}$ and $L_j^{s''}$.

The prediction of the performance of three-phase multistage separation processes is dependent on the ability to describe the thermodynamics of three-phase behavior. The prediction of column performance is only as good as the thermodynamics model. The mathematical solution of three-phase distillation columns is similar to two-phase vapor–liquid columns, the difference being in the model used to calculate the K-values. If the K-value model predicts two liquid phases, two liquid profiles must be considered in the column instead of one.

By defining a mixed K-value model, programs developed for solving vapor–liquid distillation columns have been successfully modified and used for simulating three-phase distillation (Schuil and Bool, 1985). In this method a mixed K-value is defined as the ratio of the mole fraction of a component in the vapor to its mole fraction in the mixed liquid phase. The column is solved using the mixed K-values instead of the usual vapor–liquid K-values to determine the temperatures, compositions, and flow rates of the vapor and total liquid on all the trays. The liquid phase split is then calculated on the basis of K-values for each liquid phase to determine the compositions and flow profiles of the two liquid phases.

K-Value Computations

With the molar rates of the two liquid phases designated as L' and L'' and the mole fractions of a given component i in each phase as X_i' and X_i'', the mole fraction of this component in the mixed phase is

$$X_i = \frac{X_i' L' + X_i'' L''}{L' + L''} \tag{13.58}$$

When this equation is used in the definition of the mixed K-value, the following relationship is obtained (Schuil and Bool, 1985):

$$K_i = \frac{Y_i}{X_i} = \frac{Y_i(L' + L'')}{X_i' L' + X_i'' L''} \tag{13.59}$$

The K-values for each liquid phase are given as

$$K_i' = \frac{Y_i}{X_i'}$$

$$K_i'' = \frac{Y_i}{X_i''}$$

When these equations are combined with Equation 13.59, the following is obtained for the mixed K-value:

$$K_i = \frac{(L' + L'')K_i' K_i''}{K_i'' L' + K_i' L''} \tag{13.60}$$

Defining the ratio of one liquid phase rate to the total liquid as

$$\alpha = \frac{L'}{L' + L''} \tag{13.61}$$

the mixed K-value becomes

$$K_i = \frac{K_i' K_i''}{\alpha K_i'' + (1 - \alpha)K_i'} \tag{13.62}$$

Since two liquid phases coexist at equilibrium only in systems that exhibit significant departure from ideality (Chapter 1), the K-values for three-phase distillation must be calculated using methods that are capable of predicting this nonideality. The method based on liquid activity coefficients, described in Chapter 1, is recommended for this purpose:

$$K_i = \frac{\gamma_i f_i^L}{\phi_i P}$$

(13.63)

where γ_i is the liquid phase activity coefficient, f_i^L is the standard state fugacity, ϕ_i is the fugacity coefficient, and P is the pressure. The K-values for each phase are written as

$$K_i' = \frac{\gamma_i' f_i^L}{\phi_i P}$$

(13.63a)

$$K_i'' = \frac{\gamma_i'' f_i^L}{\phi_i P}$$

(13.63b)

The only quantities that are different in these equations for the two phases are the activity coefficients, γ_i' and γ_i''. Equations suitable for calculating activity coefficients for liquid–liquid systems include the NRTL and UNIQUAC equations (Chapter 1).

By combining Equations 13.63a and 13.63b with Equation 13.62, the following is obtained for the mixed K-value:

$$K_{im} = \frac{\gamma_{im} f_i^L}{\phi_i P}$$

(13.64)

where γ_{im}, the mixed liquid phase activity coefficient, is given by

$$\gamma_{im} = \frac{\gamma_i' \gamma_i''}{\alpha \gamma_i'' + (1 - \alpha)\gamma_i'}$$

(13.65)

Hydrocarbon–Water Systems

A special case of three-phase distillation exists in hydrocarbon–water systems when both an organic liquid phase and a water phase are formed. If the hydrocarbon solubility in the water phase is neglected, it is possible to calculate a simplified mixed K-value that is not based on liquid activity coefficients.

The solubility of water in the hydrocarbon phase is given as X_s, the mole fraction of water in the water-saturated hydrocarbon phase. With the hydrocarbon phase molar rate designated as L' and the water phase as L'', the overall mole fraction of water in the mixed liquid phase is calculated as

$$X_w = \frac{L'X_s + L''}{L' + L''} \tag{13.66}$$

This equation is combined with Equation 13.61 to derive the following expression for α:

$$\alpha = \frac{1 - X_w}{1 - X_s} \tag{13.67}$$

Since the two liquid phases are in equilibrium, the component fugacities are equal in the two phases (Section 1.3). The equivalent to Equation 1.19 for two liquid phases is

$$f_i' = f_i'' \tag{13.68}$$

This equation can be written in terms of activity coefficients using the definition of fugacities as described in Section 1.3.3:

$$X_i' \gamma_i' f_i^L = X_i'' \gamma_i'' f_i^L$$

or

$$X_i' \gamma_i' = X_i'' \gamma_i'' \tag{13.69}$$

The water phase is assumed pure water; therefore,

$$X_w'' = 1$$

and

$$\gamma_w'' = 1$$

and Equation 13.69 for water in the hydrocarbon phase becomes

$$X_s \gamma_w' = 1$$

The K-value of water in the hydrocarbon phase is obtained by combining this equation with Equation 13.63a:

$$K'_w = \frac{f^L_w}{X_s \phi_w P}$$

For other components in the hydrocarbon phase, K'_i is calculated by an equation of state or any other suitable method, as described in Section 1.3.

The K-value of water in the water phase may be calculated from Equation 13.63b, with $\gamma_w'' = 1$:

$$K''_w = \frac{f^L_w}{\phi_w P}$$

For other components in the water phase, $X_i'' = 0$ since this phase is assumed pure water. From Equation 13.69

$$\gamma_w'' = \frac{X_i'}{X_i''} \gamma_i'$$

It follows that for $i \neq w$, $\gamma_i'' = \infty$, and, from Equation 13.63b, $K_i'' = \infty$.

The mixed K-values are now calculated from Equation 13.62. For water,

$$K_w = \frac{K'_w K''_w}{\alpha K''_w + (1-\alpha)K'_w}$$

And for all other components,

$$K_i = \frac{K'_i K''_i}{\alpha K''_i + (1-\alpha)K'_i}$$

Since $K_i'' = \infty$, this expression reduces to

$$K_i = \frac{K'_i}{\alpha}$$

13.2.4 Liquid–Liquid Extraction

Liquid–liquid extraction problems constitute a special case of two-phase multistage separation processes and can therefore be solved by many of the rigorous methods described in Section 13.2.1. The algorithms can be specifically adapted to extractors to take advantage of the special characteristics of these processes.

Of particular significance is the fact that in liquid–liquid extraction the transfer of components between the two liquid phases does not involve vaporization or condensation or the heats associated with these processes. The heat associated with liquid-to-liquid phase change, as well as the heat of mixing, may be neglected. Thus, if the feeds enter the extractor at the same temperature (ambient) and if the process is adiabatic, it is also isothermal. Alternatively, the column temperature may be controlled by adding or removing heat at different stages. In either case, the column temperature profile is known, and an enthalpy balance is not required as an integral part of the algorithm. The heat duties, if required, may be calculated separately once a column solution has been obtained. These assumptions are not necessary for the solution of liquid–liquid extractors but are used in the discussion that follows to illustrate how a column algorithm can be adapted to a special situation.

Another point that should be observed in extraction calculations is the nonideal nature of the system, which is responsible for the occurrence of two liquid phases in equilibrium. The liquid–liquid equilibrium distribution coefficients, or K-values, are highly composition-dependent and must be calculated by appropriate methods, namely those based on liquid activity coefficients. The NRTL and UNIQUAC liquid activity equations (Chapter 1) are among the more accurate ones for predicting liquid–liquid equilibria. The K-value is defined as the ratio of the mole fraction of a component in one liquid phase to its mole fraction in the other and is calculated as

$$K_i = \gamma_i'/\gamma_i'' \qquad (13.70)$$

where γ_i' and γ_i'' are activity coefficients of component i in L' and L''.

The modified Thiele–Geddes method described in Section 13.2.1 is adapted here to liquid–liquid extraction. Reference is made to the steps used in the Thiele–Geddes method.

In step 1 the temperatures are known if an isothermal process is considered. The flow rates of one of the liquid phases are assumed on each stage, and the flow rates of the other liquid phase are calculated by material balance (step 2). In step 3 the K-values may have to be estimated based on assumed compositions of the liquid phases on each tray rather than from a simplified correlation. Steps 4, 5, 6, and 7 are similar to the modified Thiele–Geddes method with the stipulation that liquid activity coefficients be used to calculate K-values using Equation 13.70. Steps 8 and 9 may be skipped if the special extractor model assumptions are used. Step 10 is the convergence check which needs only check for material balance and invariance of the K-values.

13.2.5 Stage Efficiencies

The column solution methods described in this chapter are based on an equilibrium stage model. Normally, the vapor and liquid leaving a tray are not at equilibrium due to imperfect mixing or insufficient residence time on the tray. Tray efficiencies, discussed in Chapter 16, are factors that relate actual performance to equilibrium stage performance. By replacing the equilibrium relationship in the model with an equation based on the Murphree tray efficiency, the actual column performance may be calculated. The accuracy of the model obviously depends on the reliability of the tray efficiencies used in the calculations.

The vapor and liquid streams leaving an equilibrium stage are both saturated at the same temperature, the stage temperature. A nonequilibrium stage implies that the vapor and liquid leaving the stage are not saturated at the same temperature. In the model they are assumed to be saturated and must therefore be at different temperatures.

The vapor phase Murphree tray efficiency for tray j is defined as (Chapter 16)

$$E_{MVj} = \frac{Y_{ji} - Y_{j+1,i}}{K_{ji}X_{ji} - Y_{j+1,i}} \qquad (13.71)$$

where Y_{ji} is the mole fraction of component i in the vapor leaving tray j. This is the actual vapor composition and is not necessarily equal to $K_{ji}X_{ji}$. The efficiency equation, rearranged as follows, is used instead of Equation 13.5 in the column calculations:

$$E_{MVj}K_{ji}X_{ji} - Y_{ji} + (1 - E_{MVj})Y_{j+1,i} = 0$$

In terms of L_{ji}, V_{ji}, and T_j, usually considered the independent variables, this equation becomes (Naphtali and Sandholm, 1971)

$$\frac{E_{MVj}K_{ji}L_{ji}}{L_j} - \frac{V_{ji}}{V_j} + \frac{\left(1 - E_{MVj}\right)V_{j+1,i}}{V_{j+1}} = 0$$

The enthalpy balance calculations for nonequilibrium stages must be modified since the temperatures of the vapor and liquid leaving a stage are different. Each phase enthalpy must be calculated based on its own temperature.

NOMENCLATURE

A_{ji}	Factor defined by Equation 13.22		F_j	Molar rate of external feed to tray j
B_{ji}	Factor defined by Equation 13.23		F_{ji}	Molar rate of component i in external feed to tray j
C	Total number of components		g	General function representing column equations
C_j	Component balance function on tray j		G	General column specification
			h	Liquid molar enthalpy
C_{pj}	Average heat capacity on tray j		H	Vapor molar enthalpy
E_j	Enthalpy balance function on tray j		K	Vapor–liquid equilibrium coefficient
E_{MVj}	Murphree tray efficiency in vapor terms		L_j	Liquid molar flow rate from tray j
			L_{ji}	Liquid molar flow rate of component i from tray j
f	Fugacity			

m_j	Total molar holdup on tray j	w	General column variable
m_{ji}	Holdup of component i on tray j	X_{ji}	Mole fraction of component i in the liquid on tray j
N	Number of theoretical stages	X_s	Solubility of water in hydrocarbon phase
P_j	Pressure on tray j	X_w	Mole fraction water
Q_j	Heat duty on tray j		
R	Reflux ratio	Y_{ji}	Mole fraction of component i in the vapor on tray j
S	Stripping factor		
S_j	Summation function on tray j	Z_{ji}	Mole fraction of component i in the feed to tray j
t	Time	α	Adjustment factor
T_j	Temperature on tray j	α	Ratio of one liquid phase to total liquid
V_j	Vapor molar flow rate from tray j	α_{ji}	Relative volatility
		γ	Activity coefficient
V_{ji}	Vapor molar flow rate of component i from tray j	θ	Correction factor
		ϕ	Fugacity coefficient

SUBSCRIPTS

b	Base component designation	r	Reference component designation
f	Feed tray designation	w	Designation of water phase or component water
i	Component designation		
j	Tray designation		

SUPERSCRIPTS

f	Designation of initial thermal conditions of external feed	k	Iteration number designation
		s	Side product designation

REFERENCES

Boston, J. F. and S. L. Sullivan, Jr., An improved algorithm for solving mass balance in multistage separation processes, *Can. J. Chem. Eng.*, 50, 663, 1972.

Brannock, N. F., V. S. Verneuil, and Y. L. Wang, Rigorous distillation simulation, *Chem. Eng. Progress*, October 1977, p. 83.

Broyden, C. G., *Mathematics of Computations*, Vol. 19, 1965, p. 577.

Carnahan, B., H. A. Luther, and J. O. Wilkes, *Applied Numerical Methods*, New York, John Wiley & Sons, 1964.

Chen, C. C., H. I. Britt, J. F. Boston, and L. B. Evans, *AIChE J.*, 28, 588, 1982.

De Leye, L. and G. F. Froment, *Computers and Chemical Engineering*, 10(5), 493, 1986.

Holland, C. D., *Fundamentals and Modeling of Separation Processes*, Englewood Cliffs, NJ, Prentice Hall, 1975.

Holland, C. D., *Fundamentals of Multicomponent Distillation*, New York, McGraw-Hill, 1981.

Holland, C. D., *Multicomponent Distillation*, Englewood Cliffs, NJ, Prentice Hall, 1963.

Khoury, F. M., Simulate absorbers by successive iteration, *Chemical Engineering*, December 29, 1980, p. 51.

Koukolik, M. and J. Marek, "Mathematical Model of HNO_3 Oxidation–Absorption Equipment," Proc. Fourth European Symp. on Chem. Reac. Eng., 1968.

Lyster, W. N., S. L. Sullivan, Jr., D. S. Billingsley, and C. D. Holland, *Petrol. Refiner*, 38(6), 221, 1959; 38(7), 151, 1959; 38(10), 139, 1959.

Naphtali, L. M. and D. P. Sandholm, Multicomponent separation calculations by linearization, *AICHE J.*, 17(1), 148, 1971.

Russell, R. A., A flexible and reliable method solves single-tower and crude-distillation-column problems, *Chemical Engineering*, October 17, 1983.

Saito, S., T. Michishita, and S. Maeda, *J. Chem. Eng. Japan*, 4, 37, 1971.

Schuil, J. A. and K. K. Bool, Three-phase flash and distillation, *Computers and Chemical Engineering*, 9(3), 295, 1985.

Thiele, E. W. and R. L. Geddes, *Ind. Eng. Chem.*, 25, 289, 1933.

Tierney, J. W. and G. D. Riquelme, *Chem. Eng. Comm.*, 16, 91, 1982.

Tomich, J. F., A new simulation method for equilibrium stage processes, *AICHE J.*, 16(2), 229, 1970.

Venkataraman, S., W. K. Chan, and J. F. Boston, *Chem. Eng. Prog.*, 86(8), 45, 1990.

Wang, J. C. and C. E. Henke, Tridiagonal matrix for distillation, *Hydrocarbon Processing*, 45(8), 155, 1966.

Wang, J. C. and Y. L. Wang, "A Review on the Modeling and Simulation of Multi-Stage Separation Processes," Proceedings of the International Conference, Foundation of Computer-Aided Chemical Process Design, July 6, 1980.

CHAPTER **14**

Tray Hydraulics

In the column model discussed in previous chapters, the internal liquid and vapor flows were considered only from the standpoint of their effect on the thermodynamic performance of the column. The performance parameters that were investigated included such quantities as compositions, temperatures, pressures, enthalpies, and K-values. The column was assumed capable of physically handling any liquid or vapor flow rate, regardless of hydraulic effects or pressure drops. The only flow limitations that were taken into account involved minimum reflux ratio and conditions under which the liquid or vapor "dried up," or approached zero flow in certain parts of the column.

The internal flow of liquid and vapor must be reevaluated from the standpoint of column capacity, both in the design and performance studies of columns. The physical dimensions of a column can handle only limited ranges of vapor and liquid flow rates. The object of this chapter is to evaluate the hydraulic aspects of fluid flow in trayed columns. The column performance is examined with regard to such factors as flooding, entrainment, and pressure drop.

The depth of coverage is, for the most part, qualitative in this area. Detailed quantitative analysis of hydraulics is outside the scope of this book since much of the subject matter relates to information that is specific to equipment manufacturers' data. The following sections are intended to provide an understanding of the factors that affect the actual performance of multistage separation processes. References are provided for more specialized reading.

A multistage tray column consists of a vertical cylinder ranging in diameter anywhere from a few inches to many feet and reaching heights of hundreds of feet. The number of trays in a column could be anywhere from a few trays to a few hundred.

Column trays come in many different types, but they all share certain features that may be described by means of a general, simplified model. Model simplification is essential in the study of tray hydraulics even when discussing one particular type of tray. The reason for this is the highly complex nature of countercurrent, two-phase fluid flow in the column and across the trays. Certain assumptions must be made in the development of flow and pressure drop correlations.

Although the overall vapor and liquid flow in the column is countercurrent, what takes place on each individual tray for most types of trays is actually crossflow, as shown in the model schematic, Figure 14.1. Counterflow tray types do exist, but their use is not as widespread as crossflow trays. Therefore, crossflow trays are only briefly mentioned here. Unlike crossflow trays, counterflow trays occupy the entire cross section of the column. Flow through the tray openings is counterflow; that is, vapor flows up and liquid flows down through the same openings. Except for the spacing between trays, counterflow-tray column hydraulics is to some degree similar to that of counterflow packed columns, discussed in Chapter 15.

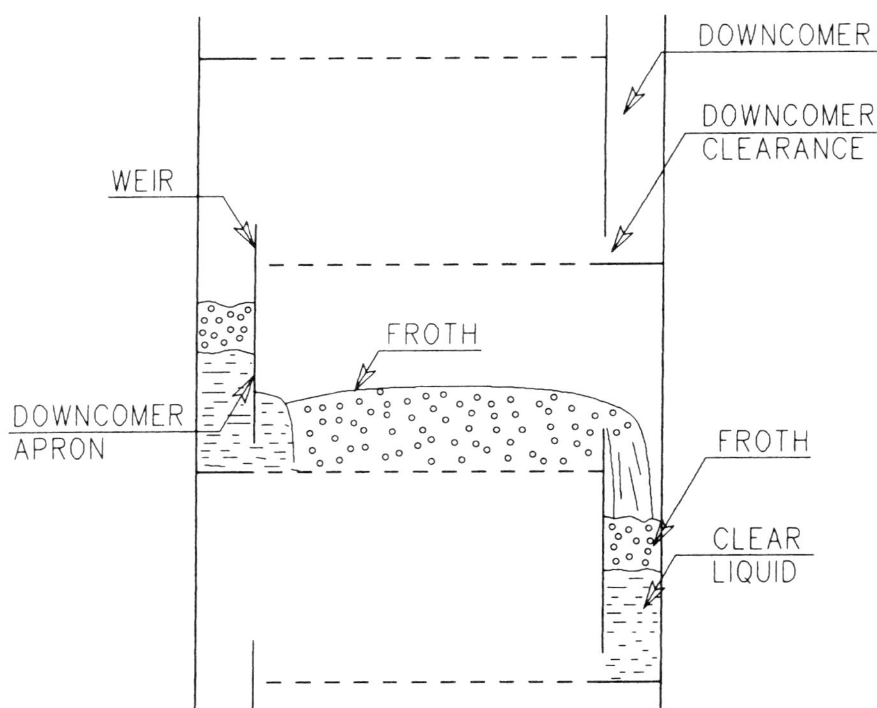

Figure 14.1 Tray schematic.

A crossflow tray consists of a horizontal metal plate that partially fills the column cross section. Figure 14.1 depicts a few single-pass trays in a column section. The other main components of each tray are the downcomer, downcomer apron, and the weir. The top view of a tray is illustrated in Figure 14.2, where the downcomer cross section appears as the segment of a circle. Opposite the downcomer is another segment of the tray where liquid flows down from the upper tray. The tray area between the two segments contains openings or valves of one type or another, through which vapor flows upward. Most of this area is the tray active area.

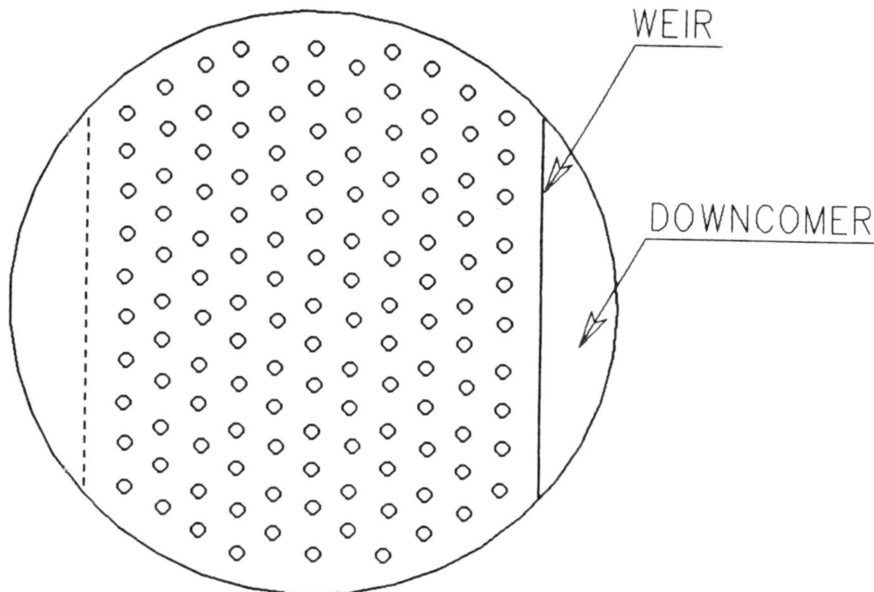

Figure 14.2 Single-pass tray.

Although the general model focuses on single-pass trays, multipass trays are also used. In multipass trays the liquid splits and flows in opposite directions so that any liquid element travels only a certain fraction of the tray width. Liquid flows down from one tray to the tray below through more than one downcomer, except on alternating trays in two-pass trays where liquid flows down one central downcomer every other tray. Figure 14.3 illustrates tray arrangements and liquid paths in two-, three-, four-, and five-pass trays.

14.1 GENERAL TRAY HYDRAULICS CHARACTERISTICS

From Figure 14.1, it is observed that liquid and froth flow down the downcomer from one tray to the tray below. As the fluid settles in the downcomer area, it separates into a layer of clear liquid and a layer of froth. The height of the froth depends on the foaming characteristics of the liquid. Clear liquid flows under the downcomer apron through the downcomer clearance and on to the active area of the tray between the downcomer apron and the weir. This part of the tray contains the openings through which vapor flows from the tray below. As a result of the vapor streaming upward through the openings and the liquid layer on the tray, froth is formed, starting a short distance from the downcomer apron. The froth flows across the tray, over the weir, and through the downcomer to the tray below. The weir height varies with tray type and in general is intended to ensure a minimum liquid height on the tray. As the froth moves from the downcomer to the weir, a liquid gradient develops across the tray to overcome the friction resistance to flow.

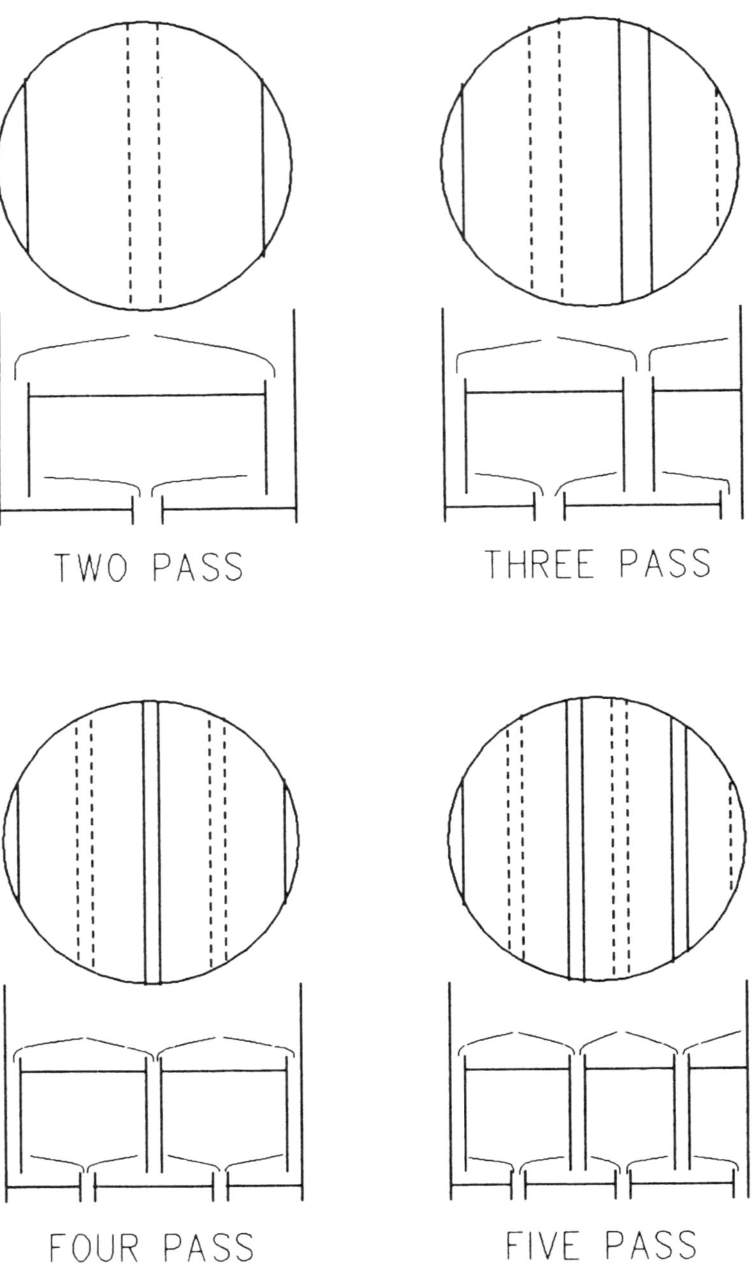

Figure 14.3 Multipass trays.

The hydraulics of liquid and vapor flow in and around the tray openings varies with tray type. One general characteristic of crossflow trays is the fact that liquid is prevented from flowing down the openings by the upward flow of the vapor. Thus, while liquid flows horizontally across the tray, vapor flows vertically through the tray and between trays. This liquid–vapor crossflow results in a composition gradient in the liquid across the tray. It is assumed that the vapor mixes thoroughly as it rises in the space between the trays and reaches a uniform composition before it penetrates the tray above. As the liquid travels across the tray, it interacts with the vapor and its composition changes progressively until it reaches the weir. Nevertheless, the simplified tray model assumes average liquid composition and steady-state conditions on the tray.

The froth on the tray is a turbulent mass of usually liquid-continuous fluid with vapor dispersed in the form of small bubbles. The various designs of tray vapor openings attempt to maximize vapor dispersion by generating the smallest possible bubbles. The froth is where mass transfer takes place between the vapor and liquid. Mass transfer, and tray efficiency along with it, is enhanced by creating the largest possible interfacial area.

Types of Tray Vapor Openings

Tray design varies with each particular manufacturer and design details are usually proprietary. Performance information is available directly from the manufacturers (Glitsch, Koch, Nutter), some of whom provide detailed design for a given set of specifications as part of the purchase. A few features of the more commonly used types of tray openings are summarized below.

Bubble Cap Trays

In this type of tray, each opening assembly consists of a cap, or inverted cup, with slots at its base, placed above an opening. The opening consists of a hole and a riser through which vapor rises from the tray below. Its flow is reversed downward by the cap, after which the vapor flows down around the riser and bubbles out through the slots and into the liquid. Bubble cap diameters are usually about 3 to 4 inches. Because of the liquid seal created by the riser, bubble caps can operate at wide ranges of vapor and liquid flows with little loss in tray efficiency.

Sieve Trays

Perhaps the simplest of crossflow column tray designs is the sieve tray or perforated tray. The tray is a flat metal plate and the vapor openings are holes drilled in the plate. The holes are usually round, ranging from 1/8- to 1/2-inch diameter. Sieve trays have no liquid seals to prevent liquid from flowing down the holes. Liquid flow down the holes is prevented only by the upward flow of the vapor.

Valve Trays

This type of tray is designed to allow for wide variations in liquid and vapor flow. One typical design of a valve vapor opening consists of an orifice in the tray plate, an orifice cover, and a travel stop. At low vapor rates, the orifice cover is settled in its lower position. In this position, slots in the orifice cover allow small amounts of vapor to be distributed evenly. At higher vapor rates, the orifice cover is elevated to its upper position set by the travel stop. In this position, large amounts of vapor can flow through the valves.

14.2 FACTORS AFFECTING TRAY HYDRAULICS

The simplified tray model assumes idealized hydraulic conditions that enhance mass transfer (high tray efficiency) and maintain a low vapor pressure drop. Proper tray design aims at limiting the effect of factors that tend to diminish "good" tray hydraulics. The adversity either may be inherent in the tray type or design for a particular situation or may be a result of operating the column outside the design conditions. A look at some of these factors follows.

Foaming

The formation of froth on the trays is desirable for maximizing the interfacial area and mass transfer between liquid and vapor. Froth is formed through agitation and turbulence brought about by the vapor flow through the liquid. The amount of froth formed (foamability) and its tendency to linger (foam stability) are related to the physical properties of the liquid. Foamability can limit the allowable vapor flow, and foam stability can limit the allowable liquid flow. The foaming properties must, therefore, be taken into account in the design and rating of column trays. In certain situations foamability may be adjusted by the addition of surfactants.

Vapor Entrainment

As liquid flows down through the downcomer, some froth flows with it and some more is formed in the downcomer by the turbulent flow. The downflowing froth should be allowed to break up in the downcomer before it reaches the tray below. Any froth flowing to the tray below entails vapor entrainment, which lowers the tray efficiency on account of vapor flowing in the wrong direction. The downcomer volume must be large enough so the residence time is sufficient to allow the froth to disintegrate.

Liquid Entrainment

When vapor disengages from the froth on a tray, liquid droplets may be entrained with it to the tray above. Liquid entrainment could be especially severe with highly foamable liquids where the froth occupies a large portion of the space between the trays. If the top of the froth is close to the tray above, the liquid droplets do not have enough time to fall back to the froth. Foamability, therefore, limits the vapor flow capacity of the trays. Handling high vapor flows with a foaming liquid requires larger tray spacing.

As liquid is entrained from one tray to the tray above, the entrained liquid, e, joins the normally flowing liquid, L, resulting in a total liquid flow, L + e. The fractional entrainment, ψ, is defined as the ratio of entrained liquid to total liquid flow:

$$\psi = \frac{e}{L + e} \tag{14.1}$$

The tray efficiency is lowered by liquid entrainment because higher-boiling liquid is carried to a tray containing lower-boiling liquid, thereby countering the fractionation process. The lowering of the Murphree tray efficiency (discussed in Chapter 15) is expressed by the Colburn equation (Colburn, 1936):

$$\frac{E_e}{E_{MV}} = \frac{1}{1 + eE_{MV}/L} \tag{14.2}$$

where E_{MV} is the Murphree tray efficiency in the absence of entrainment, and E_e is the Murphree tray efficiency with entrainment.

Liquid Gradient

As the liquid or froth moves across the tray, a liquid head develops to overcome resistance to flow caused by friction between the froth on the one hand and the tray and rising vapor on the other. The result is a liquid gradient with the liquid height at the bottom of the downcomer apron greater than that at the downflow weir.

Liquid gradient problems are understandably more severe in larger diameter trays, where the flow paths are longer. For this reason, large diameter trays are usually designed with multiple passes (Figure 14.3) in order to reduce the liquid flow path and, hence, the liquid gradient.

The main problem that may arise from the liquid gradient, unless proper design and operating practices are followed, is its effect on the vapor distribution on the tray. Vapor flowing through tray openings where the liquid level is higher must overcome a greater liquid head than in places with lower liquid level. Hence, more vapor flows through the openings next to the weir than those next to the downcomer apron. Poor vapor distribution is detrimental to the overall performance of the column because it lowers the tray efficiency.

"Weeping" is another condition that could be aggravated by a liquid gradient. Weeping is the excessive flow of liquid down through the tray openings. As long as this flow does not cause an appreciable drop in tray efficiency, it is considered normal. If the vapor flow is reduced, liquid flow through the tray increases. Below a certain vapor flow, weeping ensues. With a liquid gradient, weeping is more severe at points on the tray where the liquid head is higher, i.e., at the point of liquid entry, close to the downcomer apron.

14.3 FLOODING

At certain vapor and liquid flow rates within the column, conditions could arise where liquid backs up between the trays and the trays cease to function as distinct stages. The column becomes "flooded" and completely inoperable because of a steep decline in tray efficiency and a sharp increase in pressure drop.

Flooding can result either from the froth rising on the tray and reaching the tray above or from liquid and froth filling the downcomer up to the liquid level at the downcomer overflow weir. The rise of the froth to the tray above is caused by excessive liquid entrainment resulting from too much vapor flow. The downcomer backup is caused by liquid rates exceeding the flow capacity of the downcomer. In either case the ultimate result is flooding, with liquid and froth filling the entire column.

It is therefore clear that flooding can be caused either by the vapor flow or the liquid flow exceeding certain limitations. Usually the vapor flow is limiting and the column preliminary design is based on it. The column is then checked for its liquid handling capacity.

The vapor velocity at which vapor-induced flooding occurs is the *flooding vapor velocity*. Since flooding is usually checked in terms of vapor flow, flooding vapor velocity is the most commonly used indicator. The vapor velocity is based on the net area available for vapor flow between the trays. Normally, the net area is the column cross-sectional area less the area occupied by downcomers, baffles, etc. The actual column operation relative to its proximity to flooding is expressed as the ratio of the actual vapor velocity to the flooding vapor velocity. The ratio may be expressed as a percentage, commonly known as the percentage, commonly known as the percentage of flood. Columns are normally designed for operation at 75 to 85% of flood. If the correlation used for predicting the flooding vapor velocity does not take into account foaming characteristics, the design percentage of flood for foaming fluids should be about 70 to 75%.

The flooding vapor velocity correlations are empirical and specific to the type of tray. For a given tray type, the flooding vapor velocity is a function of the liquid flow, the vapor and liquid densities, and the tray spacing (Smith, 1963). Corrections for surface tension are included in certain correlations.

14.4 PRESSURE DROP

While liquid flows by gravity in vapor–liquid countercurrent columns, a pressure gradient is necessary to induce vapor flow. Pressure drops exist from tray to tray; so the lower trays must be maintained at a higher pressure than the upper trays. One consideration in tray design is to attempt to keep the pressure drop at a minimum.

For a given tray type, the pressure drop from tray to tray is a function of vapor and liquid rates and properties as well as certain tray parameters, such as tray thickness, the geometry of the openings, etc. Severe pressure drops are associated with certain operating conditions, such as too high downcomer backup, excessive liquid entrainment, and approach to flooding conditions. A minimum pressure drop is necessary to prevent liquid from flowing down the openings.

High pressure drops can cause the downcomer backup to increase to a point where it fills the entire downcomer, causing flooding. For proper column operation, the downcomer backup should not exceed 40% to 60% of the tray spacing. The downcomer backup is balanced by the friction pressure drop of liquid flowing through the downcomer clearance, the liquid head on the tray on the other side of the downcomer apron, and the friction pressure drop of the vapor flowing through the tray openings and the liquid on the tray above. For correlating purposes, the pressure drop components are broken down further as indicated in the following equation. Pressure drops are usually expressed as clear liquid head:

$$h_{dc} = h_0 + h_1 + h_c + h_w + h_{ow} + h_g$$

where h_{dc} is the downcomer backup, h_0 is the dry hole vapor pressure drop, h_1 is the vapor pressure drop through the liquid and froth on the tray, h_c is liquid pressure drop through the downcomer clearance, h_w is the weir height, h_{ow} is the height of liquid over the weir, and h_g is the hydraulic gradient across the tray. Semiempirical correlations are available for estimating these head losses (Smith, 1963) and are usually supplied by the tray vendor.

14.5 OPERABLE RANGES

Although a column may be designed for a given set of vapor and liquid flows, actual operating conditions could vary considerably. It is therefore important to determine the ranges of flow rates over which the column can operate satisfactorily with only minor losses in tray efficiency. Different tray types have different operating flexibility, and the selection of a tray type should take into account its operable ranges. The tray should be checked for maximum anticipated vapor and liquid flows, then checked again for minimum anticipated flows. It is not unusual, for instance, for a tray designed for 75 to 85% of flood to operate at ranges extending down to 20% of flood without substantial loss of tray efficiency.

A qualitative diagram showing the constraints that limit operable vapor and liquid rates for a typical tray is presented in Figure 14.4. At low liquid flow rates, the operable vapor flow range is small: a quick transition from weeping conditions to excessive liquid entrainment takes place as the vapor flow rate is increased (area around C). Curve CB represents the lowest operable vapor rates at varying liquid rates. The curve corresponds to conditions where the liquid gradient could cause problems. At lower liquid rates, insufficient vapor could result in weeping and at higher liquid rates insufficient vapor could result in poor vapor distribution.

At high liquid rates, the tray can operate over a narrow range of vapor rates. As the vapor rates are increased, the tray hydraulics changes rapidly from poor vapor distribution to flooding (area around B). At lower liquid rates, a higher vapor rate can be maintained before flooding is incurred (curve AB). Point A represents the upper

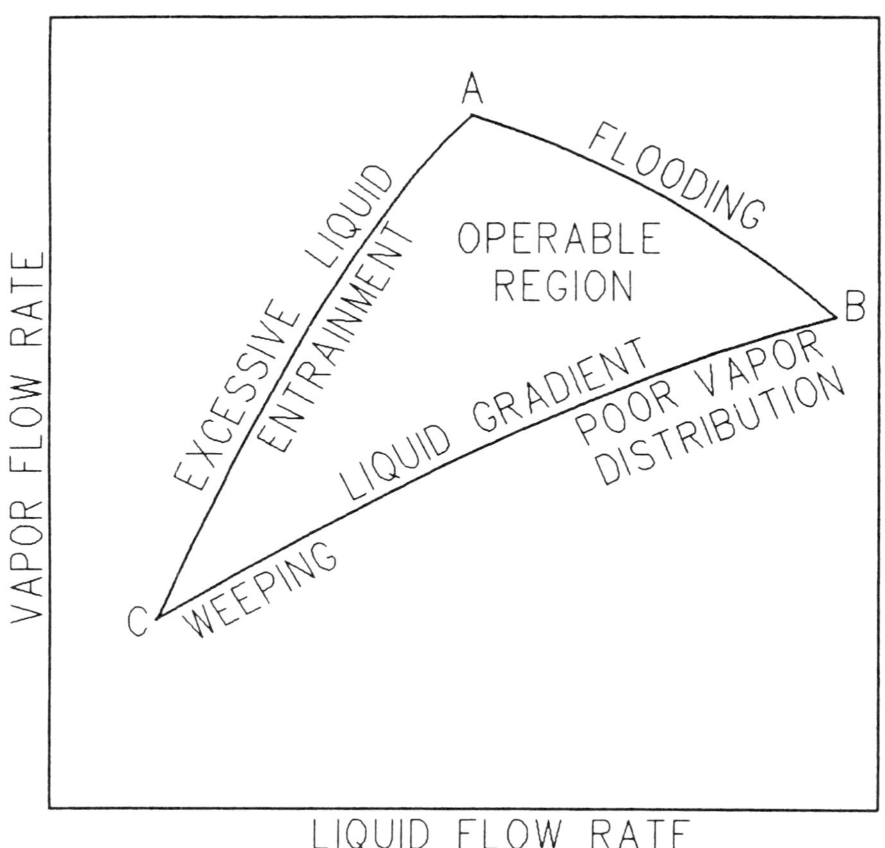

Figure 14.4 Column operable ranges.

limit of vapor flows. In that area the tray is susceptible either to excessive liquid entrainment at lower liquid rates or to flooding at higher liquid rates. Curve AC places constraints on the operable region because of excessive liquid entrainment. As the liquid rate is decreased, lower vapor rates can cause excessive liquid entrainment.

The diagram in Figure 14.4 is a qualitative one intended to demonstrate tray hydraulics at varying liquid and vapor rates. The tray can be operated satisfactorily within the area defined by ABC. The diagram should be plotted quantitatively for a given tray type. The shape and characteristics of the operable region vary from one tray type to another.

14.6 STEPS IN ANALYSIS OF TRAY HYDRAULICS

The factors influencing tray hydraulics are considered collectively in tray design or in checking tray performance under given operating conditions. Tray hydraulics is specific to tray type, and its characteristics are correlated on the basis of large amounts of operating data. Although the quantitative analysis of tray hydraulics is outside the scope of this book, the logical steps in designing a "general type" tray are described qualitatively here, relating the various hydraulics factors to tray performance.

Investigating column performance starts with heat and material balance calculations, which generate liquid and vapor flows and properties. The fluid foaming characteristics and corrosivity are also determined. Typically, the calculations are done on a computer simulator, which generates liquid and vapor flows as well as physical properties, such as densities, viscosities, and surface tension.

In the design mode, the next step is to make certain decisions and assumptions regarding the trays. Some of the assumptions may have to be refined at later stages of the design. The tray type and material of construction are selected. Tray spacing is assumed, as well as tray geometry: dimensions of openings, weir height and length, and downcomer type and size. Also, initially a single-pass crossflow tray is assumed for economy. Multiple-pass trays may be considered for large tray diameters to keep the liquid gradient in check.

The next step is to determine the flooding vapor velocity, which is a function of the vapor and liquid flow rates and their densities. The design vapor velocity is then calculated by multiplying the flooding vapor velocity by a flood factor such as 0.85. The vapor velocity is based on the vapor volumetric flow and the net flow area between the trays, that is, the column cross-sectional area minus the area blocked by the downcomers. With known vapor velocity, vapor flow, and fraction of the column cross section occupied by the downcomers, the net area can be calculated. The total area is then calculated, from which the tray diameter is determined. The actual column diameter is obtained by rounding off the tray diameter to the next larger standard size. The vapor velocity is then adjusted to the new diameter to calculate the expected flood factor.

At this point most of the column and tray parameters have been tentatively determined. Next, a performance check is carried out at various operating conditions to determine the operable ranges and whether any of the design parameters need to be adjusted. Performance checks include such items as pressure drop, liquid handling capacity, entrainment, and weeping.

If the calculated pressure drop at the vapor flow corresponding to the design flood factor is deemed excessive, consideration should be given to using larger tray openings. From another perspective, if the column is also required to operate at lower vapor rates, the weeping point should be determined, which sets the lower limit on the vapor rate. The operating range could be extended to lower vapor rates by using smaller tray openings, provided the pressure drop remains within acceptable ranges.

A liquid entrainment check is usually also made to ensure it does not exceed acceptable limits. The fractional entrainment is a function of liquid and vapor flows and densities and of the flood factor. The column should not be operated at high fractional entrainment since this reduces the tray efficiency.

Finally, the trays must be checked for their liquid-handling capacity, which is controlled by the downcomer dimensions. The liquid level in the downcomer should not be allowed to exceed a certain height, normally about half the spacing between the trays. Liquid backup in the downcomer can cause flooding, although flooding is more often initially caused by excessive vapor flow. The liquid velocity in the downcomer is obtained by dividing the liquid flow by the downcomer cross-sectional area. The residence time is then evaluated from the ratio of the assumed height in the downcomer (half the tray spacing) to the liquid velocity. The residence time provides an empirical measure of the tray liquid-handling capacity. A minimum residence time is required to ensue froth separation and breakup to prevent excessive backup. More detailed calculations specific to the tray type are carried out to determine the actual liquid level in the downcomer.

NOMENCLATURE

e	Liquid entrainment	h_{ow}	Height of liquid over the weir
E_e	Murphree tray efficiency with entrainment	h_{dc}	Downcomer backup
		h_g	Hydraulic gradient
E_{MV}	Murphree tray efficiency in vapor terms	h_l	Pressure drop through liquid and froth on the tray
h_c	Pressure drop through downcomer clearance	h_0	Dry hole pressure drop
		L	Liquid flow rate in the column
h_w	Weir height	ψ	Fractional entrainment

REFERENCES

Colburn, A. P., *Ind. Eng. Chem.*, 28, 526, 1963.

Glitsch, Inc., *Ballast Tray Design Manual,* 4th edition, Bulletin No. 4900, Dallas, TX, 1984.

Koch Engineering Company, Inc., *Design Manual — Flexitray,* Bulletin 960-1, Wichita, KS, 1982.

Nutter Engineering Company, *Float Valve Tray Design Manual,* Tulsa, OK, 1976.

Smith, B. D., *Design of Equilibrium Stage Processes*, New York, McGraw-Hill, 1963.

CHAPTER 15

Packed Columns

For the most part in this book, multistage columns were studied in terms of equilibrium stages, and a stage was assumed to represent a column tray. Multistage vapor–liquid countercurrent flow for the purpose of separation can also take place in packed columns. In these devices, instead of trays, the column is filled with an inert packing material designed for maximum mass transfer between the vapor and the liquid and for low pressure drop. The vapor and liquid compositions vary continuously with packing height rather than discretely as in trayed columns.

In trayed columns the composition changes in a stagewise manner and each stage is treated as an equilibrium stage, while lack of equilibrium is accounted for by using some form of tray efficiency. In counterflow packed columns, the passing phases are never considered to be truly at equilibrium. In trayed columns, the operating line, such as in the McCabe–Thiele diagram for binary distillation, is not truly a continuous line but a series of points, each representing the compositions of passing phases between trays. The operating line for a packed column, on the other hand, is a continuous line representing the compositions of the passing phases over the length of the column.

In contrast to trayed columns' crossflow, the flow in packed columns is vapor–liquid counterflow. While in trayed columns the construction parameter that determines separation capacity is the number of trays, in packed columns it is the packing height. A certain packing height can accomplish the same separation as an equilibrium stage. This height is known as the height equivalent to a theoretical plate, or HETP. If the total column packing height and the HETP are known, the number of equilibrium stages can be calculated by dividing the packing height by the HETP. This is analogous to determining the number of equilibrium stages in a trayed column from the number of actual trays and the overall tray efficiency. Once the number of equilibrium stages is known, a packed column can be calculated by any of the methods described in Chapters 12 and 13.

In a design situation, the height of a packed column can be determined from the number of theoretical stages and the HETP. The number of theoretical stages is determined by analytical methods as described in Chapters 12 and 13, and the HETP is estimated independently.

An alternative approach to the HETP method for calculating the column packing height, also discussed in this chapter, is based on the height of a transfer unit (HTU) and the number of transfer units (NTU). These quantities, as well as the HETP, are usually estimated from empirical correlations or based on experimental data. They depend on the type of packing, column diameter, type and flow rate of fluids, and operating conditions. The primary sources for HETP and HTU data are the packing manufacturers.

Another parameter in the design of a packed column is the column diameter, which is determined on the basis of hydraulic considerations, such as pressure drop and flooding conditions. Hydraulics information is also characteristic of the packing type and operating conditions and is usually supplied by the manufacturer, along with separation efficiency data (HETP or HTU).

The use of packing instead of trays in multistage separation columns is common for column diameters 3 to 4 feet or smaller. More recently, packing has been used for larger columns because of its low pressure drops and favorable efficiencies.

15.1 CONTINUOUS DIFFERENTIAL MASS TRANSFER

In a simplified, trayed column section, where the operating line and the equilibrium curve are assumed straight and parallel, the number of equilibrium stages required to change the vapor composition by a certain amount may be calculated simply by dividing the required composition change by the composition change per tray. Figure 15.1 represents such a process for a ternary system where one component is absorbed from an inert gas by a counterflowing liquid. The composition in the gas phase of the component being transferred is represented by Y_B at the inlet and by Y_T at the outlet. (No component subscripts are used since reference is always made to the component being transferred.) In the liquid, the compositions of the transferred component are represented by X_T at the inlet and X_B at the outlet. Points (X_T, Y_T) and (X_B, Y_B) lie on the operating line because they both represent compositions of passing phases, one at the top of the column and the other at its bottom.

If the compositions of the liquid and vapor flowing between any two stages are X and Y represented by point P on the operating line, the vapor composition at equilibrium with the liquid is Y^*, and the liquid composition at equilibrium with the vapor is X^* (Figure 15.1). The change in the vapor composition per tray is $Y - Y^*$. Since this difference is constant when the two lines are parallel, the required number of stages is calculated as

$$N = \frac{Y_B - Y_T}{Y - Y^*} \tag{15.1}$$

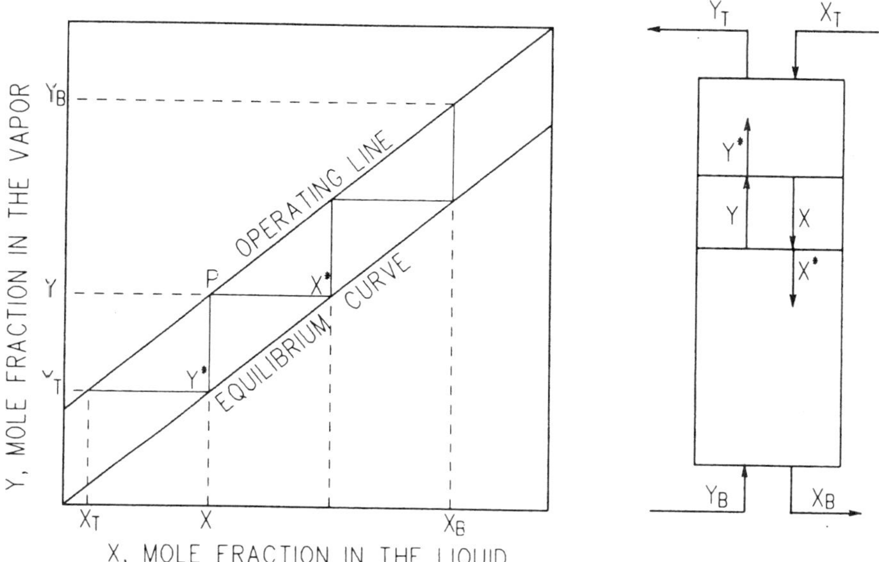

Figure 15.1 Simple absorber — parallel, straight, operating line, and equilibrium curve.

If, instead of the discrete stages, a packed column is used, an equivalent number of stages would be calculated by the same equation. In general, the operating line and the equilibrium curve are neither straight nor parallel, and Equation 15.1 would be written in differential form:

$$dN = \frac{dY}{Y - Y^*} \tag{15.2}$$

where dN is the differential number of transfer units required to bring about a change dY in the vapor composition. The difference $Y - Y^*$ is the driving force causing the composition change. The number of transfer units, which is the ratio of the composition change to the mean driving force, is obtained by integrating Equation 15.2 between the inlet and outlet vapor compositions:

$$NTU = \int_{Y_B}^{Y_T} \frac{dY}{Y - Y^*} \tag{15.3}$$

In the special case where the operating line and the equilibrium curve are straight and parallel, the difference $Y - Y^*$ is constant and the integration gives

$$NTU = \frac{Y_T - Y_B}{Y - Y^*} \tag{15.4}$$

which is the same as the number of equilibrium stages calculated by Equation 15.1. In general, the NTU is close to, but not identical to, the equivalent number of equilibrium stages. It will also be shown that the NTU may be defined in a number of different ways, giving potentially different NTU values for the same separation. The NTU could, for instance, be based on the change in the liquid composition instead of the vapor composition. Based on Figure 15.1, the liquid-based NTU is calculated as

$$NTU = \int_{X_B}^{X_T} \frac{dX}{X - X^*}$$

(15.5)

or

$$NTU = \frac{1}{X^* - X} \int_{X_B}^{X_T} dX$$

(15.5a)

or

$$NTU = \frac{X_T - X_B}{X^* - X}$$

(15.5b)

From the geometry at the $Y - X$ diagram in Figure 15.1, it can be seen that, for this simplified case, the liquid-based NTU equals the gas-based NTU (and also equals the equivalent number of equilibrium stages).

Nonparallel, Straight Operating Line and Equilibrium Curve

Figure 15.2 is a $Y - X$ diagram for a simple absorber column section, similar to the process described above, but with nonparallel, straight operating line and equilibrium curve. From the geometry of the diagram, the following relationships, based on similar triangles, may be written:

$$\frac{\left(Y - Y^*\right) - \left(Y_B - Y_B^*\right)}{\left(Y_T - Y_T^*\right) - \left(Y_B - Y_B^*\right)} = \frac{Y_B - Y}{Y_B - Y_T}$$

(15.6)

By substituting the value for $Y - Y^*$ in Equation 15.2 and integrating from $Y = Y_T$ to $Y = Y_B$, the following equation is obtained for the gas-based NTU:

$$(NTU)_G = \frac{Y_B - Y_T}{\left(Y_B - Y_B^*\right) - \left(Y_T - Y_T^*\right)} \ln \frac{Y_B - Y_B^*}{Y_T - Y_T^*}$$

(15.7)

A similar derivation for the liquid-based NTU gives

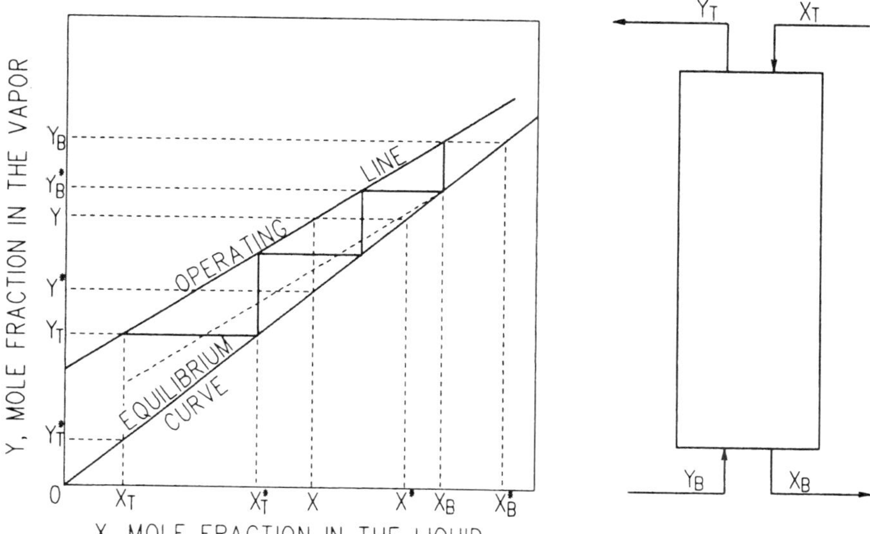

Figure 15.2 Dilute solution absorber (Example 15.1).

$$(NTU)_L = \frac{X_B - X_T}{\left(X_B^* - X_B\right) - \left(X_T^* - X_T\right)} \ln \frac{X_B^* - X_B}{X_T^* - X_T} \tag{15.8}$$

These quantities are calculated numerically in the following example and compared to the equivalent number of equilibrium stages.

Example 15.1 Dilute-Solution Absorber

Component A is absorbed from a gas by a counterflow liquid in a process similar to that depicted in Figure 15.2. The following information is given:

Inlet vapor composition:	Y_B	= 0.006
Outlet liquid composition:	X_B	= 0.0002
Equilibrium relationship:	Y^*	= 25X
Operating line slope:	m	= 20
Number of equilibrium stages:	N	= 3

It is required to calculate the outlet vapor and the inlet liquid compositions and the number of transfer units required for the separation.

The operating line, with the given slope of 20, is written as

$$Y = 20X + b$$

Since the operating line must pass through the point $X_B = 0.0002$, $Y_B = 0.006$, the intercept b is calculated as

$$b = Y_B - 20X_B$$

$$= 0.006 - (20)(0.0002)$$

$$= 0.002$$

The operating line equation is, therefore,

$$Y = 20X + 0.002$$

The equilibrium stage calculations are first performed by alternating between the operating line and the equilibrium curve. This is the analytical equivalent of the McCabe–Thiele graphical construction of equilibrium stages (Chapters 5 and 6).

$$Y_1 = Y_B = 0.006$$

$$X_1 = X_B = 0.0002$$

$$Y_2 = 25X_1 = (25)(0.0002) = 0.005$$

$$X_2 = (Y_2 - 0.002)/20 = (0.005 - 0.002)/20 = 0.00015$$

$$Y_3 = 25X_2 = (25)(0.00015) = 0.00375$$

$$X_3 = (Y_3 - 0.002)/20 = (0.00375 - 0.002)/20 = 0.0000875$$

$$Y_T = 25X_3 = (25)(0.0000875) = 0.0021875$$

$$X_T = (Y_T - 0.002)/20 = (0.0021875 - 0.002)/20 = 0.000009375$$

The calculations are stopped after three equilibrium stages as specified in the problem definition. The equilibrium compositions at the inlet and outlet are

$$Y_B^* = Y_2 = 0.005$$

$$X_B^* = Y_B/25 = 0.006/25 = 0.00024$$

$$Y_T^* = 25X_T = (25)(0.000009375) = 0.000234375$$

$$X_T^* = X_3 = 0.0000875$$

The number of transfer units is now calculated from Equations 15.7 and 15.8:

$$(NTU)_G = \frac{0.006 - 0.0021875}{(0.006 - 0.005) - (0.0021875 - 0.000234375)}$$

$$\ln\frac{0.006 - 0.005}{0.0021875 - 0.000234375} = 2.678$$

$$(NTU)_L = \frac{0.0002 - 0.000009375}{(0.00024 - 0.0002) - (0.0000875 - 0.000009375)}$$

$$\ln\frac{0.00024 - 0.0002}{0.0000875 - 0.000009375} = 3.347$$

The liquid-based number of transfer units is somewhat larger than the number of equilibrium stages, while the gas-based number of transfer units is somewhat smaller. This would be expected since the mean driving force for the vapor, $(Y - Y^*)$, is greater than that for the liquid, $(X^* - X)$.

In general, the number of transfer units must be determined by numerical or graphical integration of Equation 15.3 or 15.5 or some other equivalent form of these equations.

15.2 RATE OF MASS TRANSFER

When counter-current vapor–liquid contacting for the purpose of inducing mass transfer between the phases is carried out in a continuous differential manner as in packed columns, phase equilibrium is not, in general, achieved at any point in the column. In a trayed column, the approach to equilibrium on each tray is determined by the tray efficiency. In a packed column, the separation is controlled by the rate of mass transfer between the phases. The rate of mass transfer determines the column packing height required to achieve a specified separation. Using the concept of transfer units, the column packing height is calculated as the product of the number of transfer units (NTU) and the height of a transfer unit (HTU).

The relationship between the compositions of a given component in the liquid and vapor phases at a given column height is determined by the mass transfer rate of that component from one phase to the other. The mass transfer rate is a function of the resistance in the vicinity of the phase boundary and the composition driving force.

Several models have been proposed to represent the mechanics of mass transfer across the phases. In the two-film theory, the resistance to mass transfer is assumed to take place in a liquid film and a vapor film at the phase boundary interface, as shown schematically in Figure 15.3b. The coordinates X and Y in Figure 15.3a represent the mole fractions of a given component in the liquid and vapor, respectively. For simplicity, component subscripts are dropped and the discussions refer to one typical component. Point (X, Y) on the operating line represents a point at a given column height where the liquid bulk composition is X and the vapor bulk composition is Y. The compositions in each phase and at the phase interface are assumed constant across the column cross section since plug flow is assumed.

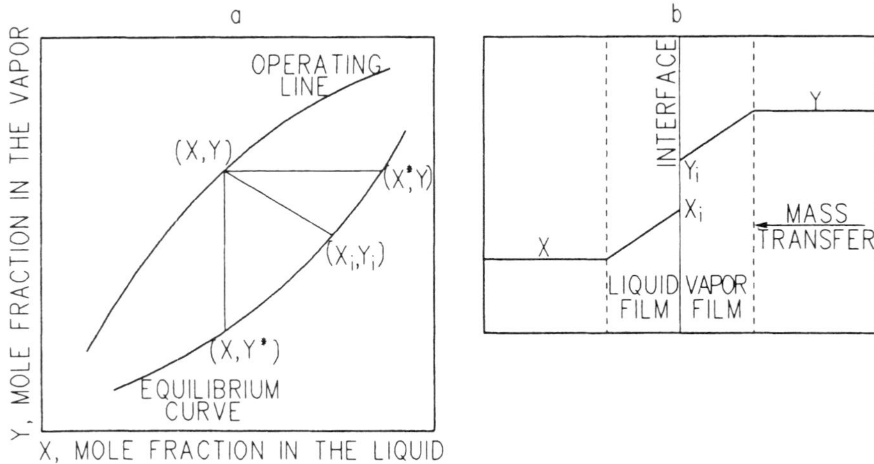

Figure 15.3 Two-film mass transfer model.

A liquid film and a vapor film are assumed to exist at the interface between the two phases, at which point the liquid and vapor compositions are X_i and Y_i. These compositions are, in general, different from the bulk liquid, and vapor compositions and are assumed to correspond to equilibrium compositions. Thus, the point (X_i, Y_i) lies on the equilibrium curve (Figure 15.3a) at the same point in the column height as point (X, Y). The equilibrium curve is obtained from vapor–liquid equilibrium data or correlations, as described in Chapter 1.

Two more points may be defined on the equilibrium curve corresponding to the same point in the column height as (X, Y). Point (X, Y^*) represents the liquid bulk composition, X, and the vapor composition, Y^*, that would be at equilibrium with it. Point (X^*, Y) represents the vapor bulk composition, Y, and the liquid composition, X^*, that would be at equilibrium with it. Note that X^* and Y^* do not exist at the same column height as (X, Y).

Figure 15.3b is a schematic of the variation of the compositions in the vicinity of the phase interface according to the two-film theory. At a given column height, the liquid and vapor compositions, X and Y, are assumed constant, except in the mass transfer films where composition gradients are hypothesized. The compositions at the interface are the equilibrium compositions X_i and Y_i. The compositions shown in Figure 15.3b correspond to a component that is being transferred from the vapor to the liquid.

The rate of mass transfer across a film per unit of interface area is determined by the composition gradient, diffusivity, and film thickness. One way of expressing this relationship is through an empirical mass transfer coefficient that includes the effects of both diffusivity and film thickness. If the compositional driving force is represented as the difference between the bulk composition and the interface composition, the rate of mass transfer in the liquid film is given as

$$N = k_X(X_i - X) \qquad (15.9)$$

and in the vapor film

$$N = k_Y(Y_i - Y) \qquad (15.10)$$

where k_X and k_Y are the mass transfer film coefficients in the liquid and vapor, respectively. The mass transfer rate, N, is the number of moles transferred across the film per unit area per unit time. The dimensions of k_X and k_Y are mole/(mole fraction × time × area). These coefficients are characteristic of the diffusing component and the mass transfer medium. Depending on the way the compositions are defined and on other considerations, the mass transfer rate equation may be expressed in different alternative forms, as presented later in this discussion. Appropriate mass transfer coefficients must be used to correspond with the particular form of the mass transfer rate equation.

Equations 15.9 and 15.10 are empirical with respect to the definition of the mass transfer coefficients, but the form of the equations is based on molecular diffusion theory. Applying the theory to a multicomponent mixture where each component has a distinct diffusivity is impractically complex and must rely on diffusivity data for all the components in the mixture. To derive usable equations from the diffusion theory, certain simplifying assumptions must be made. The basis for the derivation of Equations 15.9 and 15.10 is to assume that mass transfer takes place either as equimolar counter diffusion or as unimolar diffusion under dilute conditions.

In equimolar diffusion, a binary mixture is assumed where the two components diffuse in opposite directions at equal rates. These conditions exist in binary distillation where a mole of component 1 is vaporized for each mole of component 2 that is condensed. In unimolar diffusion, one component diffuses through a second, stagnant one. This is typical of an absorption process where one component diffuses through the gas phase to the interface boundary, is absorbed by the liquid, then diffuses to the bulk of the liquid. The other gas components are assumed to remain in the gas and the liquid components to remain in the liquid.

Note that mass is transferred in the opposite direction of the composition gradient. For instance, for the case described in Figure 15.3 and Equations 15.9 and 15.10, the component in question flows from a higher X_i to a lower X and from a higher Y to a lower Y_i.

Since the rates of mass transfer across the liquid and vapor films must be equal at steady state (by mass balance), Equations 15.9 and 15.10 may be combined to give the following:

$$N = k_X(X_i - X) = k_Y(Y - Y_i) \qquad (15.11)$$

The values of X_i and Y_i are usually unavailable and difficult to measure. It is possible, however, to calculate from vapor–liquid equilibrium correlations the vapor composition, Y^*, that would be at equilibrium with X or the liquid composition,

X^*, that would be at equilibrium with Y. The rate of mass transfer could then be defined in terms of these compositions as follows:

$$N = K_Y(Y - Y^*) = K_X(X^* - X) \qquad (15.12)$$

Since the compositions used in this equation do not represent film coefficients but rather bulk compositions, the mass transfer coefficients, K_Y and K_X, are referred to as overall mass transfer coefficients for the liquid and vapor, respectively. The dimensions of K_X and K_Y are the same as those for k_Y and k_X.

The mass transfer film coefficients (k_X and k_Y) are more directly related to physical properties, such as diffusivities and hydrodynamic conditions, than the overall mass transfer coefficients (k_X and k_Y) and are, therefore, easier to predict and correlate. On the other hand, since the interfacial compositions are difficult to measure or predict, it is more convenient to use overall mass transfer coefficients for process design and analysis. The relationship between the two sets of mass transfer coefficients may be derived by equating different groups in Equations 15.11 and 15.12:

$$K_Y(Y - Y^*) = k_Y(Y - Y_i)$$

This can be rearranged as

$$\frac{1}{K_Y} = \frac{1}{k_Y} \frac{Y - Y^*}{Y - Y_i}$$

$$= \frac{1}{k_Y} \frac{\left(Y - Y_i\right) + \left(Y_i - Y^*\right)}{Y - Y_i}$$

$$= \frac{1}{k_Y} + \frac{1}{k_Y} \frac{Y_i - Y^*}{Y - Y_i}$$

The second term in the last equation is evaluated from Equation 15.11; the resulting equation is

$$\frac{1}{K_Y} = \frac{1}{k_Y} + \frac{1}{k_X} \frac{Y_i - Y^*}{X - X_i} \qquad (15.13)$$

A similar expression is derived for K_X:

$$\frac{1}{K_X} = \frac{1}{k_X} + \frac{1}{k_Y} \frac{X^* - X_i}{Y - Y_i} \qquad (15.14)$$

From Figure 15.3a, the ratio of composition differences in Equation 15.13 and the reciprocal of the ratio of composition differences in Equation 15.14 represent an average slope of the equilibrium curve. If this curve is linear with a constant slope H, Equations 15.13 and 15.14 may be written as

$$\frac{1}{K_Y} = \frac{1}{k_Y} + \frac{H}{k_X} \tag{15.15}$$

and

$$\frac{1}{K_X} = \frac{1}{k_X} + \frac{1}{Hk_Y} \tag{15.16}$$

If the mass transfer coefficients were to be determined experimentally based on their defining equations (Equations 15.11 or 15.12), it is quite clear that the overall coefficients (K_X or K_Y) can be more conveniently determined than the film coefficients (k_X or k_Y). Whereas film coefficients involve almost impossible-to-measure interfacial compositions, overall coefficients require knowledge of bulk compositions and those at equilibrium with them.

The interfacial area is another quantity that is impractical to measure. Therefore, "capacity coefficients" are defined as $(k_X a)$, $(k_Y a)$, $(K_X a)$, and $(K_Y a)$ and are used instead of the mass transfer coefficients. The quantity a is the interfacial area per unit of active column volume. Since both a and the transfer coefficients must be determined in one way or another (usually experimentally), there is no benefit gained from determining them separately. They are, therefore, most commonly reported as groups. In terms of these coefficients, Equations 15.11 and 15.12 are rewritten as

$$Na = (k_X a)(X_i - X) = (k_Y a)(Y - Y_i) \tag{15.17}$$

and

$$Na = (K_X a)(X^* - X) = (K_Y a)(Y - Y^*) \tag{15.18}$$

The quantity Na in Equations 15.17 and 15.18 represents the number of moles transferred between the phases per unit column or packing volume per unit time.

With Equation 15.18 as the defining equation, the mass transfer coefficients in a column are conceptually determined by analyzing samples of liquid and gas for the concentration of the transferred component. The samples are drawn at a given height of the column, which is operated with the fluids (absorbent, gas, and solute) and packing material under study at the designated operating temperature and pressure. The measured quantities represent X and Y in Equation 15.18 and Figure 15.3 (the bulk compositions). The compositions at equilibrium with the bulk compositions, X^* and Y^*, are obtained from the equilibrium curve (Figure 15.3). Additional samples across a small height of packing are analyzed to determine, by material balance, the rate of mass transfer, Na, taking place in the packing volume contained in the incremental packing height.

15.3 MASS TRANSFER IN PACKED COLUMNS

The mass transfer equations discussed above are now combined with a material balance on the transferred component to calculate the column or packing height required for a given separation. The column cross-sectional area, A, is assumed known at this point although in a complete column design, A must be determined based on pressure drop considerations. The column, which is countercurrent flow, is defined as follows:

Column top:
   Liquid in:
   L_T = liquid rate, mole/hr
   X_T = mole fraction of transferred component in L_T
   Gas out:
   V_T = vapor rate, mole/hr
   Y_T = mole fraction of transferred component in V_T

Column bottom:
   Liquid out:
   L_B = liquid rate, mole/hr
   X_B = mole fraction of transferred component in L_B
   Gas in:
   V_B = vapor rate, mole/hr
   Y_B = mole fraction of transferred component in V_B

The column height, h, is measured from the column bottom. The total height, h, at the top is h_T. At any column height h, the liquid and vapor rates and the mole fractions of the transferred component in the liquid and vapor are L, V, X, and Y. In general, these variables change continuously over the height of the column. A material balance on the transferred component is, therefore, carried out over a differential column height, dh. The differential volume is Adh, and the differential mass transfer area is aAdh. The rate of mass transfer in this differential volume is NaAdh. This rate is equated to the rate of change of the transferred component in either phase:

$$NaAdh = d(LX)$$

$$NaAdh = -d(VY)$$

(15.19)

The negative sign indicates that the component is transferred from the vapor to the liquid. Substituting the value of Na from Equation 15.18 into Equations 15.19, the following equations are obtained:

$$d(LX) = (K_X a)(X^* - X)Adh \qquad (15.20)$$

$$-d(VY) = (K_Y a)(Y - Y^*)Adh \tag{15.21}$$

These equations are now rearranged to calculate the required column height, h_T, by integrating over the specified composition change, X_B to X_T or Y_B to Y_T, as follows:

$$h_T = \int_0^{h_T} dh = \int_{X_B}^{Y_T} \frac{d(LX)}{(K_Y a)A(X^* - X)} \tag{15.22}$$

and

$$h_T = \int_0^{h_T} dh = -\int_{X_B}^{Y_T} \frac{d(GY)}{(K_Y a)A(Y - Y^*)} \tag{15.23}$$

In equimolar counter diffusion or unimolar dilute diffusion, V and L are assumed constant, allowing Equations 15.22 and 15.23 to be written as

$$h_T = \frac{L}{(K_X a)A} \int_{X_B}^{X_T} \frac{dX}{X^* - X} \tag{15.24}$$

and

$$h_T = \frac{V}{(K_Y a)A} \int_{Y_T}^{Y_B} \frac{dY}{Y - Y^*} \tag{15.25}$$

The quantities before the integrals and the integrals themselves are referred to as the height of a transfer unit (HTU) and the number of transfer units (NTU), respectively. Thus, the column height calculation is reduced to evaluating the product of two quantities, each of which has a certain physical significance. The HTU has the dimension of length and expresses the efficiency of the device in facilitating mass transfer. It is a function of the vapor and liquid flow rates and the mass transfer characteristics of the system. Shorter HTUs represent higher mass transfer efficiencies. The number of transfer units (NTU), as indicated in Section 15.1, reflects the relationship between the operating line and the equilibrium curve. It is a dimensionless number, conceptually equivalent to the ratio of the required change in the vapor or liquid composition to the mean compositional driving force.

The HTU is determined from experimental data or empirical correlations, and the NTU is related to equilibrium and operating data. Since both HTU and NTU depend on the particular definition of mass transfer coefficients and compositions in the mass transfer equations, it is important to use compatible pairs of HTU and NTU. For instance, Equations 15.24 and 15.25 are rewritten as

$$h_T = (HTU)_{OG}(NTU)_{OG} \tag{15.26}$$

$$h_T = (HTU)_{OL}(NTU)_{OL} \tag{15.27}$$

where the subscripts OG and OL indicate overall gas and overall liquid, respectively. It is obviously incorrect to calculate the column height as, for example, $(HTU)_{OG}(NTU)_{OL}$. If film coefficients were used in the column height derivation instead of overall coefficients, i.e., if Equation 15.17 were used instead of Equation 15.18 as the starting point, the column height would be calculated as

$$h_T = (HTU)_G(NTU)_G \tag{15.28}$$

or

$$h_T = (HTU)_L(NTU)_L \tag{15.29}$$

where subscripts G or L refer to gas film or liquid film, respectively.

The following example is based on Example 15.1, where the NTUs were calculated from operating and equilibrium data. With additional experimental data, it is possible to determine HTUs, HETP, and mass transfer coefficients.

Example 15.2 Column Packing Characteristics from Experimental Data

In Example 15.1 and Figure 15.2, additional data are given, including a column diameter of 3 ft, a column height of 10 ft, and a total gas rate of 500 lbmole/hr. Calculate HTU, HETP, and the mass transfer coefficients.

The total liquid rate is calculated from a material balance on the transferred component:

$$V(Y_B - Y_T) = L(X_B - X_T)$$

or

$$L = V\frac{Y_B - Y_T}{X_B - X_T}$$

Using data from Example 15.1,

$$L = \frac{(500)(0.006 - 0.0021875)}{0.0002 - 0.000009375}$$

$$= 10000 \text{ lbmole / hr}$$

The number of transfer units calculated in Example 15.1 is based on overall concentration gradients, $(X^* - X)$ and $(Y - Y^*)$, and should be designated as OL and OG:

$$(NTU)_{OL} = 3.347$$

$$(NTU)_{OG} = 2.678$$

The HTUs are calculated directly as

$$(HTU)_{OL} = 10/3.347 = 2.988 \text{ ft}$$

$$(HTU)_{OG} = 10/2.678 = 3.734 \text{ ft}$$

The HETP is calculated from the number of equilibrium stages:

$$HETP = 10/3 = 3.333 \text{ ft}$$

From the definition of HTU in Equations 15.24 and 15.25:

$$(HTU)_{OL} = \frac{L}{(K_X a)A}$$

$$(HTU)_{OG} = \frac{V}{(K_Y a)A}$$

The column cross-sectional area is

$$A = \pi(3/2)^2 = 7.07 \text{ ft}^2$$

The mass transfer coefficients are

$$(K_X a) = \frac{L}{(HTU)_{OL} A}$$

$$= \frac{10000}{(2.988)(7.07)}$$

$$= 473 \text{ lbmole}/\text{ft}^3 \text{hr}$$

$$(K_Y a) = \frac{V}{(HTU)_{OG} A}$$

$$= \frac{500}{(3.734)(7.07)}$$

$$= 18.9 \text{ lbmole}/\text{ft}^3 \text{hr}$$

The mass transfer coefficients may also be calculated based on overall column performance, using Equation 15.18. The composition differences are replaced with logarithmic means of the composition differences at both ends of the column:

$$K_x a = \frac{Na}{\left(X^* - X\right)_{lm}}$$

$$K_Y a = \frac{Na}{\left(Y - Y^*\right)_{lm}}$$

where lm designates logarithmic mean.

$$X_B = 0.0002$$

$$X_B^* = 0.00024$$

$$X_B^* - X_B = 0.00024 - 0.0002 = 0.00004$$

$$X_T = 0.000009375$$

$$X_T^* = 0.0000875$$

$$X_T^* - X_T = 0.0000875 - 0.000009375 = 0.000078125$$

$$Y_B = 0.006$$

$$Y_B^* = 0.005$$

$$Y_B - Y_B^* = 0.006 - 0.005 = 0.001$$

$$Y_T = 0.0021875$$

$$Y_T^* = 0.000234375$$

$$Y_T - Y_T^* = 0.0021875 - 0.000234375 = 0.001953125$$

$$\left(X^* - X\right)_{lm} = (0.000078125 - 0.00004)/\ln(0.00078125/0.00004)$$

$$= 0.00005695$$

$$\left(Y - Y^*\right)_{lm} = (0.001953125 - 0.001)/\ln(0.001953125/0.001)$$

$$= 0.001424$$

The mass transfer rate per unit column volume, Na, is calculated from the mass transfer rate and the column volume. The mass transfer rate is calculated from the inlet and outlet compositions:

$$\text{Rate} = V\left(Y_B - Y_T\right) = L\left(X_B - X_T\right)$$

$$= 500(0.006 - 0.0021875) = 1.90625 \text{ lbmole/hr}$$

$$= 10000(0.0002 - 0.000009375) = 1.90625 \text{ lbmole/hr}$$

The column volume = (7.07)(10) = 70.7ft^3

$$\text{Na} = 1.90625/70.7 = 0.02696 \text{ lbmole/ft}^3\text{hr}$$

The mass transfer coefficients are

$$K_X a = 0.02696/0.00005695 = 473 \text{ lbmole/ft}^3\text{hr}$$

$$K_Y a = 0.02696/0.001424 = 18.9 \text{ lbmole/ft}^3\text{hr}$$

The results agree exactly with the results calculated from the HTU values because both the operating line and the equilibrium curve are straight lines.

15.4 GENERAL PACKED COLUMN DESIGN

While the mass transfer equations developed in the last section may claim roots in molecular diffusion principles, they are limited in their applicability in many ways. A basic condition for their validity is the assumption of equimolar or dilute unimolar mass transfer.

The applicability of Equations 15.22 and 15.23 could be expanded by keeping more terms inside the integral and carrying out the integration numerically. Another technique is to express liquid and gas rates on a solute-free basis, whereupon they could be assumed constant over broader operating conditions. Compositions in this case would be expressed as mole ratios instead of mole fractions.

Such modifications do not, however, address other limitations that are inherent in the NTU and HTU approach. The object of these methods is primarily absorbers, and a limited class of absorbers at that. The process model is essentially a ternary with only one component crossing the phase boundary. Although packed columns were originally designed mostly as absorbers, other multistage processes, such as vacuum distillation, often use packed columns. Under certain circumstances, packed columns have an edge over trayed columns because of lower pressure drops and lower cost. The transfer units methodology is inadequate for handling general multicomponent separation processes.

Another shortcoming of the transfer units technique is its exclusion of energy balances or temperature calculations. The processes are assumed isothermal, an assumption that could incur gross errors in calculating absorbers or other multistage processes.

Because of these drawbacks, packed columns are most commonly designed using the height equivalent to a theoretical plate (HETP) concept as described in the introduction to this chapter. With each HETP treated as an equilibrium stage, the same computational methods used for trayed columns could be used for packed columns. The problem of calculating the required column packing height, or of predicting the performance of a column with a given packing height, is thus reduced to estimating the HETP.

Having determined the column packing height on the basis of equilibrium calculations and whatever mass transfer correlations, the capacity of the column to handle the required vapor and liquid flows hydraulically must be checked. The specific items of concern in this regard are the pressure drop across the packing and the tendency of the column to flood. Based on these considerations, the column diameter is calculated.

15.4.1 Estimating the HETP

The HETP depends on the type and size of the packing, the column dimensions, the fluid properties, the operating conditions, etc. It also depends on whether the packing is dumped or ordered. The HETP is usually supplied by the packing vendors or estimated based on experimental data or empirical correlations.

Bolles and Fair (1979) estimate the HETP from vapor and liquid heights of transfer units (HTU). The vapor phase $(HTU)_G$ and liquid phase $(HTU)_L$ are first calculated separately from empirical correlations. The overall HTU is then calculated as follows:

$$(HTU)_{OG} = (HTU)_G + (mV/L)(HTU)_L \qquad (15.30)$$

where m is the slope of the equilibrium curve (Y^*/X), and V and L are the vapor and liquid flow rates. The HETP is calculated from the overall HTU using the equation

$$HETP = (HTU)_{OG} \frac{\ln(mV/L)}{(mV/L) - 1} \qquad (15.31)$$

Equations 15.30 and 15.31 are derived based on the assumption of constant operating line and equilibrium curve slopes (Vital et al., 1984).

The HETP correlations are described in detail by Vital et al. (1984), who also provide charts for estimating packing parameters to be used in these correlations. The same article contains HETP values for a variety of packing types and fluids as well as a number of rules of thumb for estimating the HETP. According to the simplest of these rules, the HETP is set equal to the column diameter. The authors recommend using the most conservative among the HETP values obtained from available data, correlations, or rules of thumb.

15.4.2 Packed Column Capacity

Multistage separation in packed columns is achieved by contacting counter-flowing vapor and liquid through the packing. Under favorable conditions, the liquid wets the packing and forms the dispersed downward flowing phase, while the continuous vapor phase flows upward around the liquid-covered packing.

As the vapor flows upward, a pressure drop develops in the vapor phase in the direction of vapor flow as a result of friction against the packing and the falling liquid. The liquid flows down by gravity against this pressure drop, resulting in some steady-state liquid-to-vapor-flow ratio, L/V.

A packed column (or a trayed column) designed for a given service is usually expected to operate over a range of L/V ratios in anticipation of variations in separation requirements, feed flows and compositions, and other factors.

If the vapor rate is increased, the pressure drop increases, which raises the resistance to the liquid flow. At a certain vapor rate, the pressure drop rises to a point where liquid flow is practically blocked and liquid fills the column, resulting in a condition known as flooding. The column ceases to function as a separation device; a larger diameter column is required to handle these flows.

Below certain vapor rates and pressure drops, channeling takes place, where the vapor and liquid pass each other with no appreciable contacting. To prevent this from happening, a smaller column diameter should be selected.

In general, columns should be designed such that pressure drops are maintained between 0.01 inch of water per foot of packing for vacuum columns and 1 inch of water per foot of packing for atmospheric fractionators. Columns should be operated above 50% and below 90% of the flooding vapor rate.

The pressure drop and flooding rate are functions of the packing type and size, the column diameter, and the vapor and liquid flow rates and properties. These rates and properties may change with the column height, resulting in different calculated diameters. The column should be designed with the most conservative (largest) diameter, although it should be checked for excessively low pressure drops that could cause channeling. The packing type and size are characterized by a packing factor, F, which has the dimension of area-per-unit volume of packing.

Figure 15.4 shows the Eckert (1975) pressure drop correlation, which may also be used to check the approach to flooding conditions. The variables in the coordinates are defined as follows: L and V are the liquid and vapor flow rates, lbmole/hr, G is the vapor mass velocity, lb/ft^{2}sec, ρ_L and ρ_V are the liquid and vapor densities, lb/ft^3, F is the packing factor, ft^2/ft^3, and μ_L is the liquid viscosity, centipoise. The column diameter is implied since G is the vapor rate per unit column cross-sectional area. The pressure drop, in inches of water per foot of packing, is reported as a parameter in this correlation. Flooding can be expected to occur at any point above the pressure drop curve of 1.5 inch of water per foot of packing.

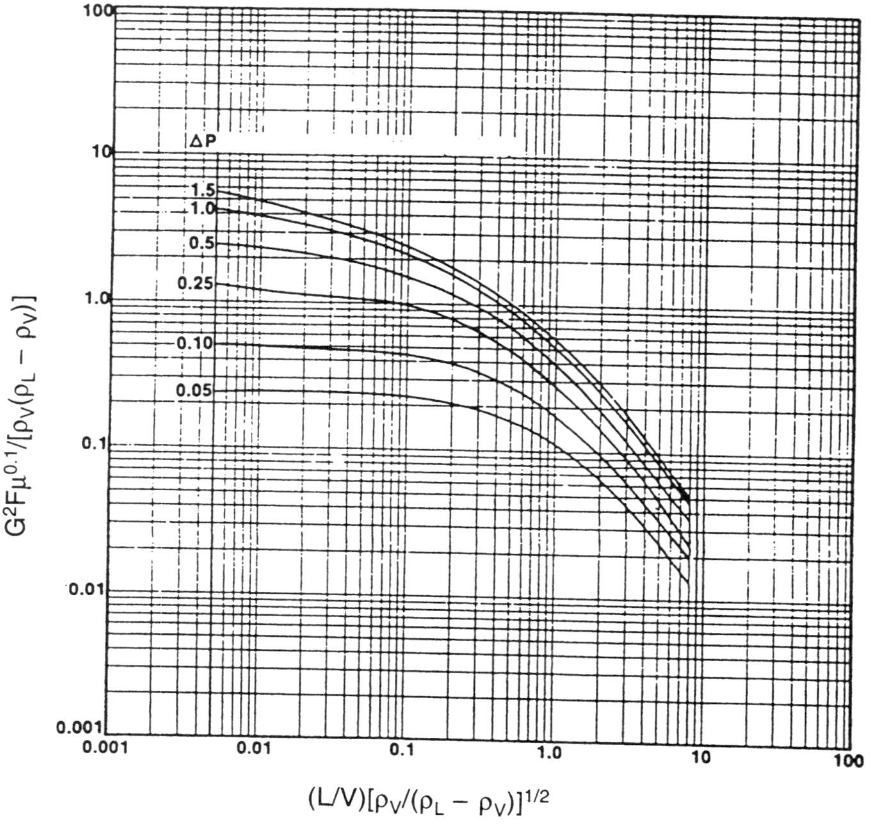

$(L/V)[\rho_V/(\rho_L - \rho_V)]^{1/2}$

Figure 15.4 Packed column pressure drop correlation. (Reprinted with permission from J. S. Eckert, *Chem. Eng.*, 82(8), 70, 1975.)

15.4.3 Packed Column Design Outline

Designing a column usually starts with heat and material balance calculations, preferably using a computer simulation program. These calculations determine the liquid and vapor flow rates and the number of equilibrium stages required to meet the designed performance specifications (separation, recovery, etc.). Fluid properties, such as densities and viscosities, may also be generated by the computer program.

Next, the packing material type and size are selected. Among the different types of packing material available, of which mesh, rings, and saddles are the most common, the type most suited for a particular application is best determined based on vendor information. Once a decision has been made regarding the packing type, its nominal size should be selected. Smaller sizes tend to give lower HETPs at the expense of higher pressure drops across the packing, while larger sizes can induce channeling. A good starting point is a packing size about one tenth the column diameter or smaller. Since the column diameter is yet to be determined, an assumed value is used for selecting the initial packing size.

The pressure drop/flooding rate correlation, Figure 15.4, is next used to determine G, the vapor mass rate per unit column cross-sectional area. The information needed includes the packing factor of the selected packing material, the L/V ratio and fluid properties (vapor and liquid densities and liquid viscosity), and a pressure drop within the acceptable range. The column cross-sectional area and diameter are then calculated from G and V. These calculations should be carried out at different column heights since the vapor and liquid rates and properties may vary.

If the resulting ratio of packing size to column diameter is reasonable, an appropriate column diameter has been determined. Otherwise, the calculations are repeated with a new packing size.

The next step is to determine the column packing height. The HETP is estimated using any of the methods described in Section 15.4.1. The column packing height is simply the number of equilibrium states multiplied by the HETP.

Packed Columns vs. Trayed Columns

Many considerations influence the choice between a packed column and a trayed column for a particular application, foremost among which are economics and performance versatility. The decision involves many variables and is best handled on a case-by-case basis. The following example compares the performance of the same column structure, using either trays or packing.

Example 15.3 Debutanizer Column — Packing vs. Trays

An existing column is to be used for separating butanes from a mixture of butanes, pentanes, and other hydrocarbons. The column is required to recover 98% of the butanes in the overhead product with a combined butanes molar purity of 94.5%. Table 15.1 shows stream compositions and conditions as obtained from simulation.

Table 15.1 Stream Data for Example 15.3

Rates, lbmole/hr	Feed	Overhead	Bottoms
C3	5.00	5.00	0.00
IC4	35.00	34.65	0.35
NC4	85.00	82.95	2.05
IC5	25.00	1.50	23.50
NC5	25.00	0.35	24.65
NC6	25.00	0.00	25.00
Total	200.00	124.45	75.55
Temperature, °F	195.00	173.00	276.00
Pressure, psia	155.00	148.00	154.00

The column is 3 feet in diameter and can fit 22 trays with 24-inch spacing between the trays. With an estimated column overall efficiency of 65%, the number of theoretical stages is

$$(22)(0.65) + 1 \text{ condenser} + 1 \text{ reboiler} = 16$$

The feed is introduced around the middle of the column. With 16 theoretical stages and in order to meet the butanes recovery and purity specifications, the required reflux ratio is computed by simulation at 2.75.

The same column could be packed to give a packing height equal to the height between the top and bottom trays:

$$\text{Packing height} = (22 - 1)(2) = 42 \text{ ft}$$

Spaces above and below the packing are reserved for the reflux and boilup distributors. The feed location is not changed from the trayed column design.

The packing material under consideration consists of 1-inch Pall rings. The estimated HETP is 18 inches and the packing factor, $F = 52$ ft^2 per cubic foot of packing (Vital et al., 1984; Eckert, 1975). The equivalent number of equilibrium stages, including the condenser and reboiler, is

$$(42)(12)/18 + 2 = 30$$

To meet the same butanes recovery and purity specifications, the required reflux ratio with 30 equilibrium stages is calculated by simulation at 1.95.

For the purpose of estimating column pressure drops, vapor and liquid flow rates are estimated from the reflux ratio. Calculations are performed at two points in the column: one in the rectifying section and the other in the stripping section.

Trayed Column

For the trayed column, the reflux ratio is 2.75 and the overhead rate is 124.45 lbmole/hr. The liquid flow in the rectifying section is calculated as

$$L_r = (2.75)(124.45) = 342.2 \text{ lbmole/hr}$$

The feed is introduced to the column as a liquid slightly below its bubble point; so the liquid flow in the stripping section is estimated as

$$L_s = L_r + F$$

$$= 342.2 + 200.00$$

$$= 542.2 \text{ lbmole / hr}$$

The boilup rate is estimated as

$$V_s = 542.2 - \text{bottoms rate}$$

$$= 542.2 - 75.55$$

$$= 466.65 \text{ lbmole / hr}$$

With the feed all liquid, the vapor rate in the rectifying section is assumed the same as in the stripping section:

$$V_r = V_S = 466.65 \text{ lbmole/hr}$$

Using the methods discussed in Chapter 14, the average pressure drop predicted by simulation is 0.015 psi per tray, giving a total column pressure drop of $(0.015)(22) = 0.33$ psi.

Packed Column

The packed column internal flows are estimated in a similar manner as the trayed column, based on a reflux ratio of 1.95:

$$L_r = (1.95)\ (124.45) = 242.7 \text{ lbmole/hr}$$

$$L_S = 242.7 + 200.0 = 442.7 \text{ lbmole/hr}$$

$$V_S = 442.7 - 75.55 = 367.15 \text{ lbmole/hr}$$

$$V_r = V_S = 367.15 \text{ lbmole/hr}$$

The following average fluid properties are given. For the rectifying section:

$$(MW)_v = 60$$

$$\rho_L = 30.1 \text{ lb/ft}^3$$

$$\rho_V = 2.12 \text{ lb/ft}^3$$

$$\mu_L = 0.24 \text{ centipoise}$$

The vapor mass velocity:

$$G = (367.15)(60)\big/\left[\pi\left(3^2/4\right)\right]$$

$$= 3116.5 \text{ lb} / \text{ft}^2\text{hr}$$

For the stripping section:

$$(MW)_v = 76$$

$$\rho_L = 34.5 \text{ lb/ft}^3$$

$$\rho_V = 2.40 \text{ lb/ft}^3$$

$$\mu_L = 0.17 \text{ centipoise}$$

The vapor mass velocity:

$$G = (367.15)(76)\big/\left[\pi\left(3^2/4\right)\right]$$

$$= 3947.5 \text{ lb} / \text{ft}^2\text{hr}$$

The pressure drop in the packed column is determined from Figure 15.4. In the rectifying section,

$$\text{Abscissa} = \frac{L}{V}\left(\frac{\rho_v}{\rho_L - \rho_V}\right)^{1/2}$$

$$= \frac{242.7}{367.15}\left(\frac{2.12}{30.1 - 2.12}\right)^{1/2}$$

$$= 0.182$$

$$\text{Ordinate} = \frac{G^2 F \mu_L^{0.1}}{\rho_V(\rho_L - \rho_V)}$$

$$= \frac{(3116.5)^2(52)(0.24)^{0.1}}{(2.12)(30.1 - 2.12)(3600)^2}$$

$$= 0.57$$

The corresponding pressure drop is 0.22 inch of water per foot of packing. In the stripping section,

$$\text{Abscissa} = \frac{442.7}{367.15}\left(\frac{2.40}{34.5 - 2.40}\right)^{1/2}$$

$$= 0.330$$

$$\text{Ordinate} = \frac{(3947.5)^2(52)(0.17)^{0.1}}{(2.40)(34.5 - 2.40)(3600)^2}$$

$$= 0.68$$

The corresponding pressure drop is 0.35 inch of water per foot of packing. These pressure drops are well below flood conditions, yet not so low as to cause channeling. The total packing pressure drop is

$$(21 \text{ ft})(0.22 \text{ in. } H_2O/\text{ft}) + (21 \text{ ft})(0.35 \text{ in. } H_2O/\text{ft}) = 11.97 \text{ in. } H_2O$$
$$= (11.97)(0.0361) = 0.432 \text{ psi}$$

This is comparable to the trayed column pressure drop.

In conclusion, this case study indicates that replacing trays with Pall rings packing in this particular column reduces by about 30% the reflux ratio required to achieve the same separation. The corresponding energy savings would have to be weighed against the cost of the packing and its installation.

It is also important to check the column performance at off-design conditions to determine the operable ranges. The above calculations could be repeated for different liquid and vapor rates to check for excessive or inadequate pressure drops.

NOMENCLATURE

a	Interfacial area per unit volume of packing	K	Overall mass transfer coefficient
A	Column cross-sectional area	L	Liquid rate in the column
F	Packing factor	MW	Molecular weight
g_C	Acceleration of gravity	N	Number of equilibrium stages
G	Vapor mass velocity in the column	N	Number of moles transferred per unit area per unit time
h	Column height	NTU	Number of transfer units
HETP	Height equivalent to a theoretical plate	V	Vapor rate in the column
		X	Mole fraction in the liquid
HTU	Height of a transfer unit	Y	Mole fraction in the vapor
		μ	Viscosity
k	Mass transfer coefficient	ρ	Density

SUBSCRIPTS

B	Column bottom	OG	Overall gas
G	Gas or vapor phase	OL	Overall liquid
i	Interfacial conditions	r	Rectifying section
L	Liquid phase	s	Stripping section
lm	Logarithmic mean	T	Column top

SUPERSCRIPTS

*   Equilibrium composition

REFERENCES

Bolles, W. L. and J. R. Fair, *I. Chem. Eng. Symp. Series, No. 56*, 1979.
Eckert, J. S., *Chem. Eng.*, 82(8), 70, 1975.
Vital, T. J., S. S. Grossel, and P. I. Olsen, *Hydrocarbon Processing*, December 1984, p. 75.

PROBLEMS

15.1 In the absorber column of Example 15.1, calculate the liquid rate per 100 lbmole/hr gas. For the same inlet gas and liquid compositions, calculate the outlet compositions if the liquid rate is halved. Determine the number of transfer units.

15.2 A 20 lbmole/hr effluent stream that is mostly air contains 0.02 mole fraction acetone. Prior to releasing it to the atmosphere the acetone content must be lowered to 100 ppm mole as a maximum. It is proposed to accomplish this by countercurrent water scrubbing in a 3-ft-diameter packed column. The scrubbing water is treated and recycled at the rate of 600 lbmole/hr with a residual acetone content of 10 ppm mole. Calculate the required packing height if the mass transfer coefficient $K_Y a = 1.75$ lbmole/ft^3-hr. The equilibrium acetone mole fraction is given by $Y^* = 1.67X$.

15.3 The sulfur dioxide content of a gas stream is lowered by 90% by scrubbing with water in a countercurrent packed column. The column is 4 ft in diameter and 16 ft in height. The gas rate is 500 lbmole/hr, and the inlet sulfur dioxide concentration is 2% mole. The column was designed such that the sulfur dioxide concentration in the bottoms does not exceed 0.04% mole. Determine the required water rate, the number of gas-phase transfer units, the height of the gas-phase transfer unit, and the mass transfer coefficient $K_Y a$. The equilibrium relationship for sulfur dioxide is given by $Y^* = 38X$.

15.4 In order to check the operable range of the debutanizer column of Example 15.3, its performance is evaluated at various reflux ratios. Check for the possibility of flooding or channeling at reflux ratios of 1.0 and 3.0.

CHAPTER 16

Tray Efficiency

Vapor–liquid multistage columns are commonly studied in terms of equilibrium stages, where the vapor and liquid phases are assumed to mix perfectly and achieve complete interaction with each other. A long-enough residence time is assumed on the stage to allow the phases to separate and leave the stage at the same temperature and with equilibrium compositions.

These assumptions must be reevaluated when applying the model to real columns. A column tray can only approach, to varying degrees, the equilibrium stage. This chapter examines tray efficiency and its role in relating actual trays to equilibrium stages.

16.1 TYPES OF TRAY EFFICIENCY

Although several different expressions are used for defining it, tray efficiency in general is a parameter that relates the actual performance of a column tray to a theoretical stage. Some of the reasons that actual trays do not achieve the separation of theoretical stages were discussed in Chapter 14. They include such factors as incomplete contact between the phases, not enough residence time, vapor entrainment, liquid entrainment, weeping, leakage, and poor vapor distribution. Tray efficiency suffers a steep decline at flood conditions.

Tray efficiency is also a function of the basic tray hydraulics model. For instance, in a crossflow tray the liquid is assumed to move across the tray in a plug flow manner with little backmixing. The vapor is assumed to mix and reach a uniform composition in the tray below before it bubbles through the liquid. As the liquid flowing across the tray interacts with the rising vapor, its composition changes progressively along the width of the tray between the downcomer and the overflow weir. Since the rising vapor interacts with a changing liquid composition, its composition as it leaves the liquid also changes across the tray width but is assumed to undergo complete mixing before it reaches the tray above. Even an ideal crossflow tray does not perform as an equilibrium stage where the vapor leaving the stage is at equilibrium with the liquid leaving the stage. In a crossflow tray all the vapor does not contact all the liquid.

Since, in general, the liquid and vapor compositions vary across the vapor–liquid interface, a point efficiency is defined in terms of compositions at a given point on the tray. The Murphree point efficiency in vapor terms is defined as (Murphree, 1925)

$$E_V = \frac{Y - Y_{j+1}}{Y^* - Y_{j+1}} \qquad (16.1)$$

where Y_{j+1} is the mole fraction of a given component in the mixed vapor coming from the tray below, designated as $j + 1$, and entering the tray in question, designated as j. Y is the actual mole fraction of the same component in the vapor at a given point on tray j. Y^* is the vapor mole fraction of that component at equilibrium with X, the mole fraction of the component in the liquid at the same point on tray j. The point efficiency is thus defined as the ratio of the actual change in the vapor concentration of a component as it passes through a tray at a given point to the change that would occur if the vapor reaches equilibrium with the liquid. Note that the efficiency is defined for a particular component, implying that different components could have different efficiencies. This reflects the different rates of mass transfer for each component.

The Murphree point efficiency may be defined in liquid terms in analogy to Equation 16.1:

$$E_L = \frac{X_{j-1} - X}{X_{j-1} - X^*} \qquad (16.2)$$

where X_{j-1} is an average mole fraction of a given component in the liquid flowing to tray j from the tray above, designated as $j - 1$. X is the actual concentration of the liquid at a given point on tray j, and X^* is the liquid concentration at equilibrium with Y, the vapor concentration at that point.

The Murphree tray efficiency relates to the tray as a whole and is defined as the ratio of the actual change in a component vapor concentration as it flows through the tray to the change that would occur at equilibrium conditions. The Murphree tray efficiency is expressed in vapor terms as

$$E_{MV} = \frac{Y_j - Y_{j+1}}{Y_j^* - Y_{j+1}} \qquad (16.3)$$

where Y_{j+1} is the mole fraction of a component in the vapor coming from the tray below, designated as $j + 1$. Y_j is the actual mole fraction of the same component in the vapor leaving tray j. Y_j^* is the vapor mole fraction of that component at equilibrium with X_j, the mole fraction of the component in the liquid leaving tray j. The Murphree tray efficiency may also be expressed in liquid terms as

$$E_{ML} = \frac{X_{j-1} - X_j}{X_{j-1} - X_j^*} \qquad (16.4)$$

with similar definitions for the liquid mole fractions. As with point efficiencies, the Murphree tray efficiency is defined specifically for a given component and may have different values for different components. The Murphree tray efficiency is applicable under the assumption that the liquid phase and the vapor phase are completely mixed on each tray. That is, the exiting vapor and liquid phases, although not at equilibrium with each other, are each assumed to have a uniform composition.

Most commonly used in engineering practice is the overall column tray efficiency. It relates the number of theoretical stages in a column or column section to the number of actual trays that would be required to achieve a similar separation. The overall efficiency is used to translate actual trays to theoretical stages and vice versa in the analysis of the performance of existing columns and in the design of new columns. Since the degree of separation depends on reflux ratio or liquid-to-vapor flows in the column as well as on the number of stages, it is important when applying overall efficiency to ensure that other factors are comparable. The overall column efficiency is defined as

$$E_0 = \frac{N}{N_A} \qquad (16.5)$$

where N is the number of theoretical stages, and N_A is the number of actual stages.

16.2 THEORETICAL MODEL

The theoretical derivation of the overall efficiency on the basis of fluid properties and tray dimensions consists of the following steps (A.I.Ch.E Tray Efficiency Research Program):

1. Predict Murphree point efficiencies from mass transfer rates in the vapor and liquid, taking tray dimensions into account.
2. Assume a mathematical model to represent liquid mixing on the tray and use the model to predict Murphree tray efficiency from the point efficiency.
3. Correct the Murphree tray efficiency to account for entrainment (Equation 14.2).
4. Use the corrected Murphree tray efficiency to calculate overall column efficiency.

This theoretical derivation is quite involved and outside the scope of this book. The reader is referred to the original reference for details. Due to the complexity and somewhat erratic nature of tray hydraulics, the accuracy of the resulting equations is not universal, and therefore their usefulness is limited to systems for which the equations were tested. The methods may be used for scaling up small column laboratory data to large columns.

The more common approach for estimating overall column efficiencies is to use empirical methods based on large amounts of experimental data.

16.3 EMPIRICAL METHODS

The overall efficiency of an existing column is determined directly from experimental data obtained from the column. The data collected include the actual number of trays, rates and compositions of the feeds and products, the column pressure, and reflux ratio or internal liquid-to vapor-flow ratios. The theoretical number of stages required to achieve the measured separation of two key components at the column conditions is then calculated using a rigorous method (Chapter 13). Since most of these methods assume a known number of theoretical stages, this number must be determined by trial and error using multiple calculations, each with a different number of theoretical stages. The calculations can be conveniently carried out on a computer simulator where the number of stages and other parameters may be varied to generate the desired cases. Once the experimental data are matched with a particular number of theoretical stages, the overall efficiency, E_O, is calculated from Equation 16.5.

The efficiency may depend on the two components selected as the keys because, although the rigorous calculations can closely match measured key component compositions, other components may be off due to inaccuracies in the theoretical column model. Also, since the results of the column calculations depend on the thermodynamic properties prediction methods, the efficiency may also depend on the selection of these methods.

Column efficiency experimental data are usually compiled and saved and are used for estimating efficiencies of similar columns or for developing empirical correlations for predicting column tray efficiencies. When attempting to estimate one tray efficiency based on data available for another column, it is important to be aware of the factors that influence tray efficiency the most and take them into consideration. For instance, the tray dimensions should be comparable. Of particular importance are the tray diameter, tray spacing, number of passes, and liquid level on the tray.

The fluid properties should also be considered, among which are the viscosity, relative volatility, and surface tension. Lower viscosities are usually associated with higher efficiencies. It follows that columns operated at higher pressures can be expected to have higher efficiencies since higher pressures imply higher temperatures and lower viscosities. Also, columns for distilling narrow-boiling mixtures usually have higher efficiencies because most of the components are close to their boiling point and therefore have low viscosities. By contrast, columns used in absorption, stripping, and vacuum distillation have low efficiencies because many of the components in the liquid phase are far removed from their boiling points and therefore tend to have higher viscosities.

Relative volatilities also have a considerable effect on column efficiencies. A component with a high relative volatility has a low solubility in the liquid phase. This means high resistance to mass transfer in the liquid, which tends to lower the efficiency in general.

Surface tension influences efficiency through its effect on foaming. Foamability becomes a factor when it reaches levels that could increase entrainment, which in turn leads to a reduced efficiency.

Trends such as these have led to the development of empirical correlations for predicting column efficiencies. The Drickamer–Bradford (1943) correlation is based on data from numerous refinery fractionation columns. It correlates the overall efficiency E_O to a single parameter: the liquid average viscosity, μ_L, evaluated for the feed at average column temperature. The correlation is reproduced in Figure 16.1. Although it was developed primarily for hydrocarbon systems, it was also found to predict nonhydrocarbon column efficiencies fairly well. However, care should be exercised when using the correlation for fluid types, operating conditions, and tray types or dimensions that are different from those for which the correlation was developed (Drickamer and Bradford, 1943). Also, the correlation should be applied only to relatively narrow-boiling mixtures because it does not account for the effect of relative volatilities.

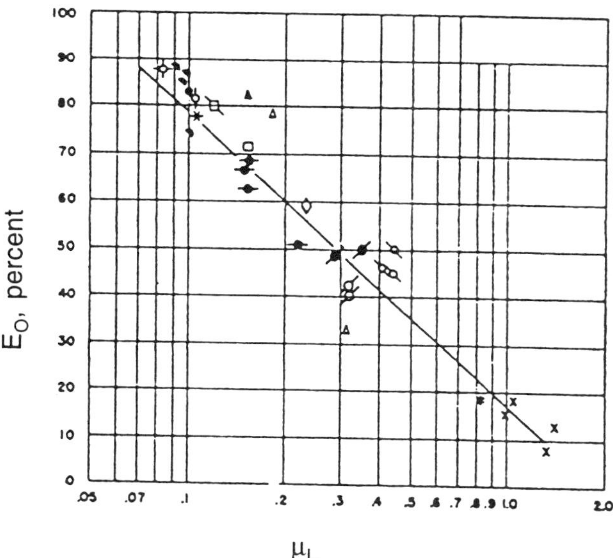

Figure 16.1 Drickamer–Bradford correlation. (Reprinted with permission from H. G. Drickamer and J. R. Bradford, *Trans. A.I.Ch.E.*, 39, 319, 1943.)

To include the effect of relative volatilities, O'Connell (1946) used the product of viscosity and relative volatility as the correlating parameter for predicting overall efficiencies. As in the Drickamer–Bradford correlation, the correlating viscosity, μ_L, is defined as the average liquid viscosity of the feed, calculated at the average column

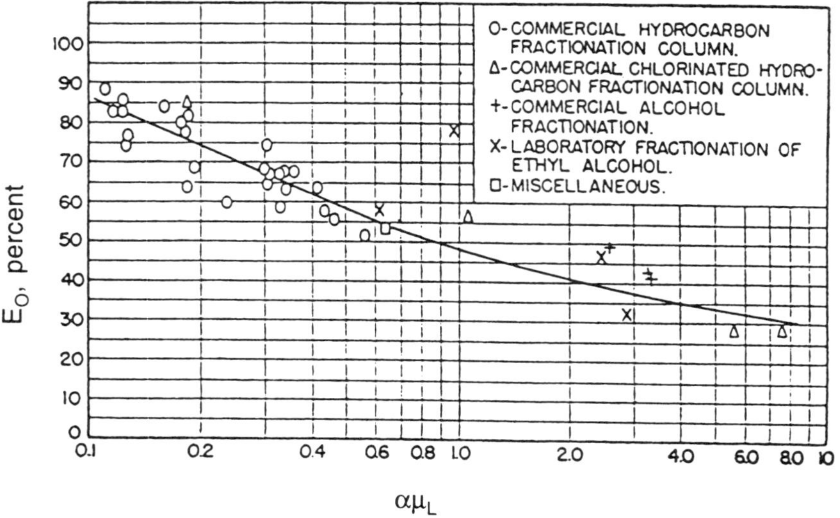

Figure 16.2 O'Connell correlation. (Reprinted with permission from H. E. O'Connell, *Trans. A.I.Ch.E.*, 42, 741, 1946.)

temperature. The correlating relative volatility, α, is that of the light key relative to the heavy key, calculated at the average column temperature. The overall efficiency, E_O, is correlated as a function of the product $\alpha\mu_L$ as shown in Figure 16.2, which is reproduced from the source (O'Connell, 1946). Although the O'Connell correlation somewhat extends the Drickamer–Bradford correlation to mixtures containing higher volatility components, it should still be used with care, especially outside the applications described in the source.

NOMENCLATURE

E_{MV}	Murphree tray efficiency in vapor terms	E_O	Overall tray efficiency
E_{ML}	Murphree tray efficiency in liquid terms	N	Number of theoretical stages
		N_A	Number of actual trays
E_V	Murphree point efficiency in vapor terms	X	Mole fraction in the liquid
		Y	Mole fraction in the vapor
E_L	Murphree point efficiency in liquid terms	α	Relative volatility
		μ_L	Liquid viscosity

SUBSCRIPTS

j Tray designation

SUPERSCRIPTS

*   Equilibrium composition

REFERENCES

Bubble Tray Design Manual, Tray Efficiency Research Program, American Institute of Chemical Engineers, New York, 1960.

Drickamer, H. G. and J. R. Bradford, *Trans. A.I.Ch.E.*, 39, 319, 1943.

Murphree, E. V., *Ind. Eng. Chem.*, 17, 747, 1925.

O'Connell, H. E., *Trans. A.I.Ch.E.*, 42, 741, 1946.

CHAPTER 17

Batch Distillation

The processes discussed so far in this book are all treated as continuous, steady-state operations, i.e., the properties or state variables at any given point in the process are considered time invariant. An important separation process that differs in this respect is batch distillation, a time-dependent process.

A batch distillation apparatus consists of a reboiler attached to the bottom of a trayed or packed column with a condenser at the top. A batch of feedstock is charged to the reboiler; then the mixture in the reboiler is gradually vaporized. The vapor flows up the column, is condensed in the condenser, and part of the condensate is returned to the column as reflux, flowing countercurrent to the vapor. The rest of the condensate is collected over a period of time as a single distillate product or a series of distillate fractions. Generally there are no side feeds or side products and the entire column acts as a rectifier.

The same principles of continuous fractionation apply in the batch rectifier, except in batch distillation the compositions and column conditions change continuously with time.

The initial distillate cut is the lightest and, as the distillation progresses, the liquid remaining in the reboiler becomes continuously richer in the heavier components, and subsequent distillate cuts become increasingly heavier. The residue remaining in the reboiler after the last distillate cut is the heaviest cut. A multicomponent feed mixture may be separated in one batch distillation column into a number of products with specified purities. Given the required number of trays and reflux ratio, a batch distillation column could, in principle, separate a normal feed mixture (one that is not azeotrope forming or reacting) into its pure constituents.

Batch distillation is commonly used for separating and purifying materials of high value that are generally handled in relatively small quantities, such as pharmaceuticals and specialty chemicals. A continuous process usually requires long periods of time to reach steady state, during which large quantities of the products are off specification. The batch still also adds versatility to the plant since the same unit may be used to process different feeds and generate multiple products.

Column hydraulics calculations for the design and operation of batch distillation rectifiers, trayed or packed, are similar to those of continuous columns (Chapters 14 and 15). Since the same column may be used for different mixtures at different operating conditions and since the operating conditions change during a given distillation task, it is important to check the column hydraulics for all the expected operating conditions. The reboiler and condenser must be designed to handle the expected rates of vaporization and condensation.

The economical design and operation of batch distillation systems must take into account many factors, most of which are time dependent. The mathematical complexity of the process provides a strong incentive for using computer stimulation to solve the problem.

17.1 PRINCIPLES OF BATCH DISTILLATION

The batch distillation system schematic shown in Figure 17.1 consists of a reboiler, a rectifying column, a condenser, and a series of receivers. The plant is piped so that each product may be directed to any one of the receivers.

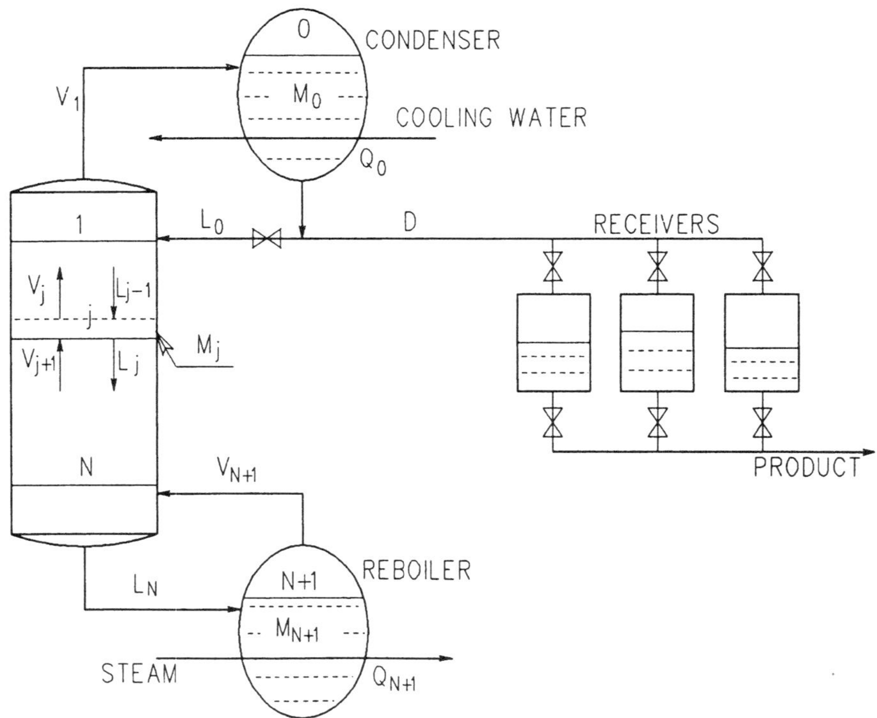

Figure 17.1 Batch distillation system schematic.

A batch distillation task is started up by boiling the liquid in the reboiler and vaporizing it at a rate determined by the heat transfer rate. The vapor generated is sent through the column and into the condenser, where it is totally condensed. Initially the entire condensate is refluxed back to the column with no product drawn. This is a condition of total reflux, similar to the total reflux operation of a continuous column. Under these conditions the batch column operates at steady state, i.e., the column operating parameters at any given point do not change with time. At total reflux, the maximum separation is achieved with the given number of trays or packing height.

Depending on the tray geometry (diameter, weir height, downcomer design, etc.) in trayed columns or on the type of packing in packed columns, a certain amount of liquid is retained within the column as liquid holdup. An additional amount of liquid holdup is associated with the condenser assembly. The remaining space within the distillation system contains the vapor holdup. In continuous columns, holdup is usually not a factor since it does not affect the separation as long as the column operates at steady state.

17.1.1 Effect of Holdup

As distillate starts to be drawn from the condenser following total reflux operation, the compositions throughout the column start changing and unsteady-state conditions ensue. It is generally accepted that the effect of the vapor holdup on the batch column performance may be neglected. The liquid holdup affects the column performance in different ways and to varying degrees. The effects become more noticeable with larger holdups relative to the amount of liquid in the reboiler.

One effect is that the holdup lowers the original concentration of the light components in the feed charge. Distillate production starts with a lower light component concentration in the reboiler, making the separation more difficult. Another effect, which seems to counter the first, is associated with the lag in composition changes in the column resulting from the holdup. The delayed drop of the light components' concentration in the holdup can enhance the separation.

These observations are too qualitative, but it is important to note that holdup is a factor that must be taken into consideration when evaluating batch processes, especially when holdups reach 5 to 10% or higher relative to the liquid in the reboiler. The more rigorous mathematical solution methods incorporate liquid holdup and quantify its effects.

17.1.2 Operating Strategies

As the distillate is withdrawn from the condenser, its composition changes continuously, becoming progressively heavy. The composition of a cut is an integrated average of the instantaneous distillate composition over the time period of collecting that cut. Subsequent cuts are made by redirecting the distillate to different receivers each time the required composition of the current cut has been attained. It often happens that, in order to meet certain purity cuts, other intermediate, off-specification cuts must be produced. The intermediate products are collected and usually blended with raw feed for recycling.

In batch distillation the column conditions vary throughout the distillation cycle; therefore, it is possible, in principle, to conduct the distillation in an infinite number of ways by arbitrarily changing the column conditions with time. Any two of the variables: reflux ratio, distillate rate, boilup rate, condenser duty, and reboiler duty could be controlled independently during the distillation cycle (Section 17.1.3). Operating strategies are sought that would meet performance objectives such as product purity specifications. Additionally, the operation may be driven by optimization criteria, such as minimizing the distillation time or maximizing profit. A discussion of some of the possible operating strategies follows.

Constant Reflux

Operating the column at a constant reflux rate results in a distillate with a continuously changing composition. The composition profile over time is determined by the reflux rate and the distillate rate. The lightest component starts at its highest value, followed by concentration peaks of components with descending volatility. The sharpness of the separation is determined by the reflux ratio. Since the reflux rate is limited by hydraulics considerations (Chapters 14 and 15), higher reflux ratios would require lowering the distillate rate, thus lengthening the distillation cycle. The separation sharpness must therefore be weighed against the time required to complete the distillation.

Constant Distillate Composition

A constant composition distillate may be produced by progressively increasing the reflux rate. When it reaches the maximum allowable from a hydraulic or economic standpoint, the distillate is redirected to a new receiver, and the reflux rate is lowered to a new starting point. The procedure is repeated for subsequent cuts, some of which may be off specification. Here again, lower distillate rates generally result in better separation at the expense of longer distillation times.

Cycling Operation

The column is initially operated at total reflux until equilibrium is reached. The condensate is then totally diverted to a receiver for a short period of time, during which the column operates at zero reflux. Before the composition of the liquid in the receiver goes off specification, the condensate is diverted back to the column, and the unit is operated again at total reflux. The cycle is repeated for all the remaining cuts.

Optimized Operation

Given an objective function to be maximized or minimized, it is possible, in principle, to compute an optimum profile of reflux ratio and distillate rate for the entire distillation cycle. The principles of such approaches are introduced in Section 17.2.4.

17.1.3 Conceptual Control and Degrees of Freedom

Reference again is made to Figure 17.1 for a degrees of freedom analysis and a study of possible control strategies. The description rule (Section 5.2.1) is applied here to an existing system. The variables that can be set by external independent means include the steam rate to the reboiler, the cooling water rate to the condenser, and the reflux and product valve positions. At a given time, only one product receiver is active, and the valves of all the other receivers are closed.

At total reflux the reboiler duty is balanced by the condenser duty. If, at this point, a distillate product is drawn at a certain rate without changing the reflux rate, both reboiler and condenser duties must be increased to handle the higher rate of vaporization and condensation. In general, any set of reboiler and condenser duties determines the reflux and distillate rates. The reflux and product valve positions must be controlled to maintain reboiler and condenser liquid levels. The boilup rate is also determined from the other variables by material balance. Thus of the five variables — reboiler duty, condenser duty, boilup rate, reflux rate, and product rate — two are independent and the system has 2 degrees of freedom.

The overall control strategy depends on the reflux policy selected and can vary from simple manual control to elaborate systems designed to control the reflux based on some optimization scheme. Aside from the overall control strategy, the primary controllers most commonly used include the following: the reboiler steam rate is controlled by the pressure drop across the column, the reflux rate by a temperature point in the column, the condenser cooling water rate by the condensate temperature, and the product rate by the liquid level in the accumulator.

17.2 SOLUTION METHODS

The complexity of batch distillation calculations is due to the unsteady-state, dynamic nature of the process. Most of the variables describing the column are constantly changing with time; hence, an accurate mathematical model must include differential as well as algebraic equations. With the ever-increasing computing power available, however, rigorous simulation using reasonable computing time is now possible for the design and performance evaluation of batch distillation operations. As is always the case, the use of simulators is most effective when based on a qualitative understanding of the problem, aided by graphical or shortcut preliminary calculations. These methods may themselves be adequate for small applications where the economic penalty of overdesign is insignificant.

17.2.1 Graphical and Shortcut Methods: Binary Systems

Binary batch distillation can be represented graphically on a Y–X diagram using the McCabe–Thiele method as in continuous binary distillation. Since the compositions throughout the system change continuously with time, one plot represents one instant in the distillation cycle. In addition to the McCabe–Thiele assumptions in continuous distillation, the graphical method for batch distillation also assumes negligible column and condenser holdups compared to the amount of liquid in the reboiler. With low percentage holdups (0 to 10% of the liquid in the reboiler), the accumulation terms in the mass balance equations (discussed in following sections) can be neglected, and instantaneous pseudo-steady-state conditions can be assumed.

At a given point in time, t^1, the amount of liquid in the reboiler is M_{N+1}^1, and its composition in terms of the mole fraction of the lighter component is X_{N+1}^1 . The distillate rate and composition at time t^1 are D^1 moles/hr and X_D^1 , respectively. The internal liquid and vapor flows at the same instant are L^1 and V^1 moles/hr, assumed constant from tray to tray. A material balance at t^1 around an envelope that includes the condenser and a column section from the condenser to any tray gives the following operating line equation if holdups are neglected:

$$Y^1 = \left(L^1/V^1\right)X^1 + \left(D^1/V^1\right)X_D^1 \tag{17.1}$$

The operating line equation at time t^2, when the reboiler liquid amount and composition are M_{N+1}^2 and X_{N+1}^2 , is given by

$$Y^2 = \left(L^2/V^2\right)X^2 + \left(D^2/V^2\right)X_D^2 \tag{17.2}$$

A material balance around the condenser relates the L/V ratio to the reflux ratio at any instant:

$$\frac{L}{V} = \frac{R}{R+1} \tag{17.3}$$

Instantaneous operating lines may be constructed based on given operating conditions. Figure 17.2 shows two operating lines drawn for a constant L/V ratio (or constant reflux ratio) at two different points in time. From the intersection point of the vertical line at X_D^1 and the 45° diagonal, an operating line is drawn with slope L/V. Next, the number of theoretical stages (three in this case) is stepped off between the operating line and the equilibrium curve. The equilibrium reboiler composition is X_{N+1}^1 . At a later time the distillate composition X_D^2 is less than X_D^1 since this is the mole fraction of the lighter component, which is decreasing as the distillation progresses. The operating line is drawn parallel to the first one with the same slope, L/V, and the three stages are stepped off, giving a reboiler composition of X_{N+1}^2 .

For an operation at constant distillate composition X_D, the reflux ratio is constantly changing. The operating lines at different times all pass through the X_D intersection with the diagonal, each with a different slope. In order to maintain a constant distillate composition, the reflux ratio, and hence the operating line slope, must increase with time. Figure 17.3 shows two operating lines at two different points in time, giving reboiler compositions X_{N+1}^1 and X_{N+1}^2 . As time progresses, the concentration of the lighter component in the reboiler decreases, i.e., $X_{N+1}^2 < X_{N+1}^1$.

The graphical procedure is applicable to any binary batch distillation process and is not limited to operations at constant reflux ratio or constant distillate com-

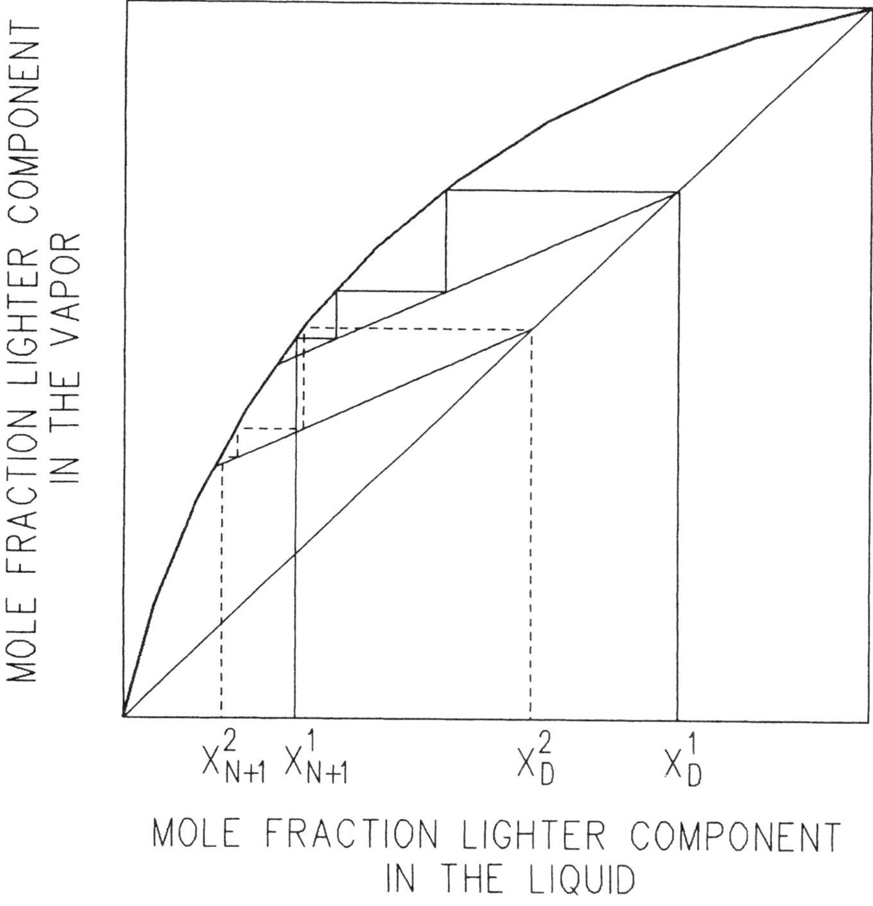

Figure 17.2 Graphical method for constant reflux ratio.

position. The result is a tabular relationship between the distillate composition, the reboiler composition, and the L/V ratio:

$$X_D = f(X_{N+1}, L/V) \tag{17.4}$$

Other variables may be calculated based on this relationship. Since the holdup is neglected, the rate of change in the reboiler liquid equals the distillate rate. By total mass balance,

$$-\frac{dM_{N+1}}{dt} = D \tag{17.5}$$

and by mass balance on the lighter component,

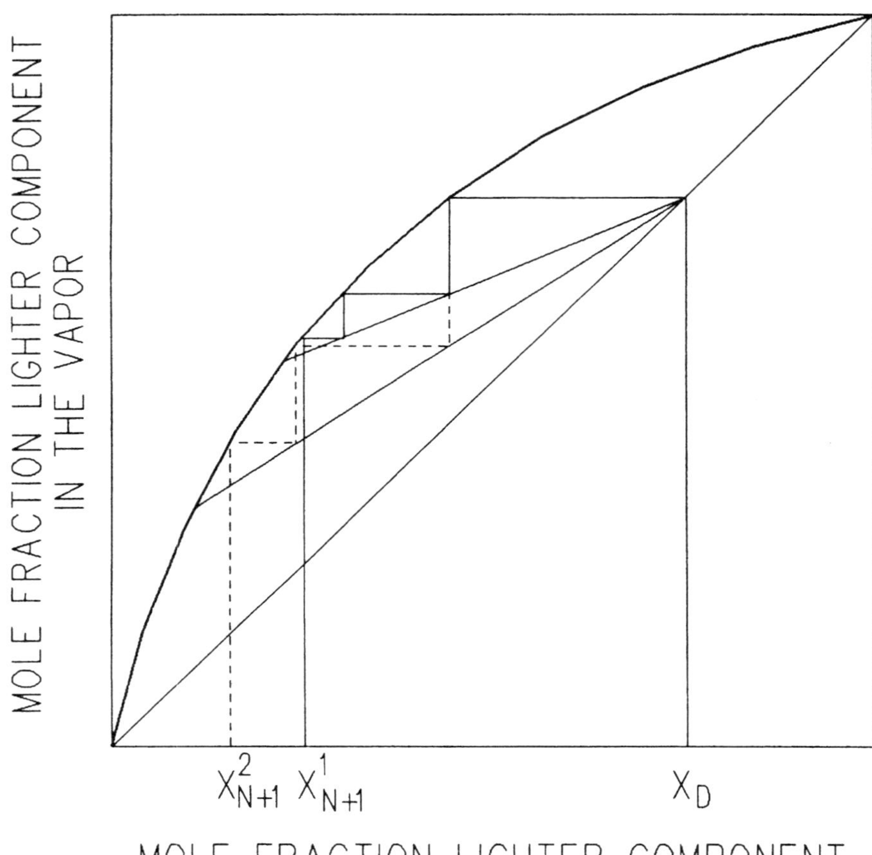

Figure 17.3 Graphical method for constant distillate composition.

$$-\frac{d\left(M_{N+1}X_{N+1}\right)}{dt} = DX_D \qquad (17.6)$$

By expanding the derivative in Equation 17.6, combining with Equation 17.5 and rearranging, the following is obtained:

$$\frac{dM_{N+1}}{M_{N+1}} = \frac{dX_{N+1}}{X_D - X_{N+1}} \qquad (17.7)$$

Integrating this equation from initial conditions (t = 0) to current conditions gives (Smoker et al., 1940)

$$\ln\left(\frac{M_{N+1}}{M_{N+1}^0}\right) = \int_{X_{N+1}^0}^{X_{N+1}} \frac{dX_{N+1}}{X_D - X_{N+1}} \tag{17.8}$$

The right-hand side is evaluated by graphical integration using the relationship 17.4. If this integral is denoted by I, then

$$M_{N+1} = M_{N+1}^0 e^{-I} \tag{17.9}$$

The amount of distillate cut, C, and its composition are calculated by total and lighter component mass balances:

$$C = M_{N+1}^0 - M_{N+1}$$

$$CX_C = M_{N+1}^0 X_{N+1}^0 - M_{N+1} X_{N+1}$$

where X_C is the cumulative mole fraction of the lighter component in C. The two mass balance equations are combined to give

$$X_C = \frac{M_{N+1}^0 X_{N+1}^0 - M_{N+1} X_{N+1}}{M_{N+1}^0 - M_{N+1}} \tag{17.10}$$

The time required to distill an amount C at constant reflux and distillate rates is calculated as

$$t = \frac{C}{D} = \frac{M_{N+1}^0 - M_{N+1}}{D} \tag{17.11}$$

The distillate rate may be expressed in terms of the boilup rate and reflux ratio:

$$D = \frac{V}{R+1} \tag{17.12}$$

Substituting Equations 17.9 and 17.12 in Equation 17.11 gives

$$t = (R+1)\frac{M_{N+1}^0}{V}\left(1 - e^{-I}\right) \tag{17.13}$$

In general, the distillate rate may vary, and the distillation time is written in differential form:

$$dt = \frac{dC}{D} = -\frac{dM_{N+1}}{D}$$

This is combined with Equation 17.7 to give

$$dt = -\frac{M_{N+1}dX_{N+1}}{D(X_D - X_{N+1})} \tag{17.14}$$

The distillation time is obtained by graphical integration of this equation. If the process is at constant distillate composition, the instantaneous and cumulative distillate compositions are identical:

$$X_D = X_C$$

Thus, Equation 17.10 may be rearranged to give

$$M_{N+1} = \frac{M_{N+1}^0\left(X_D - X_{N+1}^0\right)}{X_D - X_{N+1}} \tag{17.15}$$

The distillate rate is given in terms of the boilup rate and the L/V ratio:

$$D = V - L = V(1 - L/V) \tag{17.16}$$

Equations 17.15 and 17.16 are combined with 17.14 to give

$$dt = -\frac{M_{N+1}^0\left(X_D - X_{N+1}^0\right)dX_{N+1}}{V(1 - L/V)\left(X_D - X_{N+1}\right)^2} \tag{17.17}$$

The time required to change the reboiler composition from X_{N+1}^0 to X_{N+1} is obtained by integration:

$$t = \int_{X_{N+1}}^{X_{N+1}^0} \frac{M_{N+1}^0\left(X_D - X_{N+1}^0\right)dX_{N+1}}{V(1 - L/V)\left(X_D - X_{N+1}\right)^2}$$

The integration is carried out graphically using data as expressed by the relationship 17.4.

17.2.2 Shortcut Methods: Multicomponent Distillation

Multicomponent batch distillation computation time can be excessive with rigorous methods, especially for problems that require multiple runs, such as optimization, and for on-line applications. It is desirable to have shortcut alternatives for such situations.

The graphical-based shortcut methods for binary batch distillation may be applied to multicomponent distillation only when the separation is between two key components to produce one distillate product and the residue. In this case the calculations may be approximated by lumping the other components with either of the key components and treating the system as a pseudobinary.

Truly multicomponent solutions based on continuous distillation shortcut methods have been proposed for batch distillation. The Fenske, Underwood, and Gilliland equations or correlations are commonly used in conjunction with each other to solve continuous distillation problems as described in Section 12.3. Diwekar et al. (1991) describe how these techniques may be modified for the design of batch distillation columns for variable and constant reflux cases.

The unsteady-state operation is approximated by a pseudo-steady-state model using a series of time steps, during each of which steady-state conditions are assumed. At the end of each time step, the bottoms product becomes the feed for the next time step.

For a variable reflux case, the procedure initially assumes constant product (distillate) compositions X_{Di} for all the components. The minimum reflux ratio, R_{min}, required to produce a distillate with composition X_{Di} from a feed with composition $X_{N+1,i}^0$ (which is the reboiler initial composition) is calculated by the Underwood method (Equations 12.29 and 12.30). The ratio of refluxes is

$$ r = \frac{R}{R_{min}} $$

where the starting reflux ratio, R, is given. The minimum number of trays is calculated by the Fenske equation from the distillate composition X_{Di} and the bottoms composition, assumed initially the same as $X_{N+1,i}^0$. Equation 12.14 may be used for this purpose, rewritten in terms of the ratio of light key to heavy key mole fractions and relative volatilities:

$$ N_{min} = \frac{\ln\left(X_{Dl}X_{N+1,h}^0 \big/ X_{N+1,l}^0 X_{Dh}\right)}{\ln\left(\alpha_l / \alpha_h\right)} $$

The Gilliland correlation is used to determine the ratio N/N_{min} corresponding to r. This determines the number of trays, N, held constant in subsequent time steps in which the reflux ratio becomes the calculated variable.

During a small time step, a small amount of liquid in the reboiler is vaporized. As a result, the composition of the liquid remaining in the reboiler changes. The amount vaporized and the change in the reboiler composition during the time step are determined by the rate of heat transfer in the reboiler. The computations proceed by incrementing the composition of a reference component in the reboiler, which is equivalent to incrementing the time. By material balance, the new reboiler composition for any component i is given in terms of the new composition of the reference component, $X_{N+1,r}$ (Diwekar et al., 1991):

$$X_{N+1,i} = X_{Di} - \left[\frac{X_{Di} - X_{N+1,i}^0}{X_{Dr} - X_{N+1,r}^0} \left(X_{Dr} - X_{N+1,r} \right) \right]$$

The Fenske–Underwood–Gilliland methods are again applied to the distillate composition, X_{Di} (assumed constant); the current reboiler composition, $X_{N+1,i}$, and the number of trays, N, to determine r and hence the reflux ratio, R. The procedure is repeated by further incrementing the reference component composition for each time step until a target composition, $X_{N+1,r}^F$, is reached.

This simplified procedure is based on the assumption that the distillate compositions of all the components are constant if the reference component distillate composition is constant. If the compositions of the nonreference components change, the authors (Diwekar et al., 1991) recalculate the compositions based on the Hengstebeck–Geddes equation (Hengstebeck, 1946; Geddes, 1958). The authors also describe a procedure for the constant reflux case, where all the component compositions in the distillate change as the distillation progresses.

17.2.3 Rigorous Methods

The rigorous solution of batch distillation columns carries an extra dimension of complexity over continuous distillation because it is inherently a transient operation. The basic assumption of steady-state operation in the continuous column model obviously does not apply for batch distillation. The only possible steady-state operation in batch distillation is at total reflux, which is commonly used as the initial condition for the dynamic solution of the column.

As in continuous columns, the rigorous solution of batch columns is obtained by solving the heat and material balance and phase equilibrium equations. While holdup terms are not considered in continuous operations, they must be included in batch distillation. The column schematic of Figure 17.1 is the problem model with nomenclature similar to continuous columns. The condenser is designated as stage 0, the reboiler as stage N + 1, and any of the N column trays (or equivalent packing stages) is designated as stage j. The liquid molar holdup on stage j is designated as M_j.

In addition to the basic continuous column model assumptions of equilibrium stages and adiabatic operation, dynamics-related assumptions are made for the batch model. Distefano (1968) assumed constant volume of liquid holdup, negligible vapor holdup, and negligible fluid dynamic lag. Although different solution strategies may be employed, the fundamental model equations are the same. Condenser total mass balance:

$$\frac{dM_0}{dt} = V_1 - \left(L_0 + D \right) \tag{17.18}$$

Condenser component mass balances:

$$\frac{d\left(M_0 X_{0i} \right)}{dt} = V_1 Y_{1i} - \left(L_0 + D \right) X_{0i} \tag{17.19}$$

for i = 1, ..., C.
Condenser enthalpy balance:

$$\frac{d\left(M_0 h_0\right)}{dt} = V_1 H_1 - \left(L_0 + D\right)h_0 - Q_0 \tag{17.20}$$

Total mass balance on any tray j (1 ≤ j ≤ N):

$$\frac{dM_j}{dt} = V_{j+1} + L_{j-1} - V_j - L_j \tag{17.21}$$

Component mass balances on tray j:

$$\frac{d\left(M_j X_{ji}\right)}{dt} = V_{j+1} Y_{j+1,i} + L_{j-1} X_{j-1,i} - V_j Y_{ji} - L_j X_{ji} \tag{17.22}$$

for i = 1, ..., C.
Enthalpy balance on tray j:

$$\frac{d\left(M_j h_j\right)}{dt} = V_{j+1} H_{j+1} + L_{j-1} h_{j-1} - V_j H_j - L_j h_j \tag{17.23}$$

Reboiler total mass balance:

$$\frac{dM_{N+1}}{dt} = L_N - V_{N+1} \tag{17.24}$$

Reboiler component mass balances:

$$\frac{d\left(M_{N+1} X_{N+1,i}\right)}{dt} = L_N X_{Ni} - V_{N+1} Y_{N+1,i} \tag{17.25}$$

for i = 1, ..., C.
Reboiler enthalpy balance:

$$\frac{d\left(M_{N+1} h_{N+1}\right)}{dt} = L_N h_N - V_{N+1} H_{N+1} + Q_{N+1} \tag{17.26}$$

Phase equilibrium relations:

$$Y_{ji} = K_{ji} X_{ji} \tag{17.27}$$

for i = 1, ..., C; j = 1, ..., N + 1.

Mole fraction summation equations:

$$\sum_i X_{ji} = 1 \tag{17.28a}$$

for $j = 0, 1, \ldots, N + 1$, and

$$\sum_i Y_{ji} = 1 \tag{17.28b}$$

for $j = 1, \ldots, N + 1$.

Equations 17.18 through 17.28 encompass a coupled set of $2NC + 4N + 3C + 7$ individual equations to be solved for the time-dependent, unknown variables. These variables are listed as follows:

X_{ji}	$(N+2)C$
Y_{ji}	$(N+1)C$
L_j	$N+1$
V_j	$N+1$
D	1
T_j	$N+2$
M_j	$N+2$
Q_0	1
Q_{N+1}	1
Total	$2NC + 4N + 3C + 9$

Thus, in order to define the column operation uniquely, two specifications are required, as already concluded using the description rule (Section 17.1.3). These could be the reflux rate and distillate rate, L_0 and D. Note that a subcooled condenser is assumed so that no phase equilibrium equation is written for stage 0 and no Y_{0i} variables exist. The column pressure profile is assumed fixed or determined independently from hydraulics calculations and is not included in the column variables. Also, the enthalpies and phase equilibrium coefficients are, in general, functions of the temperature, pressure, and composition (Chapter 1) and are therefore not considered as additional unknown variables.

Equations 17.18 through 17.28 may be solved using different strategies, including simultaneous or iterative methods. Distefano (1968) describes a calculational procedure that solves the equations one at a time. Boston et al. (1981) describe a more flexible and efficient method that can handle a variety of specifications and column configurations, including multiple feeds, side draws, and side heaters. The method uses advanced numerical procedures to handle nonlinearity and stability problems. The object here is not to describe the details of the methods, which may be found in the references. The Distefano method is outlined here to illustrate briefly some of the potential numerical computational problems.

The column is initially solved at total reflux, constant overhead vapor rate, steady-state conditions. At time t = 0, transient conditions develop as distillate drawoff begins. Numerical integration of the differential equations is performed for finite amounts of distillate drawoff, using initial values of the variables from the steady state solution.

Equations 17.18 through 17.26 are first manipulated and rearranged to express composition derivatives, liquid and vapor flow rates, and heat duties explicitly:

$$\frac{dX_{0i}}{dt} = \frac{V_1}{M_0}\left(Y_{1i} - X_{0i}\right) \tag{17.29}$$

$$\frac{dX_{ji}}{dt} = \frac{V_{j+1}}{M_j}\left(Y_{j+1,i} - X_{ji}\right) + \frac{L_{j-1}}{M_j}\left(X_{j-1,i} - X_{ji}\right) - \frac{V_j}{M_j}\left(Y_{ji} - X_{ji}\right) \tag{17.30}$$

$$\frac{dX_{N+1,i}}{dt} = -\frac{V_{N+1}}{M_{N+1}}\left(Y_{N+1,i} - X_{N+1,i}\right) + \frac{L_N}{M_{N+1}}\left(X_{Ni} - X_{N+1,i}\right) \tag{17.31}$$

$$V_1 = L_0 + D + \frac{dM_0}{dt} \tag{17.32}$$

$$L_j = V_{j+1} + L_{j-1} - V_j - \frac{dM_j}{dt} \tag{17.33}$$

$$V_{j+1} = V_j \frac{H_j - h_j}{H_{j+1} - h_j} - L_{j-1}\frac{h_{j-1} - h_j}{H_{j+1} - h_j} + \frac{M_j}{H_{j+1} - h_j}\frac{dh_j}{dt} \tag{17.34}$$

$$Q_0 = V_1\left(H_1 - h_0\right) - M_0\frac{dh_0}{dt} \tag{17.35}$$

$$Q_{N+1} = V_{N+1}\left(H_{N+1} - h_{N+1}\right) - L_N\left(h_N - h_{N+1}\right) + M_{N+1}\frac{dh_{N+1}}{dt} \tag{17.36}$$

Starting from total reflux conditions, the distillate rate is incremented from 0 to D, thereby lowering the reflux rate to $L_0 - D$. Since negligible fluid dynamic lags are assumed, all the liquid rates are instantly lowered to $L_j - D$. The vapor rates are maintained at V_j. These flow rates and the steady-state total reflux mole fractions are used to calculate the mole fraction derivatives by Equations 17.29 to 17.31. The molar holdups in these equations are calculated from the assumed constant volume holdups multiplied by calculated molar densities.

The mole fractions at the end of the time increment (equivalent to the distillate increment) are calculated by numerical integration. The magnitude of the time increment is determined by stability and truncation considerations. The batch distillation model contains tray holdups with time constants much smaller than the reboiler time constant. These conditions ("stiff systems") can cause computational instability unless very small time increments are used. The penalty is excessive computing time and the likelihood of incurring truncation errors. Distefano (1968) provides values for the maximum time increment size consistent with stability for a number of integration schemes. The same time increment is used to determine the incremental distillate rate for the first step.

The calculated mole fractions on a given stage generally may not sum to 1 due to inaccuracies, truncation errors, etc., and must therefore be normalized. With the normalized compositions, a bubble point calculation is performed on each stage to determine the temperatures, T_j, and vapor mole fractions, Y_{ji}. This is equivalent to solving Equations 17.27 and 17.28 (Chapter 2).

Based on current temperatures and compositions, the liquid molar densities and the liquid and vapor enthalpies are now calculated. The molar holdups are calculated, and the holdup and enthalpy derivatives with respect to time are computed numerically from current and latest values of these variables.

Next, Equations 17.32 through 17.34 are used to update the vapor and liquid flow rates, and Equations 17.35 and 17.36 are used to calculate the condenser and reboiler duties. The reboiler molar holdup is calculated from a material balance:

$$M_{N+1} = M_{N+1}^0 - \sum_{j=0}^{N} M_j - \int_0^t D\,dt \qquad (17.37)$$

The computational cycle is repeated until the required amount of distillate has been produced.

17.2.4   Optimization

Optimizing batch distillation operations can have a significant economic impact especially when the separation of high-value chemicals is involved. The control variable for optimizing a batch distillation product is the reflux ratio policy, i.e., the variation of the reflux ratio with time. As mentioned in Section 17.1.2, two reflux ratio policies are commonly in use — a constant reflux ratio or a variable reflux ratio that produces a constant composition distillate. The alternative is to optimize the reflux ratio policy to achieve certain objectives, such as maximizing the amount of product with a specified composition in a given length of time or minimizing the distillation time required to produce a given amount of product with a specified composition.

The optimum operation of a multiproduct column is one that maximizes an overall objective function, such as the annual profit. The optimization decision variables are the duration and reflux ratio policy of each product.

The problem of extensive computing time typical of batch distillation is compounded with the superimposition of an optimization routine. For this reason batch distillation optimization algorithms are usually built around shortcut batch distillation methods. Possible approaches to formulating the optimization problem are presented, although a detailed mathematical discussion is outside the scope of this book.

With negligible liquid holdup, negligible pressure drop, constant vapor and liquid flow rates from tray to tray, time-invariant vapor flow rate, and ideal thermodynamics, the batch distillation problem for product k may be stated as (Farhat et al., 1990).

$$f_k\left[t, M_{N+1}(t), X_{ji}(t), T_j(t), M'_{N+1}(t), X'_{ji}(t), T'_j(t), R(t), v\right] = 0 \qquad (17.38)$$

where $j = 1, \ldots, N$, $i = 1, \ldots, C$, $t_{k-1} \le t \le t_k$, $R(t)$ is the reflux ratio function, and v is a vector of time-invariant parameters: the initial compositions, operating pressure, number of trays, and vapor flow rate. The *prime* indicates derivatives with respect to time. Since

$$D = \frac{V}{R(t)+1}$$

the amount of product k collected between times t_{k-1} and t_k is given by

$$C_k = \int_{t_{k-1}}^{t_k} \frac{V}{R(t)+1}\, dt \qquad (17.39)$$

The principal component in product k is also designated by *k*, and its amount in the product is given as

$$C_{kk} = \int_{t_{k-1}}^{t_k} \frac{VX_{Dk}}{R(t)+1}\, dt \qquad (17.40)$$

The purity of component k in product k is, therefore,

$$X_{Ck} = \frac{C_{kk}}{C_k} \qquad (17.41)$$

The optimization problem for a given product is to maximize C_k subject to the constraints

$$X_{Ck} = X_{Ck}^*$$

and

$$f_k = 0$$

for a given time interval t_{k-1} to t_k. X^*_{Ck} is a specified purity. This optimization problem may be formulated using a linear programming approach and solved by sequential quadratic programming.

NOMENCLATURE

C	Amount of distillate, moles	N_{min}	Minimum number of stages
C	Number of components	Q	Heat duty
D	Distillate molar rate	R	Reflux ratio
h	Liquid molar enthalpy	R_{min}	Minimum reflux ratio
H	Vapor molar enthalpy	r	R/R_{min}
I	Integral in Equation 17.7	t	Time
K	Vapor–liquid equilibrium coefficient	V	Vapor molar flow rate in the column
L	Liquid molar flow rate in the column	v	Time-invariant parameters
		X	Mole fraction in the liquid
M	Liquid molar holdup	Y	Mole fraction in the vapor
N	Number of stages in the column	α	Relative volatility

SUBSCRIPTS

C	Condenser or accumulator	k	Product number
D	Distillate	l	Light key component
h	Heavy key component	r	Reference component
j	Stage number		

SUPERSCRIPTS

0,1,2	Points in time	*	Specified value
F	Final conditions		

REFERENCES

Bogart, M. J. P., *Trans. Am. Inst. Chem. Eng.*, 33, 139, 1937.

Boston, J. F., H. I. Britt, S. Jirapongphan, and V. B. Shah, *Foundations of Computer-Aided Chemical Process Design, 2*, Edited by R. S. Mah and W. D. Seider, New York, American Institute of Chemical Engineers, 1981, 203.

Distefano, G. P., *AIChE J.*, 14, 190, 1968.

Diwekar, U. M. and K. P. Madhavan, *Ind. Eng. Chem. Res.*, 30, 713, 1991.
Farhat, S., M. Czernicki, L. Pibouleau, and S. Domenech, *AIChE J.*, 36, 1349, 1990.
Geddes, R. L., *AIChE J.*, 4, 389, 1958.
Hengstebeck, R. J., *Trans. Am. Inst. Chem. Eng.*, 42, 309, 1946.
Smoker, E. H. and A. Rose, *Trans. Am. Inst. Chem. Eng.*, 36, 285, 1940.

PROBLEMS

17.1 For a binary batch distillation process with no reflux (differential distillation) derive an expression for the amount of liquid remaining in the reboiler in terms of its composition. Assume constant relative volatility throughout the process.

If the relative volatility $\alpha = 2$, and the initial liquid composition $X^0 = 0.4$, what is the initial distillate composition? What is the composition of the liquid remaining in the reboiler when 50% of the original liquid has been distilled? When 99% has been distilled?

17.2 In a single-stage batch distillation with no reflux (differential distillation) at 1 atm, the still is charged with 100 lbmole of an equimolar binary mixture of hexane and heptane. The K-values are given by $\ln K_i = A_i - B_i/T$, where T is in degrees Rankine.

		A_i	B_i
1.	Hexane	11.08	6821
2.	Heptane	11.29	7545

Calculate the following quantities when the reboiler temperature is 650°R: (a) the instantaneous vapor composition; (b) the cumulative distillate composition; and (c) the amount and composition of the liquid remaining in the reboiler. Neglect the holdup between the reboiler and the accumulator.

17.3 A three-stage batch distillation column is charged with a 100 lbmole mixture containing 60% mole component 1 and 40% mole component 2. The column pressure is maintained at 1 atm, and the distillate rate is 20 lbmole/hr. It is desired to produce two distillation cuts. The first cut will be produced by continuously adjusting the reflux ratio to maintain the distillate composition at 90% mole component 1. Production of the second cut starts when the L/V ratio is 0.80. The L/V ratio will be fixed at this value until the second cut cumulative composition is 75% mole component 1. Determine the amount of each cut. Assume negligible tray holdups and use vapor–liquid equilibrium data from Problem 6.1.

17.4 A hydrocarbon mixture is to be separated by batch distillation at 75 psia to produce a constant composition distillate. The starting charge composition, the relative volatilities, and the desired distillate composition are as follows:

	$X^0_{N+1,i}$	α_{i4}	X_{Di}
Ethane	0.12	13.304	0.200
Propane	0.48	5.553	0.793
n-Butane	0.25	2.315	0.007
n-Pentane	0.15	1.000	0.000

If the distillation were to be started at twice the minimum reflux ratio, determine the required number of stages. If the initial charge is 100 lbmole and the distillate rate is 10 lbmole/hr, calculate the reflux rate, the amounts of distillate and residue, and the residue composition as a function of time. Irrespective of tray hydraulics and reboiler and condenser capacity constraints, when should the distillation be stopped? Assume negligible tray holdups and use shortcut methods.

Index

C

N

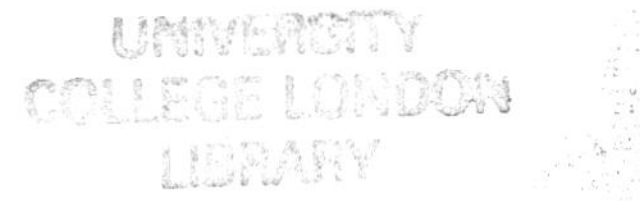